Computational
Microelectronics

Edited by
S. Selberherr

Vassil Palankovski
Rüdiger Quay

Analysis and Simulation of Heterostructure Devices

SpringerWienNewYork

Dr. techn. Dipl.-Ing. Vassil Palankovski
Institut für Mikroelektronik
Technische Universität Wien, Austria

Dr. techn. Dipl.-Phys. Rüdiger Quay
Fraunhofer Institut für Angewandte Festkörperphysik (IAF)
Freiburg, Germany

Printed in Austria

Springer-Verlag Wien New York is a part of Springer Science + Business Media
springeronline.com

Typesetting: Camera-ready by authors
Printing and binding: Druckerei Theiss GmbH, A-9431 St. Stefan im Lavanttal

Printed on acid-free and chlorine-free bleached paper
SPIN: 10909911

With 183 Figures

CIP data applied for

ISSN 0179-0307
ISBN 3-211-40537-2 Springer-Verlag Wien New York

To Dessi, Lili & Christine

Contents

List of Acronyms

AC	...	Alternating Current
ADS	...	Advanced Design System
BB	...	Band-to-Band tunneling
BCB	...	Benzocyclobutene
BGN	...	Bandgap Narrowing
BICMOS	...	Bipolar CMOS
BJT	...	Bipolar Junction Transistor
BV	...	Breakdown Voltage
CAD	...	Computer Aided Design
CML	...	Current-Mode Logic
CMOS	...	Complementary Metal Oxide Semiconductor
CQFL	...	Continuous Quasi-Fermi Level
CVD	...	Chemical Vapor Deposition
CW	...	Continuous wave
DC	...	Direct Current
DD	...	Drift-Diffusion
DHBT	...	Double Heterojunction Bipolar Transistor
DOS	...	Density of States
ECL	...	Emitter-Coupled Logic
ET	...	Energy Transport
FET	...	Field Effect Transistor
GSH	...	Global Self Heating
HBT	...	Heterojunction Bipolar Transistor
HD	...	Hydrodynamic
HEMT	...	High Electron Mobility Transistor
HFET	...	Heterostructure Field Effect Transistor
IC	...	Integrated Circuit
II	...	Impact Ionization
MAG	...	Maximum Available Gain
MBE	...	Molecular Beam Epitaxy
MC	...	Monte-Carlo
MDS	...	Microwave Design System
MESFET	...	Metal Semiconductor Field Effect Transistor

MIC	...	Microwave Integrated Circuit
MMIC	...	Monolithic Microwave Integrated Circuit
MOCVD	...	Metal Organic Chemical Vapor Deposition
MODFET	...	Modulation Doped FET
MOS	...	Metal Oxide Semiconductor
MOSFET	...	Metal Oxide Semiconductor FET
MSG	...	Maximum Stable Gain
NID	...	Non-Intentionally Doped
PAE	...	Power-Added Efficiency
PHEMT	...	Pseudomorphic HEMT
RF	...	Radio Frequency
RST	...	Real Space Transfer
SH	...	Self Heating
SHBT	...	Single Heterojunction Bipolar Transistor
SOI	...	Silicon on Insulator
SIMS	...	Secondary Ion Mass Spectroscopy
SIESTA	...	Simulation Environment for Semiconductor Technology Analysis
SRH	...	Shockley-Read-Hall
TBB	...	Trap-Assisted Band-to-Band Tunneling
TCAD	...	Technology CAD
TE	...	Thermionic Emission
TFE	...	Thermionic Field Emission
VISTA	...	Viennese Integrated System for TCAD Applications
WKB	...	Wigner-Kramers-Brillouin

List of Symbols

Δ	...	step, difference, change
ΔE_{V},ΔE_{C}	...	difference of the valence/conduction band at a heterointerface between two segments
ΔE_g	...	total difference of the bandgaps at a heterointerface between two segments
Γ	...	tunneling probability
Γ_{n}, Γ_{p}	...	field enhancement factors
α,β	...	general exponent
α	...	general temperature coefficient
α	...	in HBT compact model: amplification
α,β	...	in the bandgap model: temperature coefficients
α_ν	...	auxiliary quantity in the Hänsch mobility model for carrier type ν
α_ν	...	quantity in the impact ionization model for carrier type ν
β_F	...	inverse Fermi screening length
β_{n}, β_{p}, β_ν	...	exponents in mobility models for electrons, holes, and general carrier type
γ	...	exponent in mobility models
δE_ν	...	barrier height lowering
ε	...	permittivity
ε_{s}, $\varepsilon_{\mathrm{ins}}$	...	permittivity of semiconductor, insulator
ε_{r}	...	relative permittivity
$\varepsilon_{\mathrm{ox}}$	...	relative permittivity of the oxide in the polysilicon contact model
ϕ_{B}	...	barrier height
ϕ_{Fm}	...	Fermi potential in the metal
ϕ_{ms}	...	metal workfunction difference potential
ϕ_{Fn}	...	quasi-Fermi potential for electrons in the semiconductor
$\phi_{\mathrm{Fn,1}}$	...	quasi-Fermi potential for electrons in the semiconductor 1
κ_{L}	...	lattice thermal conductivity

κ_{n}, κ_{p}, κ_{ν}	...	thermal conductivity of electrons, holes, and general carrier type
κ_{300}^{A}	...	thermal conductivity at 300 K of material A
λ_0, λ	...	temperature dependent mean free path for the optical phonon
λ_B	...	De Broglie wave length
μ_{ν}	...	mobility of carrier type ν
μ_{ν}^{min}	...	minimum mobility
μ_{ν}^{sr}	...	surface mobility contribution of carrier ν at 300 K in the surface scattering model
μ_{ν}^{L}	...	mobility due to lattice scattering
μ_{ν}^{LI}	...	mobility due to lattice and impurity scattering
μ_{ν}^{LIS}	...	mobility due to lattice, impurity, and surface scattering
$\mu_{\nu}^{\mathrm{LISF}}$	...	mobility including lattice, impurity, surface scattering, and high-field reduction
$\mu_{\nu}^{\mathrm{LIST}}$	...	mobility including lattice, impurity, surface scattering, and high-temperature reduction
μ_{ν}^{LIT}	...	mobility including lattice, impurity, and high-temperature reduction
$\mu_{\mathrm{maj}}^{\mathrm{mid}}$, $\mu_{\mathrm{maj}}^{\mathrm{hi}}$	...	parameters of the majority mobility model
$\mu_{\mathrm{min}}^{\mathrm{mid}}$, $\mu_{\mathrm{min}}^{\mathrm{hi}}$	...	parameters of the minority mobility model
μ_{n}, μ_{p}	...	electron and hole mobilities
$\mu_{\nu,300}^{\mathrm{ref}}$	...	reference mobility of carrier ν at 300 K in the surface scattering model
ρ_{L}	...	mass density
$\sigma_{\mathrm{T,n}}$, $\sigma_{\mathrm{T,p}}$	...	trap capture cross sections for electrons and holes
σ_{ox}	...	oxide conductivity
σ_{s}	...	surface (interface) charge density
τ	...	phase term of the transconductance g_{m}
τ_{port}	...	port extension
$\tau_{\epsilon,\mathrm{n}}$, $\tau_{\epsilon,\mathrm{p}}$, $\tau_{\epsilon,\nu}$	...	energy relaxation times for electrons, holes, and general carrier type
$\tau_{\epsilon,0}$, $\tau_{\epsilon,1}$	...	parameters for the energy relaxation times model
$\tau_{\nu,\mathrm{dop}}$	...	doping dependent lifetime of carrier $\nu=$ n,p
$\tau_{\nu,\mathrm{max}}$	...	maximum lifetime of carrier $\nu=$ n,p
τ_{AUG}	...	carrier lifetime defined by *Auger* recombination
τ_{C}	...	bowing parameter energy relaxation time
τ_{DIR}	...	carrier lifetime defined by direct recombination
τ_{SRH}	...	carrier lifetime defined by SRH recombination
τ_{SURF}	...	surface contribution to carrier lifetime

τ_{T}	...	time constant, inverse of f_{T}
τ_{b}	...	base transit time
τ_{c}	...	collector transit time
$\tau_{\mathrm{cap},\nu}$	...	capture time constant for carrier ν
τ_{eb}	...	base emitter transit time
τ_{ec}	...	emitter collector transit time
$\tau_{\mathrm{em},\nu}$	...	emission time constant for carrier ν
τ_{ext}	...	time constant, inverse of extrinsic f_{T}
τ_{i}	...	intrinsic delay time
τ_{n}, τ_{p}	...	recombination lifetimes for electrons and holes
$\tau_{\mathrm{n},300}$, $\tau_{\mathrm{p},300}$, $\tau_{\nu,300}$	...	lifetime of carrier ν= n,p at T_{L}= 300 K
τ_{sc}	...	space charge transit time
τ_{tot}	...	total carrier lifetime
φ_{m}	...	metal quasi-Fermi level
φ_{w}	...	workfunction difference potential
ψ	...	electrostatic potential
ψ_{bi}	...	built-in potential
φ_{s}	...	semiconductor contact potential
ω	...	oscillation frequency
ω_0	...	oscillation frequency of phonon
A_ν	...	temperature coefficient saturation velocity for carrier ν= n,p
A_ν	...	coefficient for impact ionization for carrier ν= n,p
A^*	...	Richardson constant typical for thermionic emission process
A, B	...	general parameters in recombination models
B_ν	...	threshold energy in the impact ionization model
B_ν	...	parameter for the surface mobility in the Lombardi mobility model
BV_{CE0}	...	collector-emitter breakdown voltage
BV_{GD}	...	gate-drain diode breakdown voltage
BV_{i}	...	breakdown voltage
C_ε	...	permittivity bowing parameter
C_κ	...	thermal conductivity bowing parameter
C_μ	...	mobility bowing parameter
C_ν	...	parameter for the surface mobility in the Lombardi mobility model
$C^{\Gamma}_{\mathrm{vsat},\nu}$	...	bowing parameter of the saturation velocity of carrier type ν in Γ-valley
$C^{\mathrm{AUG}}_{\nu,300K}$	...	Auger coefficients for carrier ν at 300 K
C^{AUG}_{ν}	...	Auger coefficients for carrier ν
C^{DIR}	...	direct recombination coefficient

C_I	...	total impurity concentration
C_IN	...	parasitic input capacitance
C_OUT	...	parasitic output capacitance
C	...	bowing parameter in the energy relaxation time model for *alloy materials*
C_1, C_2, C_3	...	reference concentrations in the Masetti mobility model
C_0^AB	...	model parameter for *alloy materials* in the energy relaxation time model
C_bc	...	base collector capacitance
$C_\mathrm{bc,tot}$	...	total base collector capacitance
C_ds	...	drain-source capacitance
$C_\mathrm{eb,dep}$	...	emitter depletion capacitance
C_fb	...	feedback capacitance
C_fringe	...	fringe contribution to a capacitance
$C_\mathrm{g,tot}$	...	total gate capacitance
C_g	...	bandgap bowing parameter
$C_{\mathrm{g},\Gamma}$	...	bandgap bowing parameter Γ-valley
$C_\mathrm{g,X}$	...	bandgap bowing parameter X-valley
C_gd	...	gate drain capacitance
C_gs	...	gate source capacitance
C_i	...	i=1,..,4: parameter for the impact ionization model
C_i	...	i=0,..,3: parameter for the energy relaxation time model
C_i	...	i=1,2: parameter for the built-in potential
C_jc	...	collector junction capacitance
C_im	...	additional capacitance due to impact ionization
C_in	...	input capacitance
$C_{\mathrm{m},\nu}$	...	relative carrier mass bowing parameter
C_met	...	contribution to C_gd due to coupling of metal periphery
$C_\mathrm{maj}^\mathrm{mid}$, $C_\mathrm{maj}^\mathrm{hi}$	...	parameters of the majority mobility model
$C_\mathrm{min}^\mathrm{mid}$, $C_\mathrm{min}^\mathrm{hi}$	...	parameters of the minority mobility model
C_net	...	net concentration
C_pad	...	parasitic pad capacitance
C_pass	...	contribution to C_gd due to device passivation
C_pds, C_pgs, C_pgd	...	parasitic drain source, gate-source, and gate-drain capacitances
C_ref	...	reference impurity concentration
C_spaceDC	...	contribution to C_gs due to the space charge region
C_traps	...	frequency dispersive contribution to C_gs

C_{vsat}	...	bowing parameter of the saturation velocity
$\mathbf{D}$	...	dielectric flux
D_{ν}	...	parameter for the surface mobility in the Lombardi mobility model
D_{n}	...	effective diffusivity in the base
E_{ν}^{crit}	...	critical field for impact ionization for carrier ν
E_{crit}	...	critical field for the onset of saturated behaviour in compact models
E_{F}	...	Fermi energy
E_{Fm}	...	Fermi energy in the metal
E_{Fn}	...	Fermi energy for electrons in the semiconductor
E_{Fp}	...	Fermi energy for holes in the semiconductor
E_{T}	...	trap energy
$\mathbf{E}$	...	local electric field
E_{C}	...	conduction band energy
E_{g}	...	bandgap energy
E_{gate}	...	effective field at the gate
$E_{\mathrm{g},\Gamma}$	...	bandgap energy in Γ-valley
$E_{\mathrm{g,X}}$	...	bandgap energy in X-valley
$E_{\mathrm{g},0}$, $E_{\mathrm{g},300}$	...	bandgap energy at 0 K, and at 300 K
E_{i}	...	bandgap modeling parameters
E_{i}	...	threshold energy
$E_{\mathrm{i},\nu}$	...	threshold energy of carrier ν
$\mathbf{E}_{\mathrm{ins}}$	...	local electric field in insulator
E_{off}	...	energy offset
$E_{\perp 2}$	...	electric field orthogonal to the interface
$E_{\mathrm{r},0}$	...	energy loss per scattering at reference temperature
$\mathbf{E}_{s}$	...	local electric field in semiconductor
E_{V}	...	valence band energy
E_{w}	...	workfunction energy difference
E_{0}	...	minimum energy in the traps-assisted model
F	...	field in impact ionization model
F_{min}	...	minimum noise figure
F_{n}, F_{p}, F_{ν}	...	driving force for electrons and holes
$F_{c}^{\pm}$	...	critical field strengths in the direct band to band tunneling model
F_{o}	...	steady-state occupancy function
F_{0}	...	critical driving force at which overshoot in the velocity-field characteristics occur
G_{n}, G_{p}	...	electron and hole emission rate
G_{n}, G_{p}, G_{ν}	...	generation rate of carrier $\nu=$ n,p
G_{ν}^{II}	...	impact ionization generation of carrier ν

$\boldsymbol{G}_{\mathrm{tun}}$	...	local generation rate associated with the tunneling
H	...	heat generation
I_{B}, I_{C}, I_{E}	...	base, collector, emitter currents
I_{Con}	...	current through a contact
I_{D}	...	drain current
I_{Dmax}	...	maximum drain current
I_{DS}	...	drain source current
I_{G}	...	gate current
I_{II}	...	impact ionization contribution to the gate current
I_{TFE}	...	thermionic field emission contribution to the gate current
$\mathbf{J}_{\mathrm{n}}$, $\mathbf{J}_{\mathrm{p}}$	...	electron and hole current densities
J_{tun}	...	current density of tunneling currents
K_i	...	temperature coefficients for *Auger* recombination
L_{D}, L_{G}, L_{S}	...	drain, gate, and source inductances
MAG	...	maximum available gain
M_{C}	...	valley degeneracy factor of the conduction band
ME	...	modulation efficiency
MSG	...	maximum stable gain
N_{A}	...	acceptor doping concentration
N_{C}	...	effective density of states for electrons
$N_{\mathrm{C},0}$	...	effective density of states for electrons evaluated at reference temperature T_0
N_{D}	...	donor doping concentration
$N_{\mathrm{ref},\nu}$	...	reference doping concentration for carrier $\nu=$ n,p
N_{T}	...	concentration of traps
N_{V}	...	effective density of states for holes
$N_{\mathrm{V},0}$	...	effective density of states for holes evaluated at reference temperature T_0
P_ν	...	pressing forces for the carrier ν
P_ν^{ref}	...	reference pressing forces for the carrier ν
P_{L}	...	ratio of the Γ and L valley populations in the multivalley mobility model
Q_s	...	total charge in the device
R	...	net recombination rate
R^{AUG}	...	Auger recombination rate
R_{B}	...	parasitic base resistance
R^{BB}	...	band to band tunneling recombination rate
R_{Con}	...	contact resistance

R_{C}	...	parasitic collector resistance
R^{DIR}	...	direct (radiative) recombination rate
R_{E}	...	parasitic emitter resistance
R_{H}	...	Huang-Rhys factor
R^{II}	...	contribution from impact ionization to the net generation/recombination rate
R_{SM}, R_{DM}	...	metal contribution to R_{S}, R_{D}
R^{SRH}	...	SRH net recombination rate
R_{S}, R_{G}, R_{D}	...	parasitic source, gate, drain resistances
$R_{\mathrm{Ssemi/alloy}}$, $R_{\mathrm{Dsemi/alloy}}$	...	semiconductor or alloy contribution to R_{S}, R_{D}
R_{TH}	...	thermal resistance
R_{bb}	...	base resistance
R_{ds}	...	drain-source resistance
$R_{\mathrm{eff},\nu}$	...	effective (net) recombination rate of carrier ν
R_{gap}	...	contribution to base resistance
R_{gd}	...	gate-drain resistance
R_{gs}	...	gate-source resistance
$R_{\mathrm{TH,glob}}$	...	global thermal resistance
R_{im}	...	resistance describing impact ionization in compact models
R_{maj},R_{min}	...	majority and minority Rydberg energies
R_{n}, R_{p}	...	recombination rate of carrier n,p
R_{pgd}	...	gate leakage resistance drain side
R_{pgs}	...	gate leakage resistance source side
R_{spread}	...	contribution to base resistance
$\mathbf{S}_{\mathrm{L}}$	...	lattice heat flux density
S_{ij}	...	scattering (S-) parameter, i,j=1,2
s_{n}, s_{p}	...	surface recombination velocities for electrons and holes
$\mathbf{S}_{\mathrm{n}}$, $\mathbf{S}_{\mathrm{p}}$	...	electron and hole energy flux density
T_{C}	...	contact temperature
T_{L}	...	local lattice temperature
T_{n}, T_{p}	...	electron and hole temperatures
T_{ν}	...	carrier temperature of carrier ν
$T_{\nu,i}$	...	carrier temperature of type ν in segment i
T_{sub}	...	substrate temperature
U	...	unilateral gain
V_{BE}, V_{CE}, V_{BC}	...	base-emitter, collector-emitter, and collector-base voltages
V_{Con}	...	voltage at a contact
V_{DS}, V_{GD}, V_{GS}	...	drain-source, gate-drain, and gate-source voltages
V_{bar}	...	equivalent potential barrier height

V_{ox}	... voltage drop over the oxide at the polysilicon contact
V_{thr}	... threshold voltage
W_{g}	... gate width
X_{b}	... effective base width
Y_{ij}	... Y-parameter for i,j=1,2
Z_{ij}	... Z-parameter for i,j=1,2
c_{L}	... lattice specific heat
$c_{\mathrm{L,300}}$	... lattice specific heat at T_{L}= 300 K
c_{eff}	... effective capacitance
c_{n}, c_{p}	... heat capacity of electron gas and hole gas
d	... thickness, length
d_{eff}	... effective gate-to-channel separation
d_{ox}	... oxide thickness
d_{tun}	... effective tunneling length
d_{R}	... recess length
f_{T}	... current gain cut-off frequency
$f_{\mathrm{T,ext}}$	... extrinsic current gain cut-off frequency
f	... frequency
f_{c}	... frequency for k = 1
f_{max}	... maximum frequency of oscillation
g_{ds}	... output conductance
$g_{\mathrm{ds,ext}}$	... extrinsic output conductance
g_{m}	... transconductance
g_{mi}	... intrinsic transconductance
g_{mim}	... transconductance due to impact ionization
$g_{\mathrm{m,max}}$	... maximum transconductance
h	... Planck constant
$\hbar$	... reduced Planck constant
k_{B}	... Boltzmann constant
k	... stability factor
l_{Co}	... length contact to recess
l_{g}	... gate length
m_{Γ}	... relative carrier masses in Γ-valley
m_{ν}^{*}	... effective tunneling mass band-to-band model
m_{L}	... relative carrier masses in L valley
m_{0}	... free electron mass
$m_{0,\mathrm{n}}$, $m_{0,\mathrm{p}}$	... relative masses of electrons and holes at 0 K
m_{X}	... relative carrier masses in X valley
m_{n}, m_{p}	... relative masses of electrons and holes
$m_{1,\mathrm{n}}$, $m_{1,\mathrm{p}}$, $m_{2,\mathrm{p}}$	... parameters for the temperature dependence of the relative masses of electrons and holes
m_{nt}	... transversal electron mass

m_{nl}	...	longitudinal electron mass
m_{ph}	...	heavy hole mass
m_{pl}	...	light hole mass
m_t	...	effective tunneling mass heterojunction tunneling
n_0	...	equilibrium concentration at contact
$\mathbf{n}$	...	a normal vector
n	...	electron concentration
n_1	...	auxiliary concentration in the SRH model
n_{c}	...	reference charge
n_i	...	intrinsic electron concentration
n_{s}	...	surface concentration of electrons
n_{sheet}	...	channel sheet charge density
n_{t}	...	auxiliary concentration in the trap-assisted model
n_{T}	...	concentration of occupied traps
p	...	hole concentration
p_0	...	equilibrium concentration
p_1	...	auxiliary concentration in SRH model
p_i	...	intrinsic hole concentration
p_{t}	...	auxiliary concentration in the trap-assisted model
p_{s}	...	surface concentration of holes
q	...	elementary charge
r_s	...	density parameter for the BGN model
t	...	time
u	...	carrier energy relative to the threshold energy
v_{eff}	...	effective carrier velocity
v_{n}, v_{p}	...	velocity for electrons and holes
$v_{\mathrm{em,n}}$, $v_{\mathrm{em,p}}$	...	emission velocities for electrons and holes
$v_{\mathrm{sat,n}}$, $v_{\mathrm{sat,p}}$	...	saturation velocities for electrons and holes
$v_{\mathrm{sat,n,300}}$, $v_{\mathrm{sat,p,300}}$	...	saturation velocities for electrons and holes at 300 K
$v_{\mathrm{sat,n}}^{\mathit{\Gamma}}$, $v_{\mathrm{sat,n}}^{\mathrm{L}}$	...	saturation velocities for electrons in valley $\mathit{\Gamma}$, L
$v_{\mathrm{th,n}}$, $v_{\mathrm{th,p}}$	...	average thermal velocities for electrons and holes
$v_{\mathrm{th,n,300}}$, $v_{\mathrm{th,p,300}}$	...	thermal velocities for electrons and holes at 300 K
w_{n}, w_{p}	...	average electron/hole energy
x	...	material composition
y	...	distance
y^{ref}	...	surface reference distance

1. Introduction

Communication and information systems are subject to rapid and highly sophisticated changes. Currently semiconductor heterostructure devices, such as Heterojunction Bipolar Transistors (HBTs) and High Electron Mobility Transistors (HEMTs), are among the fastest and most advanced high-frequency devices. They satisfy the requirements for low power consumption, medium integration, low cost in large quantities, and high-speed operation capabilities in circuits. In the very high-frequency range, cut-off frequencies up to 500 GHz [557] have been reported on the device level. HEMTs and HBTs are very suitable for high-efficiency power amplifiers at 900 MHz as well as for data rates higher than 100 Gbit/s for long-range communication and thus cover a broad range of applications.

To cope with explosive development costs and the competition of today's semiconductor industry, Technology Computer-Aided Design (TCAD) methodologies are used extensively in development and production. As of 2003, III-V semiconductor HEMT and HBT micrometer and millimeter-wave integrated circuits (MICs and MMICs) are available on six-inch GaAs wafers. SiGe HBT circuits, as part of the CMOS technology on eight-inch wafers, are in volume production. Simulation tools for technology, devices, and circuits reduce expensive technological efforts. This book focuses on the application of simulation software to heterostructure devices with respect to industrial applications. In particular, a detailed discussion of physical modeling for a great variety of materials is presented.

Chapter 2 discusses the status of research regarding the most important electronic heterostructure devices, HBTs and HEMTs. It includes a review of state-of-the-art materials, devices, and driver applications. Device-specific optimization potentials are discussed systematically. Device simulators are reviewed and a discussion of the materials and material systems on which heterostructure devices are based is given.

Chapter 3 is dedicated to the derivation and application of physics-based analytical models, compiled in a comprehensive and carefully systemized collection. It contains models for the lattice, thermal, band-structure, and transport properties of various semiconductor materials, as well as models for important high-field and high-doping effects which occur in the devices. The quality of these models, in terms of their accuracy and simplicity, is decisive for the predictive capabilities

of the simulation tools, which must capture the process as well as the physics of the device.
Chapter 4 provides an overview of RF extraction with the help of TCAD for HEMTs and HBTs and a comparison of the TCAD approaches with simplified analytical approaches.
In Chapter 5, simulation results for several different types of Si-based, GaAs-based, and InP-based HBTs are shown. All examples are based on industrial requirements and the simulation results are compared to experimental data. Two-dimensional and three-dimensional DC and AC simulations of high-power AlGaAs/GaAs and InGaP/GaAs HBTs are provided in very good agreement with measured data in a wide range of application temperatures. Reliability investigations confirm the usefulness of device simulation for practical applications. A methodology for the characterization and optimization of SiGe HBTs in an integrated BiCMOS technology is proposed with emphasis on high-frequency performance and reliability. Simulation results which are in excellent agreement with measured DC and RF data are presented for devices from different manufacturers. A full set of material models for InGaAs/InP HBTs is provided and the high-speed optimization potential for InP-based HBTs is analyzed.
Chapter 6 gives a detailed DC, small-signal, and large-signal analysis of AlGaAs/InGaAs and InAlAs/InGaAs HEMTs with gate lengths down to 100 nm, based on GaAs and InP substrates. Simulation results and experimental data from a large number of industrially available processes are presented. A recent driver application for the HEMT device concept employs high-power amplifiers using the AlGaN/GaN material system on SiC and sapphire substrates. This work offers a complete simulation approach for these devices, which is useful for addressing unsolved device and optimization issues. In Chapter 7, the simulation of novel device concepts, such as double heterostructure GaAsSb/InP HBTs and GaN-based HBTs, is discussed briefly.
This book is of interest for device and circuit designers in semiconductor development and industry. It is strongly recommended for advanced undergraduate and graduate students, for researchers in the field of electrical engineering and solid-state physics, for TCAD users and developers, and for researchers who are looking for practical applications of their scientific work.
The continuous industrial demand for optimized devices and the challenge of analyzing advanced heterostructure devices through numerical simulations have been our highest source of motivation for performing the work described in this book.
We would like to cordially thank several people who made this work possible:
First of all, we would like to thank the editor, Prof. Dr. Siegfried Selberherr, for the unique chance to publish this book in the Springer series and for his careful and patient support.
We express our thanks to the authors of MINIMOS-NT and the colleagues at the Technische Universität Wien, Austria, especially to Prof. Dr. Tibor Grasser and

Prof. Dr. Hans Kosina. We deeply thank Dr. Klaus Dragosits for his irreplaceable support and for his careful proofreading. We acknowledge Dipl.Ing. Stefan Wagner for his work on the S-parameter extraction and Dr. Robert Klima for his work on three-dimensional simulations.
We would like to acknowledge Prof. Dr. Günter Weimann and Dr. Michael Schlechtweg at the Fraunhofer Institut für Angewandte Festkörperphysik (IAF) in Freiburg, Germany, for their continuous support and encouragement. Further we thank the colleagues at Fraunhofer-IAF, especially Dr. Arnulf Leuther, Dr. Rudolf Kiefer, Dr. Klaus Köhler, Dr. Rachid Driad, Dr. Wolfang Bronner, Dipl.Phys. Stefan Müller, and Dr. Friedbert van Raay, for valuable discussions. We also thank Dr. Wolfang Bronner for supplying the SEM image on the cover page.

We thank the following people for their help in the completion of this work: Prof. Dr. Erich Kasper at the Universität Stuttgart, Germany, for helpful discussions and providing SiGe HBT examples; Dipl.Ing. Georg Röhrer at Austriamicrosystems AG in Unterpremstätten, Austria, for the supply of several SiGe HBT examples; Dr. Rüdiger Schultheis from Infineon Technologies in Munich, Germany, for continuous support and for supplying several sets of measurement data and examples on GaAs HBTs; Dr. Walther Kellner from Infineon Technologies in Munich, Germany, for his dedication to device physics in an industrial context.
We deeply thank Joyce Flynt Visne for her very thorough and dedicated proofreading of the manuscript.

We would like to acknowledge the Springer editor team and would like to thank Edwin Schwarz at Springer Verlag for his kind suppport.

Those to whom we dedicate this book, knowing us as they do, understand that their names do not appear here.

Freiburg i.Br. and Wien, November 2003

2. State-of-the-Art of Materials, Device Modeling, and RF Devices

2.1 State-of-the-Art of Heterostructure RF Device Modeling

2.1.1 Semiconductor Material Aspects

Based on their electrical properties, solids can be classified as conductors, insulators, or semiconductors. Amorphous solids have little or no regular geometric arrangement of their atoms in space and therefore cannot be easily studied. Crystalline solids have a perfect periodic arrangement of atoms, which allows them to be easily analyzed. Polycrystalline solids have atomic arrangements between these two extremes. Semiconductor materials are nearly perfect crystalline solids with a small amount of imperfections, such as impurity atoms, lattice vacancies, or dislocations, which can be intentionally introduced to alter their electrical characteristics [305].

Fig. 2.1 is a summary of the chemical elements involved in the formation of semiconductors.

Semiconductors can be elemental, such as silicon, germanium, or another chemical element from Group IV; or they can be compounds, usually a combination between elements from Group III and Group V or from Group II and Group VI. Examples for such combinations are the binary compounds $Ga^{III}As^{V}$ and $Zn^{II}S^{VI}$. There are also several combinations of practical importance which involve two or more elements from the same chemical group. Such alloy semiconductors can be binary (e.g. $Si^{IV}Ge^{IV}$), ternary (e.g. $Al^{III}Ga^{III}As^{V}$), quaternary (e.g. $In^{III}Ga^{III}As^{V}P^{V}$), and even pentanary ($Ga^{III}In^{III}P^{V}Sb^{V}As^{V}$) materials.

New materials or material combinations have always stimulated the development of both silicon and III-V semiconductor devices. Table 2.1 gives an overview of the materials and the combinations suitable for the development of today's heterojunction RF devices.

Strained SiGe is already an established material for Si-based heterostructure devices. Carbon is often added in small quantities ($< 3\%$) to SiGe for technological reasons [134]. Arsenides from the III-V chemical group, such as GaAs, $Al_xGa_{1-x}As$, and strained InGaAs, are well established in GaAs-based devices. Other arsenides, such as $In_{0.53}Ga_{0.47}As$, $In_{0.52}Al_{0.48}As$, and quaternary

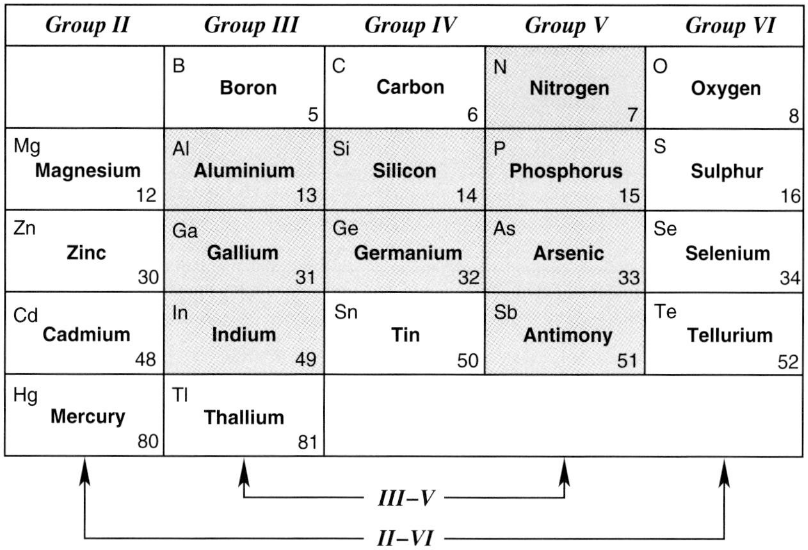

Fig. 2.1. Part of the Periodic Table of Elements: The elements involved in the formation of semiconductors are given. The elements considered in this work are highlighted by a gray background.

$(Al_xGa_{1-x})_{0.47}In_{0.53}As$, are common materials for devices on InP substrates. Phosphides, such as InGaP, AlGaP, and $(Al_xGa_{1-x})_{0.5}In_{0.5}P$, are used for example in GaAs HBTs. The quaternary $In_xGa_{1-x}As_yP_{1-y}$ can also match the lattice constant of InP. Thus it is applied in various electronic and optoelectronic InP-based devices. Nitrides with a wurtzite (hexagonal) crystal structure, such as GaN, AlN, and most recently InN, are very promising materials for wide bandgap devices. Adding minor amounts of nitrogen ($< 3\%$) to InGaAs yields a low bandgap quaternary material lattice-matched to GaAs in HBTs and optoelectronic devices. Antimonides have been restricted to optoelectronic applications for a long time. However, GaAsSb and AlInAsSb have made a significant breakthrough for HBTs and extremely high-speed HEMTs respectively, both grown on InP substrates.

A strong feature of a generic device simulator is its capability of accounting for various semiconductor materials, such as III-V binary, ternary, and quaternary compounds, and Si, SiGe, and SiGe:C. A large number of published theoretical and experimental reports have been reviewed to include the physical parameters for Si and Ge from chemical Group IV and GaAs, AlAs, InAs, InP, GaP, GaSb, AlSb, InSb, GaN, AlN, and InN, which are binary compounds from the III-V chemical group. All of these materials are referred to as *basic materials* later in this work. The combination of two III-V binary materials results in a ternary or a quaternary material. SiGe, a combination of Si and Ge, and the ternary

Table 2.1. Materials and material combinations under consideration for heterojunction RF devices: ^{C}-cubic, ^{W}-wurtzite lattice structure

Elements	Binary	Ternary	Quaternary	Substrate
Si,Ge,(C)	SiGe,SiC^{C}	SiGe:C		Si
Al,Ga,In,As	AlAs,GaAs,InAs	AlGaAs InGaAs InAlAs	AlGaInAs	GaAs,InP
(Al),Ga,In,P	(AlP),GaP,InP	InGaP (InAlP) (AlGaP)	(AlInGaP)	GaAs
Ga,In,As,P	GaAs,InAs,GaP,InP	InGaP InGaAs (InAsP) (GaAsP)	InGaAsP	GaAs,InP
Al,Ga,In,N	GaN,AlN,InN	AlGaN^{W} InGaN^{W} (AlInN)	(AlInGaN)	SiC,sapphire,Si (GaN,AlN)
In,Ga,As,N	GaAs,InAs,GaN,InN	InGaAs InGaN^{C} (InAsN) (GaAsN)	InGaAs:N	GaAs
Al,Ga,As,Sb	AlAs,GaAs,AlSb,GaSb	GaAsSb (AlGaAs) AlAsSb AlGaSb	AlGaAsSb	InP
Al,In,As,Sb	AlAs,InAs,AlSb,InSb	AlAsSb InAsSb (InAlSb) InAlAs	AlInAsSb	InP,GaSb

III-V materials, a combination of the respective binary materials, are named *alloy materials* later in this work.
An attempt has been made to allow the user to choose arbitrary material composition for the *alloy materials*. This not only gives the designer a good degree of freedom for the choice of material, but also allows a direct comparison between various devices, such as AlGaAs/GaAs, InGaP/GaAs, InP/InGaAs, InAlAs/InGaAs, and SiGe/Si HBTs, or AlGaAs/InGaAs, InAlAs/InGaAs, and AlGaN/GaN HEMTs. However, due to limited experimental data on some compound materials, such as InGaP, InAlAs, InAsP, and GaAsP, one has to consider

interpolation schemes as the only available option to model the variation of some parameters in a continuous range of material compositions. In these cases, various interpolation schemes are studied to find the best fit to the sometimes limited reported data in the literature.

2.1.2 Modeling Concept

For all models in this work, the general approach is to employ universal models, i.e. the same functional form to be used for all materials but with different parameter sets. In models for *alloy materials*, the respective models for the two *basic materials* are employed first and then combined as a function of the material composition x. Additionally, full consistency between the *alloy materials* and the *basic materials* is obtained by having all the models for *alloy materials* inherit their model parameters from the models for *basic materials.* Although it is arbitrary which of the two *basic materials* will correspond to a material composition $x = 0$ and which to a mole fraction $x = 1$, a choice has to be made, e.g. InGaAs is used for $In_xGa_{1-x}As$ and not for $In_{1-x}Ga_xAs$. The *alloy materials* ($A_{1-x}B_x$) formed by the *basic materials* (A and B) are summarized in Table 2.2.

Table 2.2. Notation of the material composition x in *alloy materials*

	A	B	$A_{1-x}B_x$
SiGe	Si	Ge	$Si_{1-x}Ge_x$
AlGaAs	GaAs	AlAs	$Al_xGa_{1-x}As$
InGaAs	GaAs	InAs	$In_xGa_{1-x}As$
InAlAs	AlAs	InAs	$In_xAl_{1-x}As$
InAsP	InP	InAs	$InAs_xP_{1-x}$
GaAsP	GaAs	GaP	$GaAs_{1-x}P_x$
InGaP	InP	GaP	$Ga_xIn_{1-x}P$
AlGaN	GaN	AlN	$Al_xGa_{1-x}N$
InGaN	GaN	InN	$In_xGa_{1-x}N$
GaAsSb	GaAs	GaSb	$GaAs_{1-x}Sb_x$
AlAsSb	AlAs	AlSb	$AlAs_{1-x}Sb_x$
AlGaSb	GaSb	AlSb	$Ga_{1-x}Al_xSb$
InAlSb	InSb	AlSb	$In_{1-x}Al_xSb$
InGaSb	GaSb	InSb	$Ga_{1-x}In_xSb$

Quaternary Materials. Quaternary materials have gained attention for advanced electronic devices, mostly for bandgap engineering purposes (see Table 2.1). General methods for deriving parameters of quaternary materials have been summarized by several authors [3, 275, 532]. To avoid the inappropriate effort of modeling quaternary materials in the complete range of material compositions x and

y, only those which are lattice-matched to common substrate materials, such as GaAs or InP, are considered. They are modeled as a combination $A_{1-x}B_x$ of two lattice-matched materials A and B, where A is a ternary and B is a binary or ternary material [532] (see Table 2.3). The parameters used for these ternaries are consistent with those modeled from the respective binaries.

Table 2.3. Notation of the material composition x in quaternary *alloy materials*

	A	B	$A_{1-x}B_x$
InGaAsP/InP	$In_{0.53}Ga_{0.47}As$	InP	$(InP)_{1-x}(In_{0.53}Ga_{0.47}As)_x$
AlGaInAs/InP	$In_{0.52}Al_{0.48}As$	$In_{0.53}Ga_{0.47}As$	$(Al_xGa_{1-x})_{0.47}In_{0.53}As$
AlGaInP/GaAs	$In_{0.49}Ga_{0.51}P$	$In_{0.49}Al_{0.51}P$	$(Al_xGa_{1-x})_{0.51}In_{0.49}P$
InGaAsP/GaAs	$In_{0.49}Ga_{0.51}P$	GaAs	$(GaAs)_{1-x}(In_{0.49}Ga_{0.51}P)_x$

2.1.3 Interpolation Schemes

Material parameters of Si, Ge, GaAs, InP, GaP, InAs, and InGaAs, as well as SiGe and AlGaAs, have received considerable attention in the past, and much experimental data and a lot of theoretical studies for these parameters can be found in the literature. On the other hand, the band structure and transport-related parameters of many III-V ternary and nearly all quaternary materials has been the topic of few or no experimental/theoretical publications. These facts necessitate the use of some interpolation scheme, essentially based on known values of the physical parameters for the related *basic materials* and *alloy materials.* In cases where experimental data scatter, the most consistent or most recent published data is adopted. In cases where experimental data are inconsistent or missing, Monte Carlo simulation data are used. Although the interpolation scheme is still open to experimental or Monte Carlo verifications, it provides more useful and reliable material parameters for numerical device simulation over the entire range of alloy compositions.

For many parameters, such as various lattice parameters, a linear interpolation is sufficient. Some parameters, like the electronic bandgap, exhibit a strong nonlinear dependence on the material composition. This arises from the effects of alloy disorder. In such cases, a quadratic interpolation is used and a so-called bowing parameter is introduced. For other parameters, such as carrier mobility, a linear interpolation of the inverse values is used, i.e. the so-called Mathiessen's rule. Finally, there are parameters, such as thermal conductivity, for which none of the interpolation schemes mentioned so far is sufficient, and a quadratic interpolation of the inverse values together with an inverse bowing factor is proposed.

The bandgap bowing parameters of InAsP and GaAsP are believed to be much smaller than those of InGaP and InGaAs [1, 60]. Similarly, for parameters where data are lacking, no bowing factors are assumed.
Most of the data available for nitrides are limited to GaN and to $Al_xGa_{1-x}N$ with $x \leq 35\%$, i.e. to materials suitable for electronic devices where the experimental base allows proper extraction. For InN and its compounds, the material database is still relatively poor, if available, and exists only for indium contents $\leq 10\%$. There is a relatively good database of material parameters for antimonides regarding the band structure and the thermal properties which have been of interest for optoelectronics. However, little is known about the carrier transport properties of this material system, especially about high-field and high-energy effects.
In Chapter 3, the choice of interpolation formula largely depends on factors such as required accuracy, the physical nature of the parameter, and available experimental or Monte Carlo simulation data. For example, bandgap energy is the most critical parameter in device modeling, and a slight variation in this parameter can significantly affect the terminal device characteristics.

2.1.4 The Effect of Strain and Anisotropy

A general aim when growing a device structure is to avoid lattice mismatch between the substrate and the epitaxial layers. A main concern is that the strain originating from such a mismatch can relax and lead to misfit dislocations and even to amorphous structures. Therefore only materials whose lattice constants match those of the typical substrate materials GaAs or InP are of interest. Such materials are $Al_xGa_{1-x}As$ in the whole composition range $0 \leq x \leq 1$ and $Ga_{0.51}In_{0.49}P$, all grown on GaAs, and $In_{0.53}Ga_{0.47}As$ and $In_{0.52}Al_{0.48}As$, both grown on InP. The quaternary GaInAsP can also match either GaAs or InP for certain material compositions. However, if x is sufficiently low or the layer is sufficiently thin, the substrate lattice constant and the strain can be preserved. The result can be alteration of the bandgap, which can sometimes be beneficial, e.g. for SiGe grown on Si. The effect on specific material models will be discussed in the appropriate sections.
For device modeling, strain and polarization effects can be introduced as effective quantities into parameters and models of choice, such as the mobility models. For the pseudomorphic HEMT grown on GaAs, strain is present due to the pseudomorphic growth; however the impact is mostly directly included in the model parameter calibration. For some devices, however, strain is a very important factor for achieving device functioning and optimization. This includes SiGe HBTs and FETs [243, 245]. The parameters affected are dielectric constant, bandgap energy, carrier effective masses, carrier mobilities, and saturation velocities. In the device simulators DESSIS [227] and MEDICI [336] the impact of anisotropy on the given quantities is modeled by introducing an anisotropy factor for those quantities.

Spontaneously and piezoelectrically induced charges form a major source of carriers in GaN-based heterostructure devices. These induced charges can by no means be neglected with respect to the Poisson equation, since their contribution can be an order of magnitude higher than the contribution of conventional doping. They are typically accounted for in the simulation by introduction of surface charges at the heterointerfaces.

2.1.5 Numerical Modeling

This subsection presents the state-of-the-art in simulation for industrial heterostructure devices based on SiGe/Si and III-V compound semiconductors. It includes a detailed comparison of device simulators and current transport models to be used and addresses critical modeling issues.
To cope with explosive development costs and strong competition in the semiconductor industry today, Technology Computer-Aided Design (TCAD) methodologies are extensively used in development and production. Several questions during device fabrication, such as performance optimization and process control, can be addressed by simulation. The choice of a given simulation tool or a combination of tools depends to a large extent on the complexity of the particular task, on the desired accuracy of the problem solution, and on the available human, computer, and time resources.
Optimization of geometry, doping, materials, and material compositions targets high output power, high breakdown voltage, high speed, low leakage, low noise, and low power consumption. This is a challenging task that can be significantly supported by device simulation.

Critical Modeling Issues for Heterostructure Devices. There are several problems, which are specific for the modeling and simulation of heterostructure devices.
A generic device simulator must not only be capable to account for various semiconductor materials (see the previous subsection) but also for different complex geometrical structures and material sequences in multiple dimensions.
The physical properties of SiGe and III-V compounds must be modeled for wide ranges of material compositions, temperatures, doping concentrations, etc. The model parameters have to be verified against several independent HEMT and HBT technologies to obtain a concise set used for all simulations. Reviewing simulation of HBTs and submicron HFETs with gate lengths down to 100 nm used for millimeter-wave devices, solutions of energy transport equations are necessary to account for non-local effects, such as velocity overshoot.
Modeling of strained SiGe is not a trivial task, since special attention has to be focused on the stress-induced change of the bandgap as a function of germanium content [133]. This effect must be separated from doping-induced bandgap narrowing, which in turn depends on the semiconductor material composition, the

doping concentration, and the lattice temperature. As the minority carrier mobility is of considerable importance for bipolar transistors, a distinction between majority and minority electron mobilities is required.
Heterointerface modeling is a key issue for devices which include abrupt junctions. Thermionic emission and field emission effects critically determine the current transport parallel and perpendicular to the heterointerfaces.
Another critical issue for recessed HFETs and for III-V HBTs is the description of the semiconductor/insulator interfaces, especially with respect to the treatment of the interfaces during the manufacturing process. Fermi-level pinning prevails for typical barrier materials such as AlGaAs or InAlAs and for ledge materials such as InGaP.
All the important generation/recombination effects, such as Shockley-Read-Hall recombination, surface recombination, direct recombination, Auger recombination, band-to-band tunneling processes, and impact ionization, must be taken into account. III-V semiconductors and SiGe are known to have a reduced heat conductivity in comparison with silicon (see Section 3.2). Thus self-heating effects must be accounted for.
Advanced device simulation allows a precise physics-based extraction of small-signal parameters [416, 534]. Measured bias-dependent S-parameters serve as a valuable source of information when compared at different bias points to simulated S-parameters from a device simulator. This procedure reflects the full RF information contained in the S-parameters and allows process control beyond the comparison of DC quantities.
A physics-based large-signal extraction is critical, due to the typical problems of compact large-signal models. These are: the accurate treatment of parasitic elements, e.g. inductances for multi-finger devices; the thermal problem, which is generically three-dimensional; and frequency dispersion due to fast traps in III-V semiconductors.

Three-Dimensional Device Simulation: Problems and Demands. During the past two decades, several three-dimensional device simulators have been developed [227, 465, 507]. Typical simulation examples have been studies on narrow channel effects in MOS devices or on leakage currents in memory cells. Other examples, both for III-V and silicon devices, are electro-thermal or thermal simulations which are by nature three-dimensional problems with regards to the boundary conditions [124]. The development of a three-dimensional device simulator is a challenging task. The demands on state-of-the-art device simulators are high in many respects:

- Choice of iterative solvers (direct solvers cannot be used for three-dimensional simulations [27]); capability to calculate systems with complex numbers for small-signal AC analysis
- Choice of iteration schemes for the solving process to speed up the simulation and improve the convergence
- Handling of three-dimensional data

- Generation of various three-dimensional simulation grids
- Generation of three-dimensional device geometries of arbitrary shape and complexity
- Visualization: on the surface of the device, by two-dimensional cuts through the device, or on discrete values on a specific grid
- Control system (input-deck) needed for such complex simulators

Three-dimensional device simulation is mandatory for investigations on real three-dimensional structures which cannot be described by two-dimensional cuts. The high computational cost is one of the major issues which, together with the challenges described above, is the reason that more than 95% of all device simulations today are two-dimensional.

Heterostructure Device Simulators. Numerical simulations have gained an important position in the silicon industry. Process, device, and interconnect simulation tools form a continuous row of tools ranging from material and device analysis to chip design and even system design. These capabilities have been extended to the SiGe bipolar industry with impressive results [239, 382]. III-V device simulation mainly focuses on device and circuit aspects [381]. Due to the extensive number of non-standardized technological process steps, heterostructure device simulation is mainly used for process control and inverse modeling of geometry or doping.
The continuously increasing computational power of computer systems allows the use of TCAD tools on a very large scale. Several commercial device simulators (such as APSYS [19], ATLAS [465], BIPOLE3 [42], DESSIS [227], G-PISCES [160], and MEDICI [336]), company-developed simulators (such as FIELDAY [80] and NEMO [360]), and university-developed simulators (like PISCES [403], MINIMOS 6 [146], FLOODS [151], GALENE [244, 337], DEVICE [123], nextnano3 [363], and PROSA [171]) have been successfully employed for device engineering applications. These simulators differ considerably in dimensionality (one, quasi-two, two, quasi-three, or three), in choice of carrier transport model (drift-diffusion, energy-transport, or Monte Carlo (MC) statistical solution of the Boltzmann transport equation), and in the capability of including electrothermal effects. The drift-diffusion transport model [448, 198] is by now the most popular model used for device simulation. With down-scaling of the feature sizes, non-local effects become more pronounced and must be accounted for by applying an energy-transport model or a hydrodynamic transport model [187]. During the last two decades, Monte Carlo methods for solving the Boltzmann transport equation have been developed [231, 213, 270, 357] and applied for device simulation [271, 138, 292]. However, reduction of computational time is still an issue, and therefore MC device simulation is still not feasible for industrial application on a daily basis. An approach to preserve accuracy at lower computational cost is to calibrate lower order transport parameters to MC simulation data. The MC simulation approach of Fischetti and Laux [147] has been widely used with some minor modifications

to determine high-field transport material properties, although full band MC codes are also available [150, 249].
Quantum mechanical effects gain increased importance with the scaling of the feature size. These are often neglected in device simulation or only accounted for by simple analytical models for quantum corrections [129, 200], since solving the Schrödinger equation [272] or the Wigner equation [156, 356] is extremely expensive in terms of computational resources [269].
Most device simulators focus on silicon devices, and the model parameters for SiGe are often simply inherited from the parameters for silicon. The database available for properties of III-V semiconductors has been limited for a long time, due to the large number of materials and material compositions. Moreover, a thorough approach of modeling them has been lacking. The quality of the physical models can be questioned, since modeling of the properties of AlGaAs, InGaAs, InAlAs is often restricted to slight modifications of the GaAs material properties. InGaP and other phosphides are required for advanced device modeling, together with new material systems, such as the GaN or the GaSb systems, which have entered the III-V world with impressive device results. A severe problem is the limited feedback from statistically-based technological process development to device simulator development. Therefore published application examples will also be mentioned here in the course of the discussion.
Critical issues concerning simulation of heterostructures, such as interface modeling at heterojunctions, silicon/poly-silicon interfaces for SiGe devices and insulator surfaces for III-V devices are frequently not considered. The importance of high-energy and high-field effects, such as carrier energy relaxation, impact ionization, and self-heating, is sometimes underestimated.
The two-dimensional device simulator PISCES [29, 403] developed at Stanford University incorporates modeling capabilities for Si, GaAs, and InP-based devices. One of its versions, PISCES-HB, includes harmonic balance for large-signal simulation. It has been applied to LDMOS devices [429] and to MESFETs [435]. Another version, G-PISCES from Gateway Modeling [160], has been extended by a full set of III-V models. Examples of MESFETs, HEMTs, and HBTs for several material systems, e.g. InAlAs/InGaAs, AlGaAs/InGaAs, AlGaAs/GaAs, and InGaP/GaAs, are demonstrated. However, in comparison with the original version of PISCES, this simulator lacks an energy-transport model, which is necessary to model high-field effects. G-PISCES also demonstrated the simulation of AlGaN/GaN HEMTs [17].
The device simulator MEDICI from Synopsis [336, 503], which is also based on PISCES, offers simulation capabilities for SiGe/Si HBTs and AlGaAs/InGaAs GaAs HEMTs. It has been used for the simulation of AlGaAs/GaAs HBTs [303] and for the evaluation of properties of GaN HBTs [409]. Advantages of this simulator are hydrodynamic simulation capabilities and the rigorous approach to generation/recombination processes. In addition, it includes a module which con-

siders anisotropic material properties in SiGe. In the newest version, an interface model including carrier tunneling is included in the III-V simulation module.
The two- and three-dimensional device simulator DESSIS from ISE [227] has demonstrated a rigorous approach to semiconductor physics including extensive trap modeling and a variety of mobility models. The capabilities to model Si and SiC are extended by a heterojunction framework to III-V materials [315]. Interface tunneling is included in a TFE model. The density-gradient method is used to model quantum effects in heterostructure devices [316]. At the quantum level, a one-dimensional Schrödinger-Poisson solver NEMO [360], based on non-equilibrium Green's functions, is offered, e.g. for GaAs resonant-tunneling diodes [46] and for sub-0.1 μm SiGe structures.
POSES [160] from Gateway Modeling is another Schrödinger-Poisson solver proposed for process control by charge analysis in HEMTs. In the program SIMBA [401], a link between a one-dimensional Schrödinger solver and a two-dimensional Poisson solver is demonstrated. SIMBA also provides drift-diffusion transport simulation of GaN HEMTs [481]. Recently the three-dimensional version of SIMBA has been used for thermal optimization of GaN HEMT layouts [124].
Quasi-two-dimensional approaches using a simplified one-dimensional current equation are demonstrated by several simulators, including BIPOLE3 from BIPSIM [42], which features good models for poly-silicon. A similar approach which couples a full hydrodynamic transport model with a Schrödinger solver has been developed at the University of Leeds [347, 348]. This computation time effective approach has been verified against MC simulations for a few examples for gate lengths down to 50 nm [347]. Fast Blaze from Silvaco is a commercial tool based on the code from Leeds. Simulations of S-parameters of AlGaAs/GaAs and pseudomorphic AlGaAs/InGaAs/GaAs HEMTs have been presented.
A software interface between the device model and the compact Root large-signal model within the Microwave Design System (MDS and ADS) has been offered by Agilent. Fast Blaze can be combined with the Advanced Design System (ADS) and has an interface with the microwave circuit simulator. An extraction with subsequent multitone excitation calculations have been presented in [240].
The two- and three-dimensional simulator ATLAS [465] from Silvaco has also claimed the simulation of AlGaAs/GaAs and AlGaAs/InGaAs/GaAs PHEMTs. Simulations of SiGe HBTs have also been announced, based on the simulator PROSA, developed at the University of Ilmenau [171]. However, the latter lacks heterointerface modeling. Several good optimization results for SiGe HBTs have been achieved with another university-developed simulator, SCORPIO [420].
We tried to address most of the critical modeling issues for heterostructure devices in the two- and three-dimensional simulator device simulator MINIMOS-NT [41], which we used for preparing all simulation examples in this book. Table 2.4 summarizes features of the device simulators discussed.

Table 2.4. Comparison of different device simulators[1]

Simulator	Dimension	Model	Features
POSES	1D		Schrödinger-Poisson solver
NEMO	1D		Schrödinger-Poisson solver
BIPOLE3	quasi-2D	DD	poly-silicon
Leeds	quasi-2D	HD	Schrödinger equation, thermal model
Fast Blaze	quasi-2D	HD	Schrödinger equation
ATLAS	2D	DD,ET	TE heterojunction model
APSYS	2D	HD	optical, interfaces
GALENE	2D	DD,HD	rigorous transport modeling
PISCES	2D	DD,ET	III-V models, poly-silicon
PISCES-HB	2D	DD	harmonic balance
G-PISCES	2D	DD	full set III-V models
MEDICI	2D	DD,HD	anisotropic properties, TFE model
SIMBA	2D,3D	DD	Schrödinger equation
FIELDAY	2D,3D	DD	electrothermal
MINIMOS-NT	2D,3D	DD,HD	see Chapter 3
DESSIS	2D,3D	DD,HD	trap modeling, TFE model

[1] (DD): drift-diffusion model; (ET): energy transport model;(HD): hydrodynamic model; (TE): thermionic emission; (TFE): thermionic field emission/interface tunneling

2.1.6 Mixed-Mode Simulation

Analytical device modeling of complex structures can be quite challenging, since the number of parameters to be extracted rises dramatically for advanced structures. Even compact models sometimes need about a hundred model calibration parameters to properly describe the physical effects in the device [458].
On the other hand, analytical circuit simulation is an established standard in industry. There are several approaches to linking material and device analysis to integrated circuit design [182]:

- Extraction of compact model parameters from numerical device simulation
- Coupled simulation between a numerical device simulator and a circuit simulator which includes lumped passive components such as inductances, capacitances, and resistors [139, 428]
- Numerical simulation with fully integrated circuit capabilities [335]

The third approach has been introduced into various numerical simulators, such as DESSIS and MINIMOS-NT [41, 227]. It has the advantage over the second approach that it avoids the communication between two different simulators. Both the third and the second approaches represent mixed-mode device/circuit simulation. Their strength over the first approach is that they offer a direct relation between the device physics and the circuit behavior. In addition, mixed-mode simulation is especially valuable for calibration and optimization of technologies for which the compact models fail to describe the electrical device behavior [379]. Thus mixed-mode simulation is capable of predictive evaluation of the performance of new technologies using test circuits such as ring oscillators, inverter chains, and others. It allows accurate steady-state or transient self-consistent simulations of the electrical/thermal behavior of novel semiconductor devices under operating conditions given by a circuit [185, 186].

2.2 State-of-the-Art of Heterostructure Devices and Optimization Potentials

Heterostructure devices are common contenders in the world of high-frequency devices up to at least 500 GHz. A heterostructure is an abrupt semiconductor/semiconductor transition in one direction. This abrupt change of the semiconductor material on a length not exceeding a few angstroms in one direction in space is the principle design feature of semiconductor heterostructure devices. The abrupt change of the material leads to several effects, most importantly a change in the bandgap energy and a change in the transport properties. A number of desirable effects for semiconductor devices can be achieved, such as carrier confinement in a restricted volume and the spatial separation of electrons and holes.

The idea of a heterostructure can be pursued in more than one direction. Very interesting devices using two-dimensional quantization are under industrial development, for example quantum dot lasers in optoelectronics. However, for electronic devices, the so-called low-dimensional electron and hole gases (one- or zero-dimensional) are still in a state of development, but they will enable further technological revolutions in the future. The remarkable progress of two-dimensional carrier gases within the last thirty years gives rise to this assumption. The development led from a highly sophisticated test sample in basic research to a modern consumer good for a few cents value.

2.2.1 Heterojunction Bipolar Transistors (HBTs)

Heterojunction Bipolar Transistors mainly utilize vertical carrier flows perpendicularly through heterointerfaces. The carrier confinement resulting from the introduction of heterojunctions allows separation of the major concentrations of electrons and holes and avoids parasitic currents (base currents) and unintended carrier movement at high operational voltages.

Silicon Germanium (SiGe) HBTs. SiGe HBTs are very promising RF devices stretching the intrinsic advantages of the CMOS silicon world into RF electronics. SiGe HBTs increasingly challenge the III-V devices in the highest available frequency ranges.

Table 2.5. Selected examples of process generations of Si BJTs and SiGe HBTs

Process	Year	Emitter [μm]	BV_{CE0} [V]	f_T [GHz]	f_{max} [GHz]	Reference
Atmel SiGe 1	1998	0.8	6/3	30/50	50/50	[33]
Atmel SiGe 2	2001	0.5	7/4/2.5	30/50/80	50/70/90	[33]
AMS BYS	1999	0.8	4.5	35	40	[437]
AMS S35	2002	0.4	5.5/2.7	35/60	50/70	[242]
Hitachi SiGe	1999	0.2	2.0	90	107	[543]
Hitachi SiGe	2000	0.2	2.5	76	180	[542]
Hitachi SiGe	2001	0.2	2.3	124	174	[369]
Hitachi SiGe	2002	0.2	2.3	122	178	[206]
IBM SiGe	1997	0.5	3.35	47	65	[7]
IBM SiGe	1999	0.18	2.7	90	90	[155]
IBM SiGe	2002	0.18	1.7	207	285	[233]
IBM SiGe	2002	0.12	1.4	350	170	[424]
Infineon Si BJT	1999	0.2	2.7	52	65	[48]
Infineon SiGe	2000	0.2	2.5	85	128	[49]
Infineon SiGe	2001	0.18	2.3	106	145	[51]
Infineon SiGe	2002	0.2	1.9	155	167	[50]

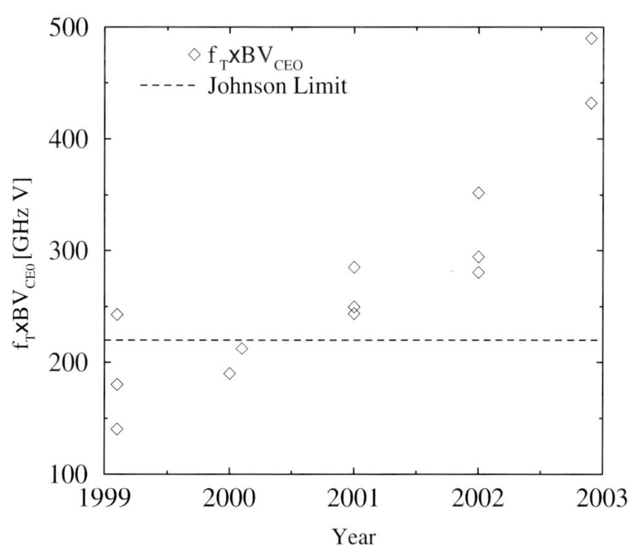

Fig. 2.2. Cut-off frequency×breakdown voltage product $f_T \times BV_{CE0}$ over time for Si BJTs and SiGe HBTs.

SiGe devices have seen a major development to an industrially mature technology in the last five years, further extending the potential of the well-established Si-bipolar RF technology towards higher frequencies and higher breakdown voltages. A strong argument in favor of these devices is their compatibility with the standard CMOS process flow, where sophisticated multilayer metalization layers and interconnects are available [7, 242]. Major contributions come from Hitachi [369], Infineon [51], and IBM [155] in processes summarized in Table 2.5.
Fig. 2.2 shows the development of the Si Bipolar Junction Transistor (BJT), the predecessor of the HBT, and SiGe HBTs over time. The constant increase in the available cut-off frequencies results in higher $f_T \times BV_{CE0}$, which is the product of current gain cut-off frequency and breakdown voltage. A limit of 170 GHz·V has been predicted by Johnson for Si BJTs [238]. The solid line in Fig. 2.2 represents the Johnson limit of $f_T \times BV_{CE0} = 220$ GHz·V [238] prevailing for SiGe HBTs in the last few years. Recent reports have demonstrated that even this limit can be overcome by advanced device technologies [50, 211, 233, 541]. The main optimization potentials of SiGe HBTs through device simulation include:

- Increase of the breakdown voltage for a given speed of operation by optimization of the implant profiling and by optimization of the insulation
- Vertical doping profile optimization towards speed and breakdown in combination with process simulation tools, as performed in [239]
- Optimization of vertical layers, i.e. reduction of base thickness without enhanced minority carrier tunneling [33]
- Optimization of horizontal doping profiles to reduce the effective emitter size in order to increase speed
- Reliability improvements for increased current densities necessary to increase device speed, such as reduction of the emitter resistance
- Reduction of parasitic elements by scaling the feature size
- Maintenance of compatibility of the SiGe HBT process module to the standard CMOS technology process flow

Based on the optimization results already achieved and documented in this section, SiGe HBTs are significantly challenging the III-V devices in RF applications, especially in low-noise and low-power, high-speed applications.

Gallium Arsenide (GaAs) HBTs. GaAs HBTs are the most popular GaAs devices, due to their use in modern cellular phone handsets. Their major advantages are the very low off-state power consumption (a few μW) and the high current amplification for low battery-driven voltage of typically 3 V. AlGaAs and InGaP are typical emitter materials [141, 387] used in combinations with GaAs bases. GaAs HBTs have been applied for 40 Gbit/s data transmission, e.g. by Hitachi [371]. Major device design issues are:

- Process-oriented design of base-collector geometries
- Increase in current densities for reliable power operation
- Improvement of power consumption by reduction of emitter resistances

- Design of the base p-type ohmic contact
- Thermal management in multi-emitter devices including emitter airbridges and substrate thinning
- Improvement of reliability with respect to doping, semiconductor/insulator interfaces, and self-heating

Major contributions come from RF Microdevices [166], based on the long experience of TRW [487]. Further contributions come from UMS [45, 281] and WIN Semiconductors [540].

Indium Phosphide (InP) HBTs. InP HBTs are most interesting for oscillator applications because of the low phase noise properties. Due to a combination of high speed and high breakdown voltages they are suitable for high-speed digital applications up to at least 100 Gbit/s for long distance communication. The principle reasons for their application include the high current amplification at low current levels, which normally is addressed as low-power capability, and the ultimate speed available in emitter-coupled logic (ECL) circuits.
Single heterojunction bipolar transistors (SHBTs) on InP have been developed by HRL for high-speed applications resulting in frequency static dividers up to 72.8 GHz [476]. Major contributions also come from TRW [260]. A low-power production process is available from Vitesse [552]. To overcome the speed×breakdown limitation of the InP SBHTs, a second heterojunction is introduced. This allows the separation of speed and breakdown considerations in the so-called double heterojunction bipolar transistors (DHBTs).
Optimization potentials using TCAD device simulation include:

- Reduction of the base-collector capacitance C_{bc}
- Reduction of base push-out effects for various bias conditions (SHBT)
- Optimization of the collector doping to increase $f_{\mathrm{T}} \times BV_{\mathrm{CE0}}$
- Reduction of the emitter access resistance to improve low-power, high-speed capabilities
- Design of the base-collector heterojunction in DHBTs by means of launcher structures or quaternary materials to avoid collector current blocking

InP DHBTs have seen a dramatic increase of attention in the recent past for two reasons: InP SHBTs fail to supply breakdown voltages suitable for many high-speed applications; InP HEMTs, the second contender, similarly deliver the high speed, but are plagued by gate current issues due to reduced gate geometries and aspect ratio considerations.
Major contributions of recent InP DHBT development come from Lucent [560], Alcatel [175, 248], NTT [224], and NEC [159]. Static frequency dividers operating beyond 100 GHz have been reported by Hughes in [345]. Fig. 2.3 shows the cut-off frequencies f_{T} and f_{max} and the product $f_{\mathrm{T}} \times BV_{\mathrm{CE0}}$. The genuine advantage of the III-V devices in terms of speed and breakdown voltage can be observed, since the product $f_{\mathrm{T}} \times BV_{\mathrm{CE0}}$ for high-speed III-V devices exceeds the Si-based values given in Fig. 2.2 by a factor of 4 to 5, allowing larger device geometries

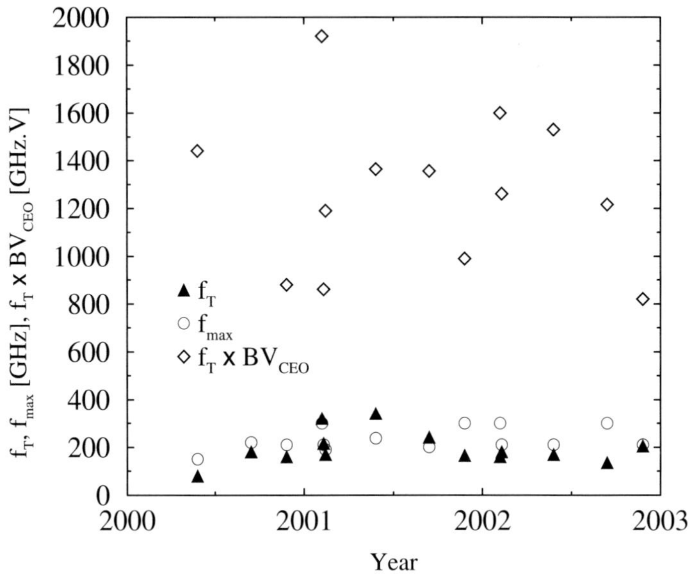

Fig. 2.3. Cut-off frequencies, f_T and f_{max}, and the $f_T \times BV_{CE0}$ product over time for InP DHBTs.

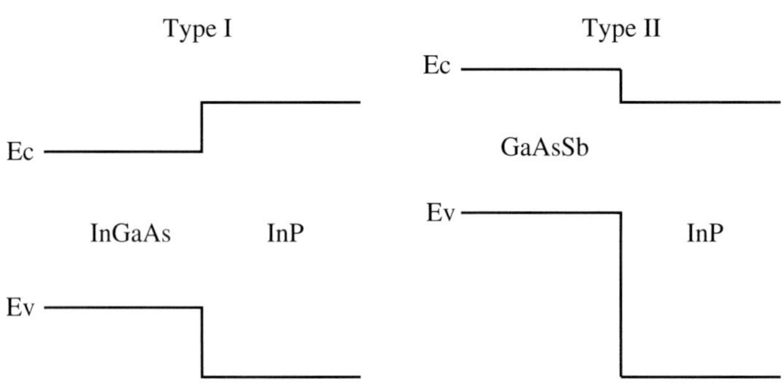

Fig. 2.4. Schematic description of type I and type II band alignments.

for the same speed and simpler emitter definition. Typical emitter lengths for high-speed III-V HBTs are about 0.5-0.8 μm, while those for SiGe HBTs are as small as 0.12 μm [233].

The base-collector current blocking in InP DHBTs is typically reduced by using staggered or linearly-graded quaternary materials for bandgap engineering. Antimonides, such as GaAsSb, have been used in DHBTs to overcome collector current blocking [55], with a different type of heterojunction replacing the type I transition by the type II transition [53] (see Fig. 2.4).

2.2.2 High Electron Mobility Transistors (HEMTs)

AlGaAs/GaAs and Pseudomorphic AlGaAs/InGaAs HEMTs. The most prominent HEMT on GaAs is currently the pseudomorphic HEMT (PHEMT) using the AlGaAs/InGaAs material system. The focus has moved from research to industrial production of these devices. Major contributions come from TriQuint, TRW, and UMS [36, 283, 518]. Although being challenged by SiGe HBTs, GaAs PHEMTs still deliver remarkable performance for low-noise applications in receiver circuits up to 100 GHz [221].
Furthermore, the combination of outstanding breakdown voltages and high-speed performance has led to considerable power performance in X-band applications [142] and in the Ka-band [36, 81]. With automotive collision avoidance applications entering large scale production, the 77 GHz W-band applications are currently based on GaAs PHEMT MMICs [190].
Typical optimization issues for PHEMTs remaining are:

- Increase of applicable bias for high-speed and high-gain applications in the Ka-frequency band (26.5-40 GHz) [263]
- Stabilization of processes with gate lengths shorter than 150 nm for industrial production [103]
- Simplification of process flows for high-performance high-volume production [142]

Next to the increase in wafer diameter to six inches, most of the optimization issues involve ultimate performance near the theoretical limit of the AlGaAs/InGaAs material system [263, 296].

InAlAs/InGaAs HEMTs. Narrow bandgap materials such as InGaAs lattice-matched to InP have pushed the cut-off frequencies beyond 400 GHz for sub-100 nm gate length HEMTs. They have been the fastest three-terminal devices for the last decade, as can be seen in Fig. 2.5. However, Fig. 2.5 also shows the continuous reduction of the product $f_T \times l_g$ from 1998 onward. Fig. 2.5 demonstrates the increase of gate leakage and short channel effects for the gate lengths between 100 nm and 30 nm given in the figure. The devices are grown either pseudomorphically or lattice-matched on InP.
Major contributions for InP-based HEMTs come from TRW, NTT, and Fujitsu [282, 491, 557]. InAlAs/InGaAs HEMTs can also be grown metamorphically on GaAs [101] with similar quality and reliability [111, 112]. Metamorphic HEMTs on GaAs make use of the larger diameters of GaAs substrate relative to InP substrates and are discussed for production on six-inch wafers by Raytheon [218].
Typical optimization issues for InAlAs/InGaAs HEMTs are:

- Stabilization of the InAlAs/passivation or InP/passivation interface to avoid dispersive or current slump effects for a given gate geometry
- Reduction of the access resistances for continuously reduced gate lengths for low-power operation

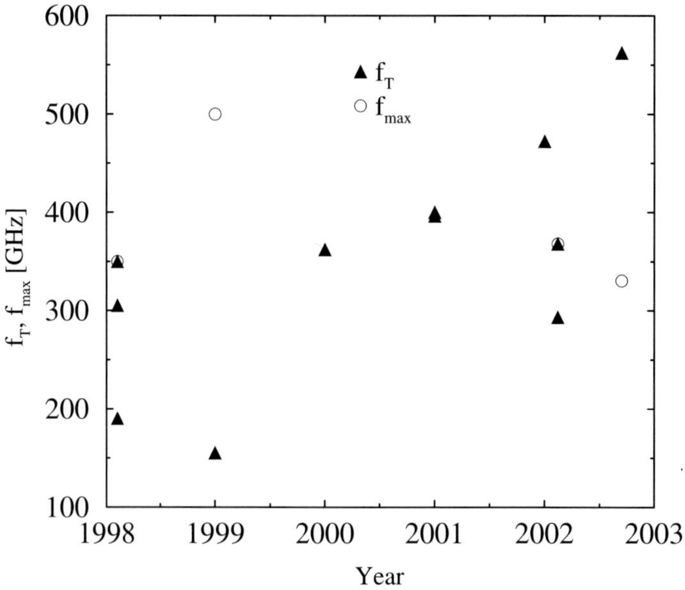

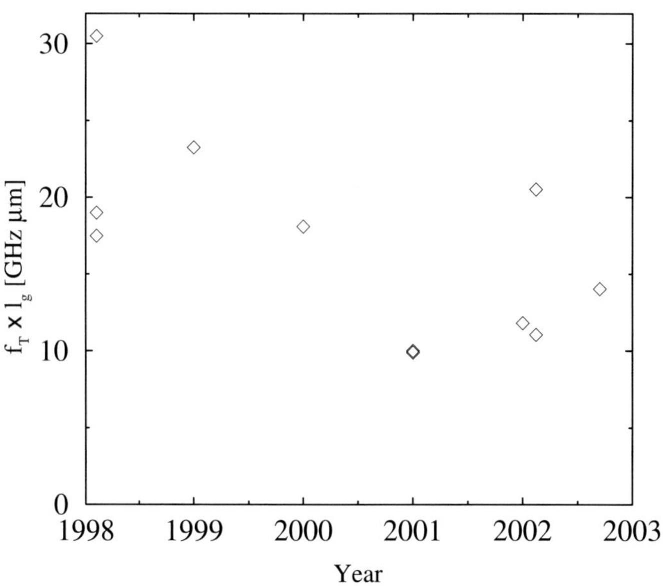

Fig. 2.5. Cut-off frequencies, f_T and f_{max}, and product current gain cut-off frequency × gate length $f_T \times l_g$ over time for InAlAs/InGaAs HEMTs.

- Reduction of gate leakage for scaled gate and barrier geometries
- Optimization of the on-state breakdown voltage for a given speed of operation, i.e. reduction of high-field effects such as impact ionization and other hot-electron effects

InAlAs/InGaAs HEMTs on InP substrates are industrially available from TRW and its spin-off Velocium [282]. Industrial metamorphic devices are announced by WIN Semiconductors [100] and Raytheon [218].

AlGaN/GaN HEMTs. The III-nitride material system has recently gained much attention. The devices are increasingly subject to device optimization issues, since a number of basic questions for material growth and device processing have been solved. III-nitride devices combine wide bandgaps resulting in high breakdown-fields with good transport properties and good thermal conductivity of the semiconductor channel layers. Moreover, barrier layers with bearable transport properties can be grown.
A major task for GaN-based devices is the development of semi-insulating substrates. Currently semi-insulating SiC, sapphire, and silicon are used. GaN and AlN substrates are still in a phase of major research efforts [386].

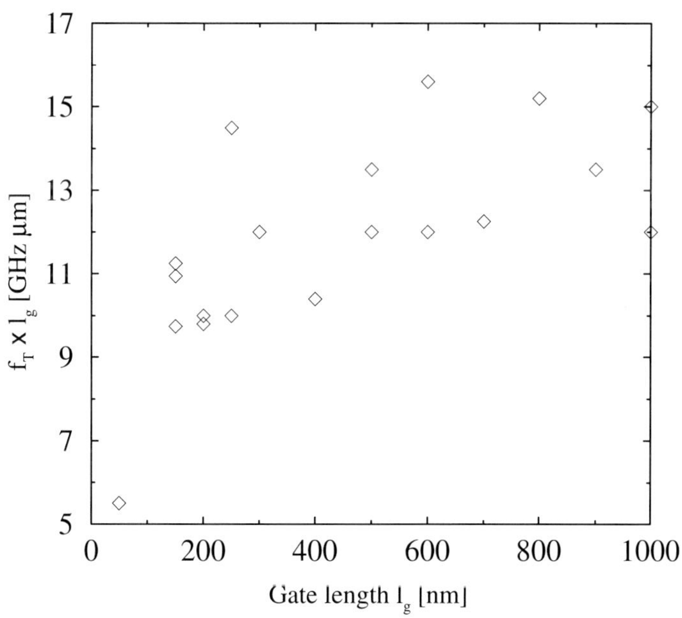

Fig. 2.6. Cut-off frequency×gate length product $f_T \times l_g$ versus gate length l_g for AlGaN/GaN HEMTs.

Fig. 2.6 shows the product $f_T \times l_g$ as a function of gate length l_g for AlGaN/GaN HEMTs on SiC collected from recent publications. It can be observed that currently only a few devices reach the $f_T \times l_g$ products of optimized GaAs PHEMTs (15-16 GHz·μm). However, such values have been theoretically predicted for AlGaN/GaN HEMTs [13]. These and other considerations demonstrate the optimization potential. Major device optimization issues include:

- Increase of breakdown voltages to fully exploit the principal wide bandgap material properties
- Full use in HFETs of the saturation velocity measured in bulk GaN material
- Reduction of contact resistances by ohmic contact design
- Improvement of Schottky contacts including metal/oxide-based concepts
- Reduction of current slump effects by properly adjusting the III-nitride/passivation interface
- Thermal management
- Charge control of the carriers available due to doping and polarization
- Introduction of more complex processing steps, such as reproducible gate recess processes

In the development of AlGaN/GaN HEMTs, fundamental contributions have come from US universities (see [132] and [342]). On the industrial side, development for both GaN HFETs and SiC MESFETs on semi-insulating SiC substrates has been performed by Cree [386]. GaN devices on silicon and sapphire substrates are promoted by RF Microdevices [456] in an industrial context. Further contributions come from NEC [14] for GaN on sapphire substrates at 2 GHz and for GaN on SiC substrates at 30 GHz [251].

2.2.3 Novel Devices

Novel devices and materials have always been a source of motivation for the designers of electronic devices and circuits [194, 277]. For example, the introduction of heterostructures in the silicon material system, which enables SiGe/Si devices, is considered to be one of the most successful innovations. New design considerations for the insulator/semiconductor or the semiconductor/semiconductor interfaces are the key to device innovation.

InP/GaAsSb/InP DHBTs. Introduction of III-antimonide materials enables the design of even faster InP HBTs, since the InP/GaAsSb heterojunction is of the type II [54]. This avoids the effects related to heterostructure current blocking in forward bias [334, 492]. Furthermore, reports suggest that the active p-doping levels in GaAsSb exceed the values typical for InGaAs bases [545]. Industrial interest for this type of material is demonstrated in [325, 565]. Major development issues include:

- Improvement of the material quality of GaAsSb to make the epitaxial layers suitable for standard processes

- Development of etch processes compatible to the InP HBT process flow

With the increasing experience in growing and processing antimonides, GaAsSb/InP HBTs are becoming strong competitors of the conventional InP HBTs.

Gallium Nitride (GaN) HBTs. This type of bipolar device is currently in a state of active development but is facing several problems:

- Lack of a genuine GaN substrate or a buffer technology with reduced defect density suitable for high-voltage HBTs
- Insufficient activation of magnesium used as p-type doping
- Processing issues, such as etch selectivity for mesa etching
- High base contact and sheet resistances reducing or nearly eliminating the current gain

A summary of the problems with GaN HBTs is given in [333]. The devices feature extremely high breakdown voltages exceeding several hundred volts but, unfortunately, very small current amplification.

InGaN-based devices. InN-based materials promise low effective masses and wide bandgap properties. Based on these expectations, InGaN-based devices have been fabricated. InGaN-based double channel HFETs are reported in [114]. The solubility of indium in GaN currently limits the amount of indium in $In_xGa_{1-x}N$ to about 7%.
GaAs HBTs with cubic InGaAsN bases have been proposed to reduce the turn-on voltage [562]. However, there are issues remaining [94] due to the quaternary material growth.

3. Physical Models

3.1 Sets of Partial Differential Equations

Transport of charged carriers in semiconductor devices can be described by the Boltzmann transport equation, which provides a semiclassical description. In this context, semiclassical means that the transport is described by the classical equations of motion, whereas the scattering events, which are assumed to happen instantaneously, are described by the laws of quantum mechanics. This formulation of carrier transport is generally considered to be valid for modern devices.
The Boltzmann transport equation can be solved numerically only at very high computational expense. Therefore an approximation of moments [214, 313] is applied. The well established drift-diffusion transport model (DD) [448, 475] is predominantly used by engineers around the world. It is the simplest macroscopic transport model and is formulated with the two lowest order moments of the expansion of the Boltzmann equation. It cannot account for non-local effects, since the electron gas is assumed to be in thermal equilibrium with the lattice temperature [337]. The hydrodynamic (HD) transport model includes higher order moments [47, 485, 499, 506]. It treats the carrier energy as an independent solution variable and is capable of describing non-local and hot-carrier effects. In addition, the lattice temperature can be treated either as a constant or as an unknown governed by the lattice heat flow equation.

3.1.1 Basic Semiconductor Equations

The basic equations solved in a device simulator are the Poisson equation and the continuity equations for electrons and holes. They read

$$\mathrm{div}(\varepsilon \cdot \mathrm{grad}\,\psi) = \mathrm{q} \cdot (n - p - C_{\mathrm{net}}), \tag{3.1}$$

$$\mathrm{div}\,\mathbf{J}_{\mathrm{n}} = \mathrm{q} \cdot \left(R + \frac{\partial n}{\partial t}\right), \tag{3.2}$$

$$\mathrm{div}\,\mathbf{J}_{\mathrm{p}} = -\mathrm{q} \cdot \left(R + \frac{\partial p}{\partial t}\right). \tag{3.3}$$

The unknown quantities in (3.1)-(3.3) are the electrostatic potential, ψ, and the electron and hole concentrations, n and p respectively. C_{net} denotes the net concentration of the ionized dopants and other charged defects, ε is the dielectric permittivity of the semiconductor or insulator, and R is the net recombination rate.

3.1.2 The Drift-Diffusion Transport Model

The drift-diffusion current relations can be derived from the Boltzmann transport equation by the method of moments [448] or from basic principles of irreversible thermodynamics [63]. The electron and hole current densities are given by

$$\mathbf{J}_\mathrm{n} = \mathrm{q} \cdot \mu_\mathrm{n} \cdot n \cdot \left(\mathrm{grad} \left(\frac{E_\mathrm{C}}{\mathrm{q}} - \psi \right) + \frac{k_\mathrm{B}}{\mathrm{q}} \cdot \frac{N_{\mathrm{C},0}}{n} \cdot \mathrm{grad} \left(\frac{n \cdot T_\mathrm{L}}{N_{\mathrm{C},0}} \right) \right) \tag{3.4}$$

$$\mathbf{J}_\mathrm{p} = \mathrm{q} \cdot \mu_\mathrm{p} \cdot p \cdot \left(\mathrm{grad} \left(\frac{E_\mathrm{V}}{\mathrm{q}} - \psi \right) - \frac{k_\mathrm{B}}{\mathrm{q}} \cdot \frac{N_{\mathrm{V},0}}{p} \cdot \mathrm{grad} \left(\frac{p \cdot T_\mathrm{L}}{N_{\mathrm{V},0}} \right) \right) . \tag{3.5}$$

These current relations account for position-dependent band edge energies, E_C and E_V, and for position-dependent effective masses which are included in the effective density of states, $N_{\mathrm{C},0}$ and $N_{\mathrm{V},0}$ [450]. The index 0 indicates that $N_{\mathrm{C},0}$ and $N_{\mathrm{V},0}$ are evaluated at some (arbitrary) reference temperature, T_0, which is constant in real space regardless of what the local values of the lattice and carrier temperatures are.

3.1.3 The Hydrodynamic Transport Model

In the hydrodynamic transport model, carrier temperatures are allowed to be different from the lattice temperature. The basic equations (3.1) through (3.3) are augmented by energy balance equations which determine the carrier temperatures. The current relations take the form

$$\mathbf{J}_\mathrm{n} = \mathrm{q} \cdot \mu_\mathrm{n} \cdot n \cdot \left(\mathrm{grad} \left(\frac{E_\mathrm{C}}{\mathrm{q}} - \psi \right) + \frac{k_\mathrm{B}}{\mathrm{q}} \cdot \frac{N_{\mathrm{C},0}}{n} \cdot \mathrm{grad} \left(\frac{n \cdot T_\mathrm{n}}{N_{\mathrm{C},0}} \right) \right) \tag{3.6}$$

$$\mathbf{J}_\mathrm{p} = \mathrm{q} \cdot \mu_\mathrm{p} \cdot p \cdot \left(\mathrm{grad} \left(\frac{E_\mathrm{V}}{\mathrm{q}} - \psi \right) - \frac{k_\mathrm{B}}{\mathrm{q}} \cdot \frac{N_{\mathrm{V},0}}{p} \cdot \mathrm{grad} \left(\frac{p \cdot T_\mathrm{p}}{N_{\mathrm{V},0}} \right) \right) . \tag{3.7}$$

The energy balance equations state conservation of the average carrier energies. In terms of the carrier temperatures T_n and T_p they can be written as

$$\begin{aligned} \mathrm{div}\, \mathbf{S}_\mathrm{n} = {} & \mathrm{grad} \left(\frac{E_\mathrm{C}}{\mathrm{q}} - \psi \right) \cdot \mathbf{J}_\mathrm{n} \\ & - \frac{3 \cdot k_\mathrm{B}}{2} \cdot \left(\frac{\partial \left(n \cdot T_\mathrm{n} \right)}{\partial t} + R \cdot T_\mathrm{n} + n \cdot \frac{T_\mathrm{n} - T_\mathrm{L}}{\tau_{\epsilon,\mathrm{n}}} \right) \end{aligned} \tag{3.8}$$

$$\begin{aligned} \mathrm{div}\, \mathbf{S}_\mathrm{p} = {} & \mathrm{grad} \left(\frac{E_\mathrm{V}}{\mathrm{q}} - \psi \right) \cdot \mathbf{J}_\mathrm{p} \\ & - \frac{3 \cdot k_\mathrm{B}}{2} \cdot \left(\frac{\partial \left(p \cdot T_\mathrm{p} \right)}{\partial t} + R \cdot T_\mathrm{p} + p \cdot \frac{T_\mathrm{p} - T_\mathrm{L}}{\tau_{\epsilon,\mathrm{p}}} \right) . \end{aligned} \tag{3.9}$$

Here $\tau_{\epsilon,\mathrm{n}}$ and $\tau_{\epsilon,\mathrm{p}}$ denote the energy relaxation times, while $\mathbf{S}_\mathrm{n}$ and $\mathbf{S}_\mathrm{p}$ are the energy fluxes.

$$\mathbf{S}_\mathrm{n} = -\kappa_\mathrm{n} \cdot \mathrm{grad}\, T_\mathrm{n} - \frac{5}{2} \cdot \frac{k_\mathrm{B} \cdot T_\mathrm{n}}{\mathrm{q}} \cdot \mathbf{J}_\mathrm{n} \tag{3.10}$$

$$\mathbf{S}_\mathrm{p} = -\kappa_\mathrm{p} \cdot \mathrm{grad}\, T_\mathrm{p} + \frac{5}{2} \cdot \frac{k_\mathrm{B} \cdot T_\mathrm{p}}{\mathrm{q}} \cdot \mathbf{J}_\mathrm{p}. \tag{3.11}$$

The carrier thermal conductivities, κ_n and κ_p, are assumed to obey a generalized Wiedemann-Franz law.

$$\kappa_\mathrm{n} = \left(\frac{5}{2} + c_\mathrm{n}\right) \cdot \frac{k_\mathrm{B}^2}{\mathrm{q}} \cdot T_\mathrm{n} \cdot \mu_\mathrm{n} \cdot n \tag{3.12}$$

$$\kappa_\mathrm{p} = \left(\frac{5}{2} + c_\mathrm{p}\right) \cdot \frac{k_\mathrm{B}^2}{\mathrm{q}} \cdot T_\mathrm{p} \cdot \mu_\mathrm{p} \cdot p. \tag{3.13}$$

The heat capacities c_n and c_p are usually neglected ($c_\mathrm{n} = c_\mathrm{p} = 0$). Strictly speaking, the model described above in this subsection represents an energy transport (ET) model. Such a model is obtained in the course of deriving the moment equations under the assumption that the average kinetic energy of the carriers is negligible in comparison to their thermal energy.

$$\frac{m_\mathrm{n} \cdot v_\mathrm{n}^2}{2} \ll \frac{3 \cdot k_\mathrm{B} \cdot T_\mathrm{n}}{2} \tag{3.14}$$

$$\frac{m_\mathrm{p} \cdot v_\mathrm{p}^2}{2} \ll \frac{3 \cdot k_\mathrm{B} \cdot T_\mathrm{p}}{2} \tag{3.15}$$

Thus the average carrier energies w_n and w_p are approximated as

$$w_\mathrm{n} = \frac{3 \cdot k_\mathrm{B} \cdot T_\mathrm{n}}{2} + \frac{m_\mathrm{n} \cdot v_\mathrm{n}^2}{2} \approx \frac{3 \cdot k_\mathrm{B} \cdot T_\mathrm{n}}{2} \tag{3.16}$$

$$w_\mathrm{p} = \frac{3 \cdot k_\mathrm{B} \cdot T_\mathrm{p}}{2} + \frac{m_\mathrm{p} \cdot v_\mathrm{p}^2}{2} \approx \frac{3 \cdot k_\mathrm{B} \cdot T_\mathrm{p}}{2}. \tag{3.17}$$

Since in the literature the terms *energy transport* and *hydrodynamic transport* are often used synonymously, the notation *hydrodynamic transport* is retained throughout this work.

3.1.4 Extensions of the Hydrodynamic Transport Model

The hydrodynamic transport model described in the previous section is available in modern commercial device simulators [227, 503]. It includes three or four moments of the distribution function in the Boltzmann transport equation and uses a so-called closure relation, as reviewed in [187].

The introduction of further moments has been considered and the parameters have been calibrated for silicon [187, 358]. A concise overview of the extensions of the energy transport model is given in [187] and the references therein.

3.1.5 The Lattice Heat Flow Equation

Self-heating effects in semiconductor devices can be accounted for by solving the lattice heat flow equation (3.18) self-consistently with the DD transport equations (3.1)–(3.3), (3.4) and (3.5), or HD transport equations (3.1)–(3.3), (3.6) and (3.7), forming a system of four, or respectively six, partial differential equations.

$$\mathrm{div}\ \mathbf{S}_\mathrm{L} = -\rho_\mathrm{L} \cdot c_\mathrm{L} \cdot \frac{\partial T_\mathrm{L}}{\partial t} + H \tag{3.18}$$

$$\mathbf{S}_\mathrm{L} = -\kappa_\mathrm{L} \cdot \mathrm{grad}\ T_\mathrm{L} \tag{3.19}$$

In (3.18), T_L denotes the lattice temperature, t the time variable, and H the heat generation term. The coefficients ρ_L, c_L, and κ_L are the mass density, the specific heat, and the thermal conductivity of the respective materials.
The model for the heat generation, H, depends on the transport model used. In drift-diffusion simulation H equals the Joule heat,

$$H = \mathrm{grad}\left(\frac{E_\mathrm{C}}{\mathrm{q}} - \psi\right) \cdot \mathbf{J}_\mathrm{n} + \mathrm{grad}\left(\frac{E_\mathrm{V}}{\mathrm{q}} - \psi\right) \cdot \mathbf{J}_\mathrm{p} + R \cdot (E_\mathrm{C} - E_\mathrm{V}), \tag{3.20}$$

whereas in the hydrodynamic simulation the relaxation terms

$$H = \frac{3 \cdot k_\mathrm{B}}{2} \cdot \left(n \cdot \frac{T_\mathrm{n} - T_\mathrm{L}}{\tau_{\epsilon,\mathrm{n}}} + p \cdot \frac{T_\mathrm{p} - T_\mathrm{L}}{\tau_{\epsilon,\mathrm{p}}}\right) + R \cdot (E_\mathrm{C} - E_\mathrm{V}) \tag{3.21}$$

are used. (3.21) is obtained from the stationary energy flux balance derived from the second moment solving the Boltzmann transport equation, the carrier continuity equations, (3.2) and (3.3), for the hydrodynamic model used [63, 259]. The first term of (3.21) represents an energy relaxation approximation of the scattering terms of the Boltzmann equation [259]. The same second (effective recombination) term is used in (3.20) and (3.21). This is sound and introduces only a small error, since the recombination energies are generally much smaller than the Joule-like contribution in (3.20) and (3.21). This is different for intensive light irradiation where additional contributions have to be added in order to model the effective recombination term.

Global Self-Heating Model. An alternative global self-heating model (GSH) has been suggested in [182]. The model calculates a spatially constant (global) lattice temperature T_L in the device from the dissipated power as

$$T_\mathrm{L} = T_\mathrm{C} + R_\mathrm{TH,glob} \cdot \sum_C I_\mathrm{Con} \cdot V_\mathrm{Con} \tag{3.22}$$

with I_Con and V_Con being the contact currents and voltages, T_C being contact temperature, and R_glob being the global thermal resistance. For the terminal characteristics, the GSH model delivers similar results to the SH model, but consumes about half the CPU time.

3.1.6 The Insulator Equations

In insulating materials, only the Poisson equation (3.1) and the lattice heat flow equation (3.18) are solved. The presence of doping and mobile carriers is neglected ($n = p = C_{\mathrm{net}} = 0$), therefore the carrier continuity equations are not solved. This assumption leads to the reduction of (3.1) to the Laplace equation (3.23) and neglects the heat generation term of (3.18), which is invoked for semiconductor segments only.

$$\mathrm{div}(\varepsilon \cdot \mathrm{grad}\, \psi) = 0 \tag{3.23}$$

$$\mathrm{div}(\kappa_{\mathrm{L}} \cdot \mathrm{grad}\, T_{\mathrm{L}}) = \rho_{\mathrm{L}} \cdot c_{\mathrm{L}} \cdot \frac{\partial T_{\mathrm{L}}}{\partial t} \tag{3.24}$$

As a result, neither currents nor charges are considered in insulators. If necessary, a workaround is to treat insulators as semiconductor materials, as has been demonstrated for Si_3N_4 in [383]. The consideration of charges at semiconductor/insulator interfaces is discussed in the next subsection.

3.1.7 Boundary and Interface Conditions

The basic semiconductor equations are solved in a closed domain. At the boundaries of this domain appropriate boundary conditions need to be specified for the unknowns ψ, n, p, T_{n}, T_{p}, and T_{L}.

Artificial Boundaries. In order to separate the simulated device from neighboring devices, artificial boundaries which are not boundaries in a physical sense must be specified. The Neumann boundary condition guarantees that the simulation domain is self-contained and that there are no fluxes across the boundary.

$$\mathbf{n} \cdot \mathbf{E} = 0, \qquad \mathbf{n} \cdot \mathbf{J}_{\mathrm{n,p}} = 0, \qquad \mathbf{n} \cdot \mathbf{S}_{\mathrm{n,p}} = 0, \qquad \mathbf{n} \cdot \mathbf{S}_{\mathrm{L}} = 0, \tag{3.25}$$

$$\mathbf{n} \cdot \mathbf{E} = 0, \qquad \mathbf{n} \cdot \mathbf{S}_{\mathrm{L}} = 0 \tag{3.26}$$

Here $\mathbf{n}$ denotes an outward-oriented vector normal to the boundary. (3.25) and (3.26) give the boundary conditions at the artificial boundaries for semiconductor and insulator segments respectively. These conditions are a good assumption for the electrical boundaries of multifinger devices. However, this approximation introduces serious errors in the solution of the heat flow equation, since the thermal interaction between fingers can be significant [320].

Semiconductor/Metal Boundaries.

Idealized Ohmic Contact. Although the microscopic situation at an ohmic contact is complicated due to alloying and diffusion processes during manufacturing, simple Dirichlet boundary conditions are normally assumed. The contact potential φ_{s}, the carrier contact concentrations n_{s} and p_{s}, and, in the HD simulation case, the contact carrier temperatures T_{n} and T_{p} are fixed. The metal quasi-Fermi level (which is specified by the contact potential φ_{m}) is equal to the semiconductor quasi-Fermi level. The contact potential at the semiconductor boundary reads

$$\varphi_{\mathrm{s}} = \varphi_{\mathrm{m}} + \psi_{\mathrm{bi}}. \tag{3.27}$$

The built-in potential ψ_{bi} is calculated after [145].

$$\begin{aligned}\psi_{\mathrm{bi}} &= \frac{k_{\mathrm{B}} \cdot T_{\mathrm{L}}}{\mathrm{q}} \cdot \ln\left(\frac{1}{2 \cdot C_1} \cdot \left(C_{\mathrm{net}} + \sqrt{C_{\mathrm{net}}^2 + 4 \cdot C_1 \cdot C_2}\right)\right) \\ &= -\frac{k_{\mathrm{B}} \cdot T_{\mathrm{L}}}{\mathrm{q}} \cdot \ln\left(\frac{1}{2 \cdot C_2} \cdot \left(-C_{\mathrm{net}} + \sqrt{C_{\mathrm{net}}^2 + 4 \cdot C_1 \cdot C_2}\right)\right)\end{aligned} \tag{3.28}$$

Here $\mathrm{C}_{\mathrm{net}} = N_{\mathrm{D}} - N_{\mathrm{A}}$ is the net concentration of dopants and other charged defects at the contact boundary. The auxiliary variables C_1 and C_2 are defined by

$$C_1 = N_{\mathrm{C}} \cdot \exp\left(\frac{-E_{\mathrm{C}}}{k_{\mathrm{B}} \cdot T_{\mathrm{L}}}\right) \tag{3.29}$$

$$C_2 = N_{\mathrm{V}} \cdot \exp\left(\frac{E_{\mathrm{V}}}{k_{\mathrm{B}} \cdot T_{\mathrm{L}}}\right). \tag{3.30}$$

The carrier concentrations in the semiconductor are pinned to the equilibrium carrier concentrations at the contact. They are expressed as

$$n_s = N_{\mathrm{C}} \cdot \exp\left(\frac{-E_{\mathrm{C}} + q \cdot \psi_{\mathrm{bi}}}{k_{\mathrm{B}} \cdot T_{\mathrm{L}}}\right) \tag{3.31}$$

$$p_s = N_{\mathrm{V}} \cdot \exp\left(\frac{E_{\mathrm{V}} - \mathrm{q} \cdot \psi_{\mathrm{bi}}}{k_{\mathrm{B}} \cdot T_{\mathrm{L}}}\right). \tag{3.32}$$

The use of such a model assumes a high doping in the semiconductor $\geq 10^{18}\ \mathrm{cm}^{-3}$. Any specific interface effects, such as dipole charges, surface charges, or semiconductor/metal alloying, are neglected.

Hydrodynamic Ohmic Contact. The boundary model in the previous paragraph assumes local thermal equilibrium at the boundary. Considering the transport at the ohmic contact a thermionic process where the band discontinuity is obtained by the charge neutrality, Schroeder obtained the following boundary conditions [443] for electrons:

$$n - p + N_{\mathrm{A}} - N_{\mathrm{D}} = 0 \tag{3.33}$$

$$\mathbf{n} \cdot \mathbf{J}_{\mathrm{n}} = q \cdot [v_{\mathrm{em,n}}(T_{\mathrm{n}}) \cdot n - v_{\mathrm{em,n}}(T_{\mathrm{C}}) \cdot n_0] \tag{3.34}$$

$$\mathbf{n} \cdot \mathbf{S}_{\mathrm{n}} = -2 \cdot k_{\mathrm{B}} \cdot [T_{\mathrm{n}} \cdot v_{\mathrm{em,n}}(T_{\mathrm{n}}) \cdot n - T_{\mathrm{C}} \cdot v_{\mathrm{em,n}}(T_{\mathrm{C}}) \cdot n_0] \tag{3.35}$$

T_{n} and T_{C} are the electron temperatures in the semiconductor and the metal. The emission velocity $v_{\mathrm{em,n}}$ and the concentration n_0 are defined by:

$$v_{\mathrm{em,n}} = \sqrt{\frac{2 \cdot k_{\mathrm{B}} \cdot T_{\mathrm{n}}}{\pi \cdot m_{\mathrm{n}}}} \tag{3.36}$$

$$n_0 = \left(\frac{T_C}{T_n}\right)^{3/2} \cdot N_C \cdot \exp\left(\frac{-E_C + q \cdot \varphi_m}{k_B \cdot T_C}\right) \tag{3.37}$$

N_C denotes the effective density of states. Another more advanced approach to model non-ideal ohmic contacts is described in [225] including the gradual change towards a Schottky contact.

Semiconductor/Metal Thermal Interface. The carrier temperatures T_n and T_p are set equal to the lattice temperature T_L.

$$T_n = T_L, \qquad T_p = T_L \tag{3.38}$$

In the case of a thermal contact, the lattice temperature T_L is calculated using a specified contact temperature T_C and thermal resistance R_{TH}. R_{TH} represents a global thermal resistance of materials surrounding the device and T_C the temperature of the heat reservoir assumed. The thermal heat flow density $\mathbf{S}_L$ at the contact boundary reads

$$\mathbf{n} \cdot \mathbf{S}_L = \frac{T_L - T_C}{R_{TH}}. \tag{3.39}$$

In case no thermal resistance is specified in a contact model, an isothermal boundary condition is assumed and the lattice temperature T_L is set equal to the contact temperature T_C.

$$T_L = T_C \tag{3.40}$$

In the case of DD simulation including self-heating effects, an additional thermal energy is accounted for. This thermal energy is produced when the carriers surmount the potential difference between the conduction or valence band and the metal quasi-Fermi level. The energy equations read

$$\mathbf{J}_n \cdot \left(\frac{E_C}{q} + \varphi_m\right) + \mathbf{J}_p \cdot \left(\frac{E_V}{q} + \varphi_m\right) = \mathrm{div}_A\ (\mathbf{S}_L) \tag{3.41}$$

The expression $\mathrm{div}_A\ (\mathbf{S}_L)$ denotes the surface divergence of the thermal heat flux at the boundary considered. In the case of HD simulation including self-heating, the thermal heat flow across the boundary is accounted for self-consistently.

Metal Contact Resistance. A finite electrical line resistance of the contact metal R_{Con} can be included using

$$\varphi_m = V_{Con} - I_{Con} \cdot R_{Con} \tag{3.42}$$

where V_{Con} denotes the applied terminal voltage and I_{Con} the current through the contact.

Polysilicon Contact. For polysilicon-emitter bipolar devices, the polysilicon contact model [569] can be used. The dielectric flux D through the oxide reads

$$\mathbf{n} \cdot \mathbf{D} = -\frac{\varepsilon_{\mathrm{ox}}}{d_{\mathrm{ox}}} \cdot V_{\mathrm{ox}} \tag{3.43}$$

where V_{ox} is the voltage drop over the thin oxide layer (which is introduced between polysilicon and silicon) and $\varepsilon_{\mathrm{ox}}$ and d_{ox} denote the permittivity and thickness of this layer. The electron and hole current densities across the contact interface read

$$\mathbf{n} \cdot \mathbf{J}_{\mathrm{n}} = \sigma_{\mathrm{ox}} \cdot V_{ox} \tag{3.44}$$

$$\mathbf{n} \cdot \mathbf{J}_{\mathrm{p}} = \mathrm{q} \cdot p \cdot S_{\mathrm{p}} \tag{3.45}$$

where σ_{ox} is the oxide conductivity, p the hole concentration in the semiconductor, and S_{p} the hole surface recombination velocity. V_{ox} depends on the quasi-Fermi level in the metal (which is specified by the contact potential φ_{m}), the potential in the semiconductor φ_{s}, and the built-in potential ψ_{bi}.

$$V_{\mathrm{ox}} = \varphi_{\mathrm{s}} - \varphi_{\mathrm{m}} - \psi_{\mathrm{bi}} \tag{3.46}$$

The polysilicon contact boundary conditions for the carrier temperatures, T_{n} and T_{p}, and the lattice temperature, T_{L}, are the same as those applied for ohmic contacts, i.e. (3.38), and (3.39) or (3.40).

Schottky Contact. At a Schottky contact, mixed boundary conditions apply. The contact potential φ_{s}, the carrier contact concentrations n_{s} and p_{s}, and, in the HD simulation case, the contact carrier temperatures T_{n} and T_{p} are fixed. The semiconductor contact potential is the difference between the metal quasi-Fermi level (which is specified by the contact voltage φ_{m}) and the metal workfunction difference potential φ_{w}

$$\varphi_{\mathrm{s}} = \varphi_{\mathrm{m}} - \varphi_{\mathrm{w}}, \qquad \text{where} \qquad \varphi_{\mathrm{w}} = -\frac{\mathrm{E_w}}{\mathrm{q}}. \tag{3.47}$$

The difference between the conduction band energy E_{C} and the metal workfunction energy gives the workfunction difference energy $\mathrm{E_w}$ which is the so-called barrier height of the Schottky contact otherwise denoted as ϕ_{B}. The applied boundary conditions are

$$\mathbf{n} \cdot \mathbf{J}_{\mathrm{n}} = -\mathrm{q} \cdot v_{\mathrm{n}} \cdot (n_0 - n_{\mathrm{s}}) \tag{3.48}$$

$$\mathbf{n} \cdot \mathbf{J}_{\mathrm{p}} = \mathrm{q} \cdot v_{\mathrm{p}} \cdot (p_0 - p_{\mathrm{s}}). \tag{3.49}$$

Here v_{n} and v_{p} denote the recombination velocities. The carrier concentrations are expressed as:

$$n_{\mathrm{s}} = N_{\mathrm{C}} \cdot \exp\left(-\frac{E_{\mathrm{C}} - E_{\mathrm{Fn}}}{k_{\mathrm{B}} \cdot T_{\mathrm{L}}}\right) \tag{3.50}$$

$$p_{\mathrm{s}} = N_{\mathrm{V}} \cdot \exp\left(\frac{E_{\mathrm{V}} - E_{\mathrm{Fp}}}{k_{\mathrm{B}} \cdot T_{\mathrm{L}}}\right) \tag{3.51}$$

where E_{Fn} denotes the Fermi-energy in the semiconductor. This implies that the currents through a Schottky contact are determined by the recombination velocities. Setting the values for v_n and v_p to 0 suppresses current flow through the Schottky contact ($\mathbf{n} \cdot \mathbf{J}_{n,p} = 0$). The Schottky contact boundary conditions for the carrier temperatures T_n and T_p and for the lattice temperature T_L are similar to those which apply for the ohmic contact, i.e. (3.38) and, respectively, (3.39) or (3.40).

Rewriting the equations above for the equilibrium situation, the quasi-equilibrium concentrations n_0 and p_0 can be expressed as

$$n_0 = N_C \cdot \exp\left(\frac{-\phi_B}{k_B \cdot T_L}\right) \tag{3.52}$$

$$p_0 = N_V \cdot \exp\left(\frac{-\phi_B}{k_B \cdot T_L}\right). \tag{3.53}$$

For the Schottky barrier height ϕ_B, typical values range between 0.61 eV and 1 eV on typical semiconductor barriers, as given in Table 3.1. For the hydrodynamic case, the carrier temperatures are fixed, thus thermal equilibrium is assumed similarly to the ohmic contact. The thermal boundary conditions are similar to those applied for ohmic contact, i.e. either by a thermal resistance or an isothermal boundary condition.

Table 3.1. Parameter values for the Schottky contact model

Material	ϕ_B [eV]	Refs.	Material	ϕ_B [eV]	Refs.
n-Si/Au	0.55-0.8	[460, 495]	n-$Al_{0.25}Ga_{0.75}As$/Ti	0.7	-
p-Si/Au	-0.25	[496]	n-$Al_{0.25}Ga_{0.75}As$/Al	0.75	-
n-GaAs/Au	0.84	[460]	$In_{0.52}Al_{0.48}As$/Ti	0.6	-
n-GaAs/Al	0.8	[442]	n-$Al_{0.15}Ga_{0.85}N$/Ni	1.26	[568]
n-$Al_{0.25}Ga_{0.75}As$/WSi	0.61	[65]	n-$Al_{0.15}Ga_{0.85}N$/Ti	1.0	[568]

The model presented so far can also be called a thermionic emission model. The recombination velocity v_ν in (3.49) is then called a thermionic emission velocity. The approach can be rewritten for electrons:

$$\mathbf{n}\cdot\mathbf{J}_n = -\mathrm{q}\cdot v_n \cdot N_C \cdot \left[\exp\left(\frac{-\phi_B}{k_B \cdot T_L}\right) - \exp\left(-\frac{E_C - E_{Fn}}{k_B \cdot T_L}\right)\right]. \tag{3.54}$$

As stated in [442], (3.54) can thus be written as

$$\mathbf{n}\cdot\mathbf{J}_n = \frac{\mathrm{q} \cdot m_n \cdot k_B^2}{2\pi^2 \cdot \hbar^3} \cdot T_L^2 \cdot \exp\left(-\frac{\phi_B}{k_B \cdot T_L}\right)\left[\exp\left(-\frac{E_{Fm} - E_{Fn}}{k_B \cdot T_L}\right) - 1\right] \tag{3.55}$$

using $\phi_B = E_C - E_{Fm}$ which again is the Schottky barrier height. Furthermore, $E_w = E_{Fm} - E_{Fn}$ expresses that the workfunction difference is equal to the difference between the Fermi levels of the metal E_{Fm} and the semiconductor E_{Fn}. The emission velocity v_{n} (given here for electrons) is rewritten as

$$v_{\mathrm{n}} = \sqrt{\frac{k_{\mathrm{B}} \cdot T_{\mathrm{L}}}{2\pi \cdot m_{\mathrm{n}}}} = \frac{\mathrm{q} \cdot m_{\mathrm{n}} \cdot k_{\mathrm{B}}^2}{2\pi^2 \cdot \hbar^3} \cdot \frac{T_{\mathrm{L}}^2}{\mathrm{q} \cdot N_{\mathrm{C}}} = A^* \cdot \frac{T_{\mathrm{L}}^2}{\mathrm{q} \cdot N_{\mathrm{C}}} \tag{3.56}$$

where A^* is the Richardson constant typical for thermionic emission processes. Thus (3.54) represents a typical thermionic emission equation, which is determined by the Schottky barrier height ϕ_{B} and the difference between the Fermi levels. Typically the Fermi level of the semiconductor E_{Fn} is chosen as the reference energy.

Schottky Barrier Tunneling. Additionally tunneling can be introduced into the Schottky contact model by rewriting the tunneling current through the contact as a local generation rate:

$$\mathbf{n} \cdot \mathbf{G}_{\mathrm{tun}} = \frac{1}{\mathrm{q}} \cdot \mathbf{n} \cdot \nabla J_{\mathrm{tun}} = \tag{3.57}$$

$$= \frac{A^* \cdot T_{\mathrm{L}}}{k_{\mathrm{B}}} \cdot E \cdot \Gamma(r) \cdot \ln \left[\frac{1 + \exp\left(- \mathrm{q} \cdot \dfrac{\psi - \phi_{\mathrm{Fn}}}{k_{\mathrm{B}} \cdot T_{\mathrm{L}}} \right)}{1 + \exp\left(- \mathrm{q} \cdot \dfrac{\psi - \phi_{\mathrm{Fm}}}{k_{\mathrm{B}} \cdot T_{\mathrm{L}}} \right)} \right] \tag{3.58}$$

and inserting an additional tunneling probability, e.g.,

$$\Gamma(r) = \exp \left[-\frac{2}{\hbar} \int_0^r \sqrt{2 \cdot m_t \cdot \left(\frac{\phi_{\mathrm{B}}}{q} + \phi_{\mathrm{Fm}} - \varphi(r) \right)} dr \right]. \tag{3.59}$$

In this case ϕ_{Fm} and ϕ_{Fn} denote the local Fermi-potentials in the metal and the semiconductor and ψ the electrostatic potential in the metal and the semiconductor. ϕ_{B} again denotes the Schottky barrier height.

This description of the tunneling in (3.58) is based on the WKB-approach of solving the Schrödinger equation. Critical to the usage of (3.59) is the implementation of the path integral, as explained further below. It can be used in both planar and non-planar structures; however, this introduces additional computational complexity. The tunneling mass m_t is used as a fitting parameter to calibrate the model and to resolve the problem of the abrupt change of the effective mass assumed in the model at the heterointerface towards the metal, i.e. to the electron gas.

Insulator/Metal Boundaries. In the case of insulator/metal boundaries, a model similar to the Schottky contact model is used to calculate the insulator contact potential. The semiconductor contact potential is the difference of the metal quasi Fermi level (which is specified by the contact potential φ_{m}) and the metal workfunction difference potential ϕ_{ms}.

$$\varphi_{\mathrm{s}} = \varphi_{\mathrm{m}} - \phi_{\mathrm{ms}}, \qquad \text{where} \qquad \phi_{\mathrm{ms}} = -\frac{\mathrm{E_w}}{\mathrm{q}}. \tag{3.60}$$

Again, $\mathrm{E_w}$ is the workfunction difference energy. The lattice temperature is set equal to the contact temperature (3.40).

Semiconductor/Insulator Interfaces. The semiconductor/insulator interface is most critical for the modeling and functioning of semiconductor devices. In a very simple approach, in the absence of surface charges at the semiconductor/insulator interface, the normal component of the dielectric displacement is continuous and so is the potential.

$$\mathbf{n}\cdot\varepsilon_{\mathrm{s}}\cdot\mathbf{E}_{\mathrm{s}}-\mathbf{n}\cdot\varepsilon_{\mathrm{ins}}\cdot\mathbf{E}_{\mathrm{ins}}=0, \qquad \varphi_{\mathrm{s}}=\varphi_{\mathrm{ins}}. \tag{3.61}$$

In the presence of surface charges along the interface, the dielectric displacement obeys the law of Gauß

$$\mathbf{n}\cdot\varepsilon_{\mathrm{s}}\cdot\mathbf{E}_{\mathrm{s}}-\mathbf{n}\cdot\varepsilon_{\mathrm{ins}}\cdot\mathbf{E}_{\mathrm{ins}}=\sigma_{\mathrm{s}}. \tag{3.62}$$

To describe the Fermi level pinning at the surface due to a high trap density of states, distributed surface charges are introduced. They lead to surface depletion due to band edge bending.
At the semiconductor/insulator interface the carrier current densities (or driving forces) and the carrier heat fluxes normal to the interface vanish.

$$\mathbf{n}\cdot\mathbf{J}_{\mathrm{n,p}}=0, \qquad \text{i.e.} \qquad \mathbf{n}\cdot\mathbf{F}_{\mathrm{n,p}}=0 \tag{3.63}$$

$$\mathbf{n}\cdot\mathbf{S}_{\mathrm{n,p}}=0. \tag{3.64}$$

The lattice temperature at the interface is continuous.

Insulator/Insulator Interfaces. Similarly to the semiconductor/insulator interface, (3.61) or (3.62) apply depending on the presence of surface charges. The lattice temperature is modeled to be continuous.

Semiconductor/Semiconductor Interfaces. The calculation of the electrostatic potential at interfaces between two semiconductor segments is similar to that for the semiconductor/insulator interfaces:

$$\varphi_{\mathrm{s1}}=\varphi_{\mathrm{s2}} \tag{3.65}$$

$$\mathbf{n}\cdot\varepsilon_{\mathrm{s1}}\cdot\mathbf{E}_{\mathrm{s1}}-\mathbf{n}\cdot\varepsilon_{\mathrm{s2}}\cdot\mathbf{E}_{\mathrm{s2}}=\sigma_{\mathrm{s}}. \tag{3.66}$$

Here σ_{s} is the interface charge density which can be zero or non-zero. The subscripts are used to distinguish between the two semiconductor segments on both sides of the interface.
To calculate the carrier concentrations and the carrier temperatures at the interface of two semiconductor segments, three different approaches are considered. These are:

- A model with a continuous quasi-Fermi level across the interface (CQFL)
- A thermionic emission model (TE)
- A thermionic field emission model (TFE) or tunneling-assisted emission model.

In a strict physical sense, in the third model one has to distinguish between the field-assisted current component and a tunneling component which occurs independently of the field. A derivation for basic relations of the models is given

in [466]. Each model can typically be specified separately for electrons and holes for each semiconductor/semiconductor interface.
In the following, J denotes the current density, S the energy flux density, and ΔE_ν the difference in the conduction band edges (ΔE_C) or valence band edges (ΔE_V). The effective carrier masses are denoted by $m_{\nu i}$. The subscripts denote the semiconductor segment i and the carrier type ν with n for electrons and p for holes and their concentrations.

Continuous Quasi-Fermi Level Model. The equations for the continuous quasi-Fermi level model (CQFL) read

$$\nu_2 = \nu_1 \cdot \left(\frac{m_{\nu 2}}{m_{\nu 1}}\right)^{3/2} \cdot \exp\left(-\frac{\Delta E_\nu}{k_\mathrm{B} \cdot T_{\nu 1}}\right) \tag{3.67}$$

$$T_{\nu 2} = T_{\nu 1} \tag{3.68}$$

$$J_{\nu 2} = J_{\nu 1} \tag{3.69}$$

$$S_{\nu 2} = S_{\nu 1} - \frac{1}{q} \cdot \Delta E_\nu \cdot J_{\nu 2}. \tag{3.70}$$

When the CQFL model is used, a Dirichlet interface condition is applied. The carrier concentrations are directly determined in a way that the quasi-Fermi level remains continuous across the interface. The model is suitable for use at homojunctions. However, it is erroneous to assume such continuous quasi-Fermi levels at abrupt heterojunctions, since the bandgap alignment of the adjacent semiconductors is ignored when such a continuous condition is enforced.

Local Thermionic Emission and Thermionic Field Emission Model. In order to consider the bandgap alignment, interface models such as the TE model or the TFE model must be used. These models determine the current flux across heterointerfaces by applying a Neumann interface condition. For the modeling of the electron and hole currents, as well as for the energy flux across the interfaces, several models for different types of interfaces have been proposed [193, 349, 551, 559]. As a computationally efficient approach, the TE model is commonly used to model the current across heterojunctions with small impact of tunneling components. In this case, the impact of tunneling can be neglected. The respective interface equations read

$$J_{\nu 2} = J_{\nu 1} \tag{3.71}$$

$$J_{\nu 2} = \mathrm{q} \cdot v_{\nu 2} \cdot \nu_2 - \mathrm{q} \cdot v_{\nu 1} \cdot \nu_1 \cdot \frac{m_{\nu 2}}{m_{\nu 1}} \cdot \exp\left(-\frac{\Delta E_\nu}{k_\mathrm{B} \cdot T_{\nu 1}}\right) \cdot \Gamma \tag{3.72}$$

$$S_{\nu 2} = S_{\nu 1} - \frac{1}{\mathrm{q}} \cdot (\Delta E_\nu) \cdot J_{\nu 2} \tag{3.73}$$

$$S_{\nu 2} = -2k_\mathrm{B} \left[T_{\nu 2} \cdot v_{\nu 2} \cdot \nu_2 + T_{\nu 1} \cdot v_{\nu 1} \cdot \nu_1 \cdot \frac{m_{\nu 2}}{m_{\nu 1}} \cdot \exp\left(-\frac{\Delta E_\nu}{k_\mathrm{B} \cdot T_{\nu 1}}\right)\right]. \tag{3.74}$$

As a derivation of the thermionic emission model, a tunneling component is introduced in (3.72). The modeling of the tunneling probability Γ determines whether the TE model becomes a thermionic field emission (TFE) model, as discussed in the next paragraph. Tunneling is a non-local process introduced into the semi-classical transport model discussed above. The introduction of tunneling into a modern simulator thus requires assumptions about how the available quantities in a semiclassical transport model are to be used to calculate the tunneling probability.

Heterojunction Carrier Tunneling. Tunneling can be a major mechanism for carrier transport, for example on the drain side of the channel in HEMTs or through abrupt or staggered heterojunctions in HBTs. Several approaches have been proposed to model the tunneling probability in (3.73).

Field-Dependent Approach with Constant Tunneling Length. A first approach for modeling the tunneling is to assume a local field-dependent model. The thermionic emission velocity $v_{\nu i}$ for the segment i and the barrier height lowering δE_ν read

$$v_{\nu i} = \sqrt{\frac{2 \cdot k_\mathrm{B} \cdot T_{\nu i}}{\pi \cdot m_{\nu i}}} \tag{3.75}$$

$$\delta E_\nu = \begin{cases} \mathrm{q} \cdot E_{\perp 2} \cdot d_\mathrm{tun}, & E_{\perp 2} > 0 \\ 0, & E_{\perp 2} \leq 0 \end{cases} \tag{3.76}$$

The barrier height lowering depends on the electric field orthogonal to the interface $E_{\perp 2}$ and the effective tunneling length d_tun. A fit of the barrier reduction δE_ν for the $Al_{0.2}Ga_{0.8}As/In_{0.25}Ga_{0.75}As$ interface has been performed in [64] and is extended in this work to different material systems, such as AlGaAs/GaAs, InAlAs/InGaAs, and AlGaN/GaN. The tunneling parameters are found in Table 3.2. The values in Table 3.2 are effective fitting parameters; however, they scale with the tunneling probabilities found. Thus this tunneling model extends the TE model by accounting for tunneling effects through the heterojunction barrier by introducing a field-dependent barrier height lowering.

Table 3.2. Tunneling parameter values for various materials

Interface	Material Composition, x	d_tun [nm]
$Al_xGa_{1-x}As/GaAs$	0.2	3
$Al_xGa_{1-x}As/In_{0.25}Ga_{0.75}As$	0.2	7
$In_xAl_{1-x}As/In_xGa_{1-x}As$	$0.33 \leq x \leq 0.66$	7-8
$Al_xGa_{1-x}N/GaN$	0.25	3

Wigner-Kramers-Brillouin (WKB) Tunneling Approximation. A non-local methodology for unified treatment of interface models has been presented in [468]. It allows a change of the interface condition from the Neumann to the Dirichlet type in the limiting case of very strong barrier reduction due to tunneling.

The equivalence of a generation process and the local tunneling current can be used as in the non-local Schottky contact model. Thus, a second model for the metal/semiconductor and semiconductor/semiconductor interfaces including a tunneling current component has been described [225, 329] which is similar to the one described for the Schottky contact. The tunneling current transport between the semiconductor segments 1 and 2 can be described. The thermionic current component is additionally evaluated.

$$\mathbf{n} \cdot \mathbf{G}_{\mathrm{tun}} = \frac{1}{\mathrm{q}} \cdot \mathbf{n} \cdot \nabla J_{\mathrm{tun}} \tag{3.77}$$

$$= \frac{A^* \cdot T_{\mathrm{L}}}{k_{\mathrm{B}}} \cdot E \cdot \Gamma(r) \cdot \ln \left[\frac{1 + \exp\left(-\mathrm{q} \cdot \dfrac{\psi_1 - \phi_{\mathrm{Fn},1}}{k_{\mathrm{B}} \cdot T_{\mathrm{L}}}\right)}{1 + \exp\left(-\mathrm{q} \cdot \dfrac{\psi_2 - \phi_{\mathrm{Fn},2}}{k_{\mathrm{B}} \cdot T_{\mathrm{L}}}\right)} \right] . \tag{3.78}$$

The probability for tunneling in (3.73) is calculated in a similar fashion as it is calculated in (3.59). Such a non-local model is physically sound and useful. The model implies the calculation of a path integral on the given mesh.

Its usefulness depends critically on the actual two- or three-dimensional implementation of the integration path for the tunneling probability in general-purpose simulators. Therefore the problem is shifted from finding a sound model to providing a sound and computationally efficient implementation in the two- or three-dimensional model. In DESSIS [227] a suggestion is made to form a submesh for the simulation for the tunneling, and the assumption is made that tunneling occurs in a planar process. In [227] a planar situation is constructed from the neighboring boundaries if a non-planar situation is encountered. Tunneling of this kind is restricted to the semiconductor/metal or semiconductor/semiconductor interface. Non-ideal insulators, such as SiN, cannot be modeled. Similar to the field-dependent model given in (3.76), (3.59) contains a fitting parameter, the so-called tunneling mass, which can be used to calibrate the model. This resolves the principal problem that the effective masses at a heterointerface change abruptly, although in a strict sense the electron wave function cannot be localized to either material. Additional effort is necessary in (3.78) in order to ensure zero current at thermal equilibrium.

Semiconductor/Semiconductor Thermal Interface. The lattice temperature is assumed to be continuous across semiconductor/semiconductor interfaces. In the case of DD simulation with self-heating, an additional thermal energy is accounted for at heterojunction interfaces. This thermal energy is produced when the carriers have to surmount the energy difference in the conduction and valence bands, ΔE_{C} and ΔE_{V} respectively. The energy equation reads

$$\frac{\mathbf{J}_{\mathrm{n}}}{\mathrm{q}} \cdot \Delta E_{\mathrm{C}} + \frac{\mathbf{J}_{\mathrm{p}}}{\mathrm{q}} \cdot \Delta E_{\mathrm{V}} = \mathrm{div}_{\mathrm{A}} \left(\mathbf{S}_{\mathrm{L}} \right) . \tag{3.79}$$

The expression $\mathrm{div}_{\mathrm{A}} \left(\mathbf{S}_{\mathrm{L}} \right)$ denotes the surface divergence of the thermal heat flux at the boundary considered. In HD simulation with self-heating, the thermal heat flow across the boundary is accounted for self-consistently.

3.2 Lattice and Thermal Properties

The semiconductor equations discussed in Section 3.1 contain several physical parameters. Accurate models for these parameters are crucial for successful device simulation. Lattice properties, such as the lattice constant, relative permittivity ε_r, and mass density ρ_L, are well determined for most of the important semiconductors. Others, such as thermal conductivity and heat capacity, have been studied but not always included consistently in device simulators as a function of lattice temperature or material composition. Due to the importance of these parameters for device simulation, they are discussed in the following section in a consistent manner. These parameters are used in the Poisson equation (3.1) and the lattice heat flow equation (3.18), which are solved not only in the semiconductor, but also in the insulating materials. Since the importance of various dielectrics has risen dramatically in the recent past, models for SiO_2, Si_3N_4, and high-k and low-k dielectrics are also presented.

3.2.1 Permittivity

The relative permittivity ε_r is one of the basic properties of semiconductor and insulating materials. The data reported in the literature show minor discrepancies about the values of this parameter (see Table 3.3). The temperature dependence can be neglected. Recently, insulating materials with high dielectric constants (high-k materials) have attracted much interest as a possible replacement of the gate oxide (SiO_2) in MOSFETs, in order to avoid tunneling and gate leakage currents. Low-k materials have gained attention for reduction of capacitances in multilayer passive structures. Table 3.3 also includes some of these materials.
The permittivities of *alloy materials* $A_{1-x}B_x$ are interpolated from the values of the *basic materials* as a quadratic function of x [1].

$$\varepsilon_r^{AB} = (1-x)\cdot\varepsilon_r^{A} + x\cdot\varepsilon_r^{B} + x\cdot(1-x)\cdot C_\varepsilon \tag{3.80}$$

C_ε is referred to as a nonlinear or bowing parameter. Setting $C_\varepsilon = 0$ yields a linear interpolation, as reported in the case of SiGe [288], AlGaAs [5, 207], InGaP [35], AlGaN and InGaN [143], GaAsSb [298], or InAlGaAs [208]. The non-zero bowing parameters used in this model are summarized in Table 3.4.

3.2.2 Mass Density

The values of the mass densities of the *basic materials* are well-known and are used to model the values for *alloy materials*. A linear interpolation between the values of *basic materials* is used. For Al_2O_3 the value for sapphire is given, which is the substrate material for GaN devices.

$$\rho_L^{AB} = (1-x)\cdot\rho_L^{A} + x\cdot\rho_L^{B} \tag{3.81}$$

The parameter values used are summarized in Table 3.5.

Table 3.3. Parameter values for the relative static permittivity

Material	ε_r	Reported Range	References
Si	11.9	11.7-11.9	[147, 364, 496]
Ge	16.0	16.0-16.2	[364, 496]
GaAs	13.1	12.5-13.18	[1, 5, 44, 67, 177, 364]
AlAs	10.1	10.0-10.1	[5, 44, 286, 444, 469, 513]
InAs	14.6	14.55-15.5	[1, 4, 67, 444, 147]
InP	12.4	12.3-12.61	[4, 59, 137, 147, 178, 444]
GaP	11.1	11.1	[1, 286, 469]
GaN	9.7	10.4 $\parallel$, 9.5 $\perp$ c-axis, 8.9	[143, 513]
AlN	8.5	8.5-9.0	[143, 328]
InN	13.52	13.52-15.3	[143, 373]
GaSb	15.7	15.69-15.7	[109, 110, 297, 385, 505]
AlSb	12.04	10.3-12.04	[257, 288, 385, 513]
InSb	16.8	16.8-17.88	[257, 297, 385, 513]
SiO_2	3.9	3.75-3.9	[179, 364, 496]
Si_3N_4	7.4	6.0-8.0	[331, 364, 496]
$BaSrTiO_3$	200.0	60.0-300.0	[26, 98, 276]
Ta_2O_5	30.0	20.0-75.0	[98, 276]
TiO_2	20.0	4.0-100.0	[98, 276]
Al_2O_3	9.8	8.5-13.0	[26, 276, 328]
BCB	2.6	2.65	[86, 402]
BeO	6.5	6.5	[328]
SiC-6H	10.3	9.66 $\parallel$, 10.3 $\perp$ c-axis	[205]
Diamond	5.7	5.7	[115]

Table 3.4. Non-zero bowing parameter values for the relative static permittivity

Material	C_ε	Reported Range ε_r	References
InGaAs	-1.0	13.1-14.1	[304, 444, 4] x=0.53
InAlAs	-1.0	12.46	[304, 207] x=0.52

3.2.3 Thermal Conductivity

The temperature dependence of the thermal conductivity of the *basic materials* and insulators is modeled by a simple power law

$$\kappa_L(T_L) = \kappa_{300} \cdot \left(\frac{T_L}{300\ \mathrm{K}} \right)^{\alpha} \tag{3.82}$$

where κ_{300} is the value for the thermal conductivity at 300 K. This approximation agrees well with experimental data [253, 256, 287, 330], as presented in Fig. 3.1 and Fig. 3.2, where comparisons in the temperature range from 300 K to 800 K are given. The parameter values used are summarized in Table 3.6.

Table 3.5. Parameter values for the mass density

Material	ρ_L [g/cm^3]	Reported Range	References
Si	2.33	2.328 -2.33	[364, 469, 496]
Ge	5.327	5.3234-5.327	[364, 469]
GaAs	5.32	5.307-5.36	[2, 5, 147, 177, 364]
AlAs	3.76	3.729-3.76	[2, 5, 147]
InAs	5.667	5.667-5.70	[1, 67, 147, 513]
InP	4.81	4.79-4.81	[1, 4, 137, 147, 178]
GaP	4.138	4.13-4.14	[1, 147]
GaN	6.087	6.087-6.15	[143, 297]
AlN	3.23	3.23-3.26	[143, 402]
InN	6.81	6.78-6.81	[143, 278]
GaSb	5.613	5.61-5.63	[109, 110, 385, 469]
AlSb	4.26	4.26-4.34	[288, 385, 469]
InSb	5.77	5.77-5.80	[297, 385, 469]
SiO_2	2.203	2.2-2.27	[364, 496]
Si_3N_4	3.1	3.1	[364, 496]
Al_2O_3	3.9	3.8-3.98	[328, 402]
BCB	0.95	0.93-1.051	[86, 402]
BeO	2.85	2.85	[328]
SiC-6H	3.21	3.21-3.248	[205]
Diamond	3.515	3.515	[115]

Most of the III-V materials exhibit at most half the thermal conductivity of silicon. This is not the case for GaN and AlN, which have similar or even higher κ_L. Unfortunately many references published for the temperature dependence in GaN refer back to the measurements of [463], which may be outdated due to the progress in material growth. The value available for InN relies on measurements of [278], and the maximum value in the range is the theoretical limit given in the same work. Fig. 3.3 compares modeled and measured thermal conductivity of GaN, AlN, and SiC as a function of temperature. Natural diamond is the best heat-conducting material and is included for reference.

In the case of *alloy materials* $A_{1-x}B_x$, κ_L varies between the values of the *basic materials* (A and B). A harmonic mean is used to model κ_{300}. An additional bowing factor C_κ is introduced in order to account for the significant reduction of the thermal conductivity for material compositions x in the mid ranges between the *basic materials*. The exponent α is linearly interpolated because of the lack of experimental data at temperatures other than 300 K. There are, to the authors' knowledge, no measurements of the temperature dependence of the thermal conductivity available for III-V ternary nitrides and antimonides except of InAsSb. The parameter values used are summarized in Table 3.7.

Table 3.6. Parameter values for the thermal conductivity model

Material	κ_{300} [W/K m]	α	Reported κ_{300} [W/K m]	References
Si	148	-1.65	150-154	[364, 448, 496]
Ge	60	-1.25	60	[364, 496]
GaAs	46	-1.25	37-46	[137, 253, 364, 513]
AlAs	80	-1.37	80-91	[5, 253]
InAs	27.3	-1.1	27-48	[137, 253, 513]
InP	68	-1.4	68	[137, 253, 513]
GaP	77	-1.4	77	[253, 513]
GaN	125	-0.43	125-150	[462, 463]
AlN	285	-1.577	200-285	[462, 471]
InN	38.4	-	38.4-176	[278, 500]
GaSb	33	-1.36	32-35	[297, 330, 505, 510]
AlSb	57	-1.36	57-91	[288, 330, 505]
InSb	17	-1.26	15-18	[297, 330, 505, 510]
SiO_2	1.38	0.33	1.38-1.4	[90, 364, 496]
Si_3N_4	18.5	0.33	15-27	[90, 331, 364]
Al_2O_3	28	-1.0	15-40	[132, 331, 328]
BCB	0.2	-	0.18-0.24	[402]
BeO	248	-1.27	248-300	[320, 328]
AlN (ceramic)	150	-1.84	145.9-180	[320, 402, 470]
SiC-6H	490	-1.61	320-490	[132, 205, 470]
Diamond	2500	-1.85	500-3200	[528]

Table 3.7. Parameter values for thermal conductivity bowing factors

Material	C_κ [W/K m]	Refs.	Material	C_κ [W/K m]	Refs.
SiGe	2.8	[330]	AlGaN	–	-
AlGaAs	3.3	[4, 5]	InGaN	–	-
InGaAs	1.4	[444]	GaAsSb	–	-
InAlAs	3.3	-	AlGaSb	–	-
InAsP	3.3	[330]	AlAsSb	–	-
GaAsP	1.4	[330]	InAlSb	–	-
InGaP	1.4	[5]	InAsSb	1.6	[298]

$$\kappa_{300}^{AB} = \left(\frac{1-x}{\kappa_{300}^{A}} + \frac{x}{\kappa_{300}^{B}} + \frac{(1-x)\cdot x}{C_\kappa} \right)^{-1} \tag{3.83}$$

$$\alpha^{AB} = (1-x)\cdot\alpha^{A} + x\cdot\alpha^{B} \tag{3.84}$$

Fig. 3.4 and Fig. 3.5 compare the modeled thermal conductivity of *alloy materials* as a function of the material composition at 300 K to experimental data from [4, 5, 38, 253, 287, 330].

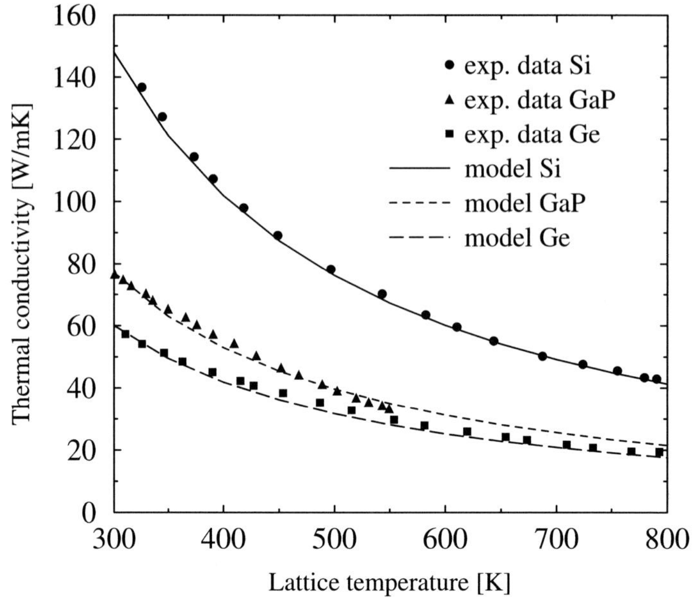

Fig. 3.1. Temperature dependence of the thermal conductivity: Comparison of the model and experimental data for Si, Ge, and GaP.

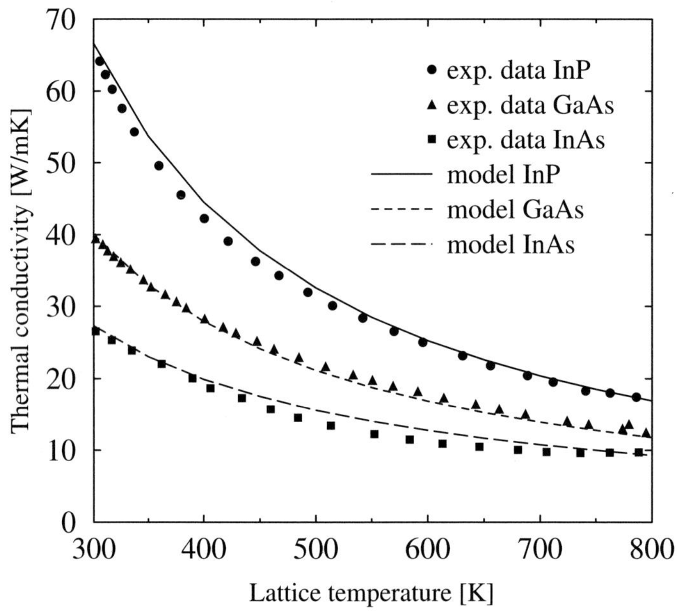

Fig. 3.2. Temperature dependence of the thermal conductivity: Comparison of the model and experimental data for InP, GaAs, and InAs.

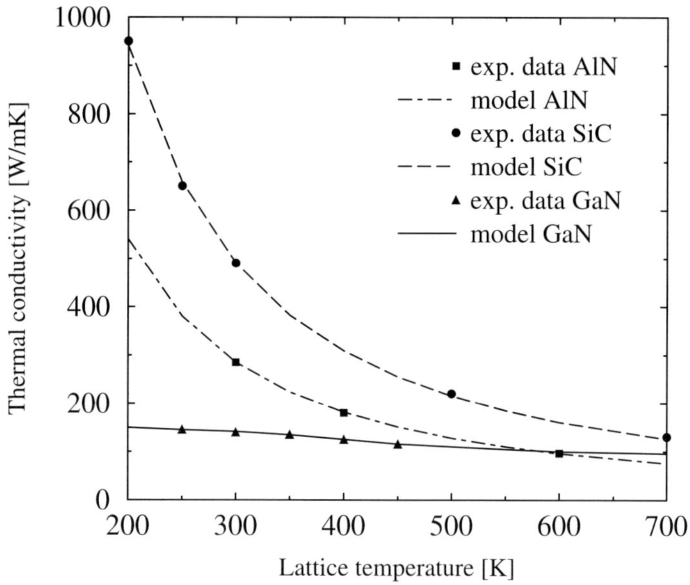

Fig. 3.3. Temperature dependence of the thermal conductivity: Comparison of the model and experimental data for GaN, AlN, and SiC.

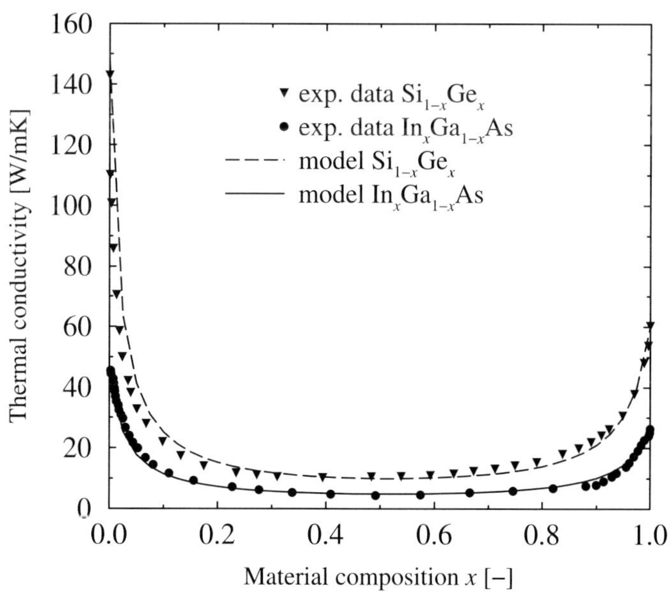

Fig. 3.4. Material composition dependence of the thermal conductivity: Comparison of the model and experimental data for SiGe and InGaAs.

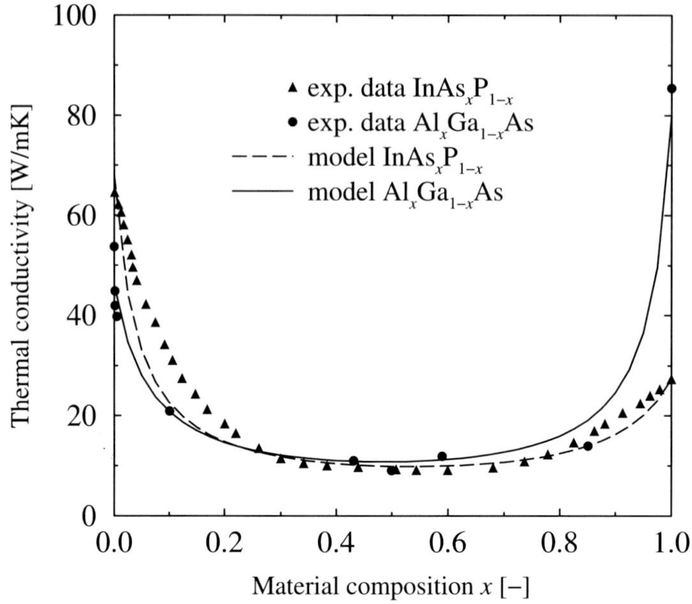

Fig. 3.5. Material composition dependence of the thermal conductivity: Comparison of experimental data and the model for $InAs_xP_{1-x}$ and $Al_xGa_{1-x}As$.

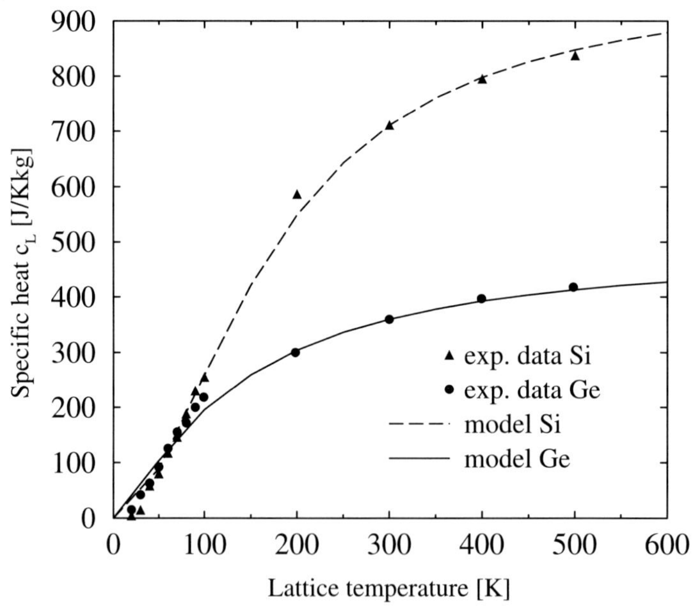

Fig. 3.6. Temperature dependence of the specific heat: Comparison of experimental data and the model for Si and Ge.

3.2.4 Specific Heat

The lattice specific heat c_L is modeled by

$$c_\mathrm{L}(T_\mathrm{L}) = c_{\mathrm{L},300} + c_1 \cdot \frac{\left(\dfrac{T_\mathrm{L}}{300\ \mathrm{K}}\right)^\beta - 1}{\left(\dfrac{T_\mathrm{L}}{300\ \mathrm{K}}\right)^\beta + \dfrac{c_1}{c_{\mathrm{L},300}}} . \tag{3.85}$$

$c_{\mathrm{L},300}$ is the value for the specific heat at 300 K and c_1 is a fitting parameter. The model is used for the *basic materials* and for insulators. Fig. 3.6, Fig. 3.7, and Fig. 3.8 show comparisons between experimental data and the results obtained with the model for the specific heat as a function of temperature. Excellent agreement is obtained in a wide temperature range (50 K – 800 K). The parameter values are summarized in Table 3.8. The specific heat coefficients for *alloy materials* are expressed by a linear interpolation between the values of the *basic materials* (A and B).

$$c_\mathrm{L}^\mathrm{AB} = (1-x) \cdot c_\mathrm{L}^\mathrm{A} + x \cdot c_\mathrm{L}^\mathrm{B} \tag{3.86}$$

The specific heat is then expressed by (3.85).

Table 3.8. Parameter values for the specific heat model

Material	c_{300} [J/K kg]	c_1 [J/K kg]	β	Reported c_{300} [J/K kg]	References
Si	711	255	1.85	700-711	[364, 496]
Ge	360	130	1.3	310-360	[364]
GaAs	322	50	1.6	322-350	[364, 496, 513]
AlAs	441	50	1.2	441-490	[513]
InAs	394	50	1.95	394	[513]
InP	410	50	2.05	410	[513]
GaP	519	50	2.6	519	[513]
GaN	491	70	1.0	491	[10]
AlN (poly)	748	482	2.29	490-748	[338, 513]
InN	325	57	1.0	325	[278, 501]
GaSb	250	40	2.0	250-404	[297, 513]
AlSb	437	11	2.0	151-437	[513]
InSb	230	20	2.0	200-370	[297, 513]
SiO_2	709	696	1.5	782-1000	[364, 448]
Si_3N_4	787	820	1.5	787-800	[331, 448]
Al_2O_3	796	-	-	795-796	[328, 402]
BCB	1267	1807	2.0	1267	[402]
SiC-6H	715	620	2.6	715	[366]
Diamond	520	1366	3.0	520	[297]

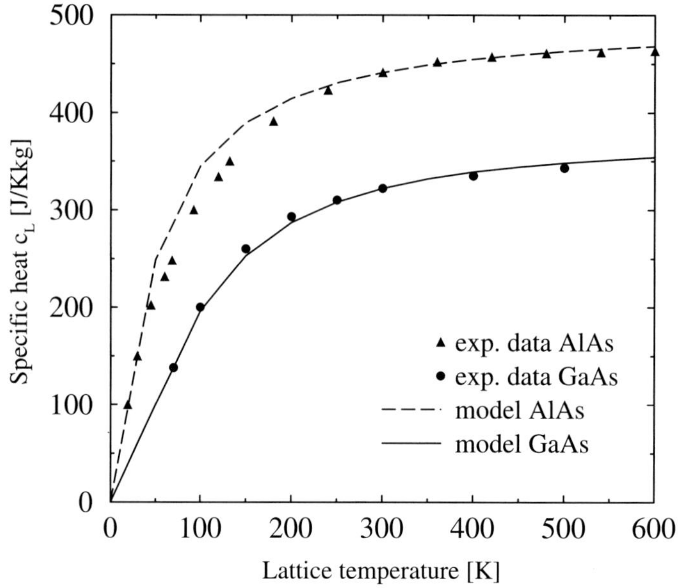

Fig. 3.7. Temperature dependence of the specific heat: Comparison of experimental data and the model for GaAs and AlAs.

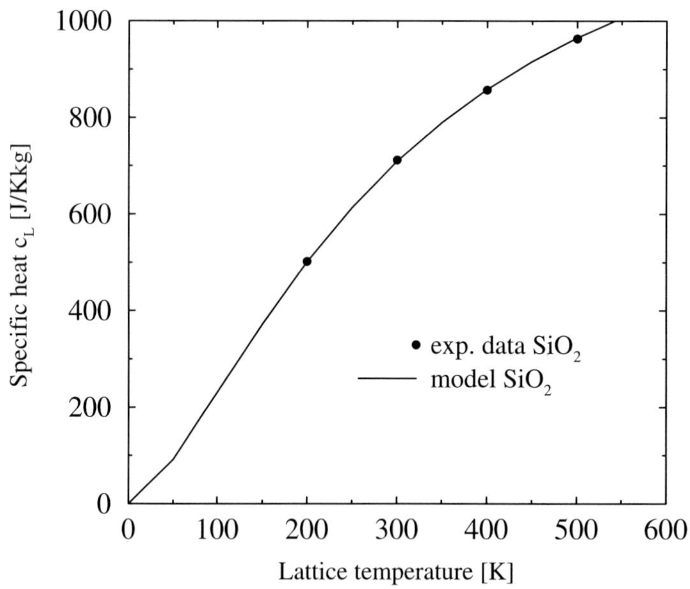

Fig. 3.8. Temperature dependence of the specific heat: Comparison of experimental data and the model for SiO_2.

3.3 Band Structure

In this section models for the bandgap energies, for the effective masses of electrons and holes, and for the effective density of states in silicon, germanium, SiGe, and III-V materials are presented.

3.3.1 Bandgap Energy

The bandgap (or forbidden energy zone) is one of the most important semiconductor parameters. Various models define the temperature dependence of the bandgap energy in semiconductors (e.g. [530]). For an alloy $A_{1-x}B_x$, the temperature-dependent bandgap energies of the constituents (A and B) are calculated first. The bandgap and the energy offset are then calculated as a function of the material composition. This is important to assure consistency between the values for *alloy materials* at $x = 0$ and $x = 1$ and the values for the respective *basic materials*. For materials where the bandgap changes between direct and indirect, the multiple valley conduction bands are considered.

Temperature Dependence of the Bandgap. In this work the model of Varshni [530] is used for *basic materials*. The temperature dependence is calculated by (3.87), where $E_{g,0}$ is the bandgap at $T_L = 0$ K.

$$E_g = E_{g,0} - \frac{\alpha \cdot T_L^2}{\beta + T_L} \tag{3.87}$$

The parameter values $E_{g,0}$, α, and β are summarized in Table 3.9. Note that for these materials, always the lowest conduction band valley minimum is taken into account. Table 3.10 summarizes the resulting values $E_{g,300}$ for the bandgap energies at $T_L = 300$ K and compares them to data reported in the literature. A good review for the III-V materials is given in [532].
For silicon, two additional models can be chosen which are based on second and third order polynomial fits, as given in (3.88). The first one has been proposed by Gaensslen [161, 162]. The second model is based on data from Green [188].

$$E_g = E_0 + E_1 \cdot \left(\frac{T_L}{300\text{ K}}\right) + E_2 \cdot \left(\frac{T_L}{300\text{ K}}\right)^2 + E_3 \cdot \left(\frac{T_L}{300\text{ K}}\right)^3 \tag{3.88}$$

The parameter values for these models are summarized in Table 3.11.
In Fig. 3.9, the results obtained with the three different models for silicon are compared to data from [188]. Fig. 3.10 compares different models for the temperature dependence of the direct gap in GaAs [457, 91, 191] to experimental data from [88, 389]. Fig. 3.11 shows the modeled temperature dependence for InP [197, 529] and compares it to data summarized in [290].

Table 3.9. Parameter values of the lowest valleys for the bandgap energy model

Material	Min.	$E_{g,0}$ [eV]	α [eV/K]	β [K]	Rep. $E_{g,0}$ [eV]	References
Si	X	1.1695	4.73×10^{-4}	636	1.17	[364, 469, 496]
Ge	L	0.7437	4.774×10^{-4}	235	0.74	[496]
GaAs	Γ	1.521	5.58×10^{-4}	220	1.51-1.55	[354, 406, 538, 549]
AlAs	X	2.239	6.0×10^{-4}	408	2.22-2.239	[192, 253, 469]
InAs	Γ	0.420	2.5×10^{-4}	75	0.414-0.43	[275, 513, 549]
InP	Γ	1.421	3.63×10^{-4}	162	1.42-1.432	[253, 496, 538]
GaP	X	2.338	5.771×10^{-4}	372	2.338-2.346	[253, 469, 496]
GaN	Γ	3.50	7.32×10^{-4}	700	3.50-3.51	[409, 487, 532, 533]
AlN	Γ	6.312	18.0×10^{-4}	1462	6.118-6.312	[409, 487, 532]
InN	Γ	2.01	12.0×10^{-4}	627	1.994-2.01	[378, 501, 532]
GaSb	Γ	0.816	3.78×10^{-4}	94	0.81-0.813	[288, 297, 339, 469]
AlSb	X	1.687	4.97×10^{-4}	213	1.66-1.715	[288, 469, 87]
InSb	Γ	0.237	3.2×10^{-4}	170	0.236-0.24	[288, 469, 532]

Table 3.10. Bandgap energies of the lowest valleys at 300 K compared to reported experimental data

Material	Minimum	$E_{g,300}$ [eV]	Reported $E_{g,300}$ [eV]	References
Si	X	1.124	1.12-1.1242	[364, 496, 469]
Ge	L	0.663	0.66-0.67	[364, 469, 496]
GaAs	Γ	1.424	1.42-1.43	[2, 59, 275, 364]
AlAs	X	2.163	2.14-2.168	[192, 275, 469]
InAs	Γ	0.360	0.354-0.37	[237, 275, 354, 444, 513]
InP	Γ	1.350	1.34-1.351	[354, 469, 513]
GaP	X	2.261	2.26-2.272	[513, 364, 469]
GaN	Γ	3.434	3.2-3.43	[409, 487, 533]
AlN	Γ	6.22	6.02-6.23	[409, 487, 532]
InN	Γ	1.894	1.89-1.95	[378, 501]
InN	Γ	0.8	0.8	[326]
GaSb	Γ	0.73	0.67-0.75	[56, 275, 297, 339, 469]
AlSb	X	1.60	1.58-1.615	[84, 275, 288, 469, 505]
InSb	Γ	0.176	0.165-0.23	[84, 275, 297, 469, 505]

Table 3.11. Parameter values for the bandgap energy model in silicon

Model	E_0 [eV]	E_1 [eV]	E_2 [eV]	E_3 [eV]	$E_{g,300}$ [eV]
Gaensslen	1.1785	-0.02708	-0.02745	0.0	1.124
Green	1.17	0.00572	-0.06948	0.018	1.124

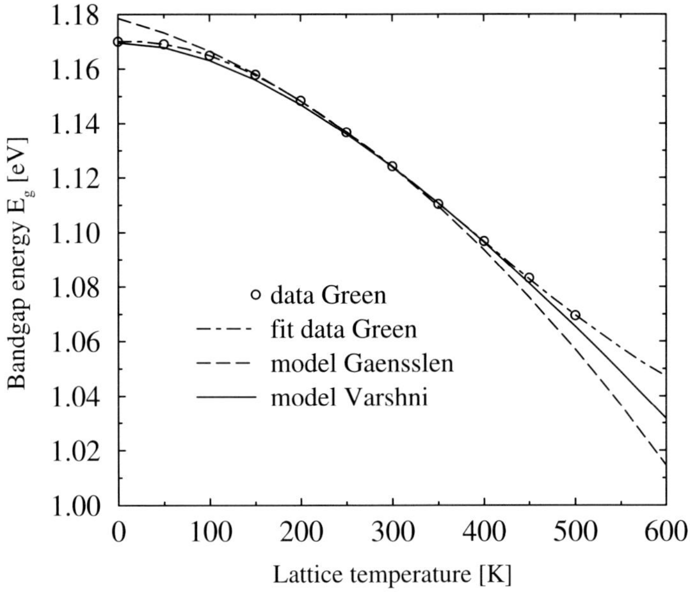

Fig. 3.9. Comparison of experimental data and different models for the temperature dependence of the bandgap in Si.

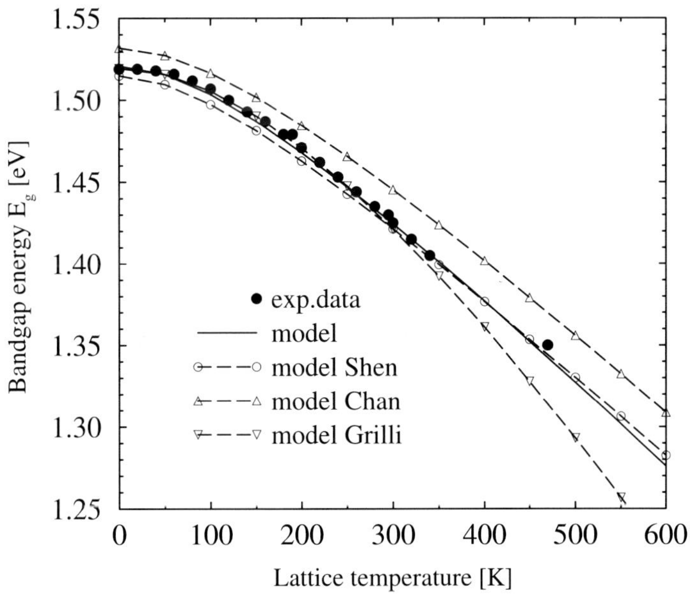

Fig. 3.10. Comparison of experimental data and different models for the temperature dependence of the bandgap in GaAs.

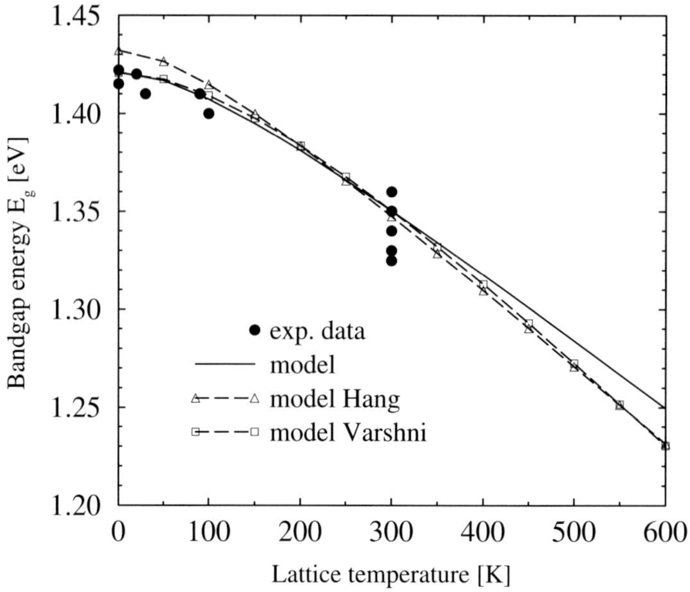

Fig. 3.11. Comparison of experimental data and different models for the temperature dependence of the bandgap in InP.

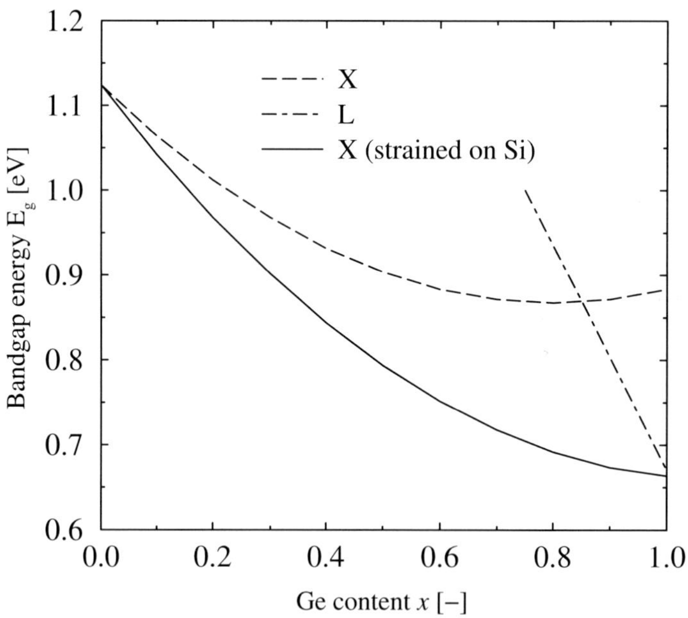

Fig. 3.12. Material composition dependence of the X and L bandgaps in $Si_{1-x}Ge_x$ at 300 K.

Table 3.12. Parameter values of higher valleys for the bandgap energy model

Material	Min.	$E_{g,0}$ [eV]	α [eV/K]	β [K]	Rep. $E_{g,0}$ [eV]	References
GaAs	X	1.981	4.6×10^{-4}	204	1.981-1988	[2, 275, 391, 532]
AlAs	Γ	2.891	8.78×10^{-4}	332	2.907-3.02	[2, 59, 391, 467]
InAs	X	2.278	5.78×10^{-4}	83	1.43-2.27	[275, 460, 532]
InP	X	2.32	7.66×10^{-4}	327	2.32-2.38	[4, 147, 459]
GaP	Γ	2.88	8.0×10^{-4}	300	2.869-2.895	[147, 255, 286]
AlN	Γ	7.0	18.0×10^{-4}	1462	6.9-7.0	[154, 462, 297]
GaSb	X	1.141	4.75×10^{-4}	94	1.12-1.242	[297, 532]
AlSb	Γ	2.386	4.2×10^{-4}	140	2.35-2.386	[532]

Table 3.13. Bandgap energies of the higher valleys at 300 K compared to reported data

Material	Minimum	$E_{g,300}$ [eV]	Reported $E_{g,300}$ [eV]	References
GaAs	X	1.899	1.899-1.91	[2, 275, 391]
AlAs	Γ	2.766	2.671-2.766	[391, 444, 467]
InAs	X	2.142	1.37-2.14	[253, 275, 460]
InP	X	2.21	2.21-2.30	[3, 275, 459, 460]
GaP	Γ	2.76	2.73-2.85	[147, 255, 286, 532]
AlN	Γ	6.9	6.9	[154]
GaSb	X	1.033	1.03-1.05	[3, 297]
AlSb	Γ	2.3	2.2-2.3	[3, 87, 275]
InSb	X	1.0	1.0-1.63	[3, 275, 297]

Semiconductor Alloys. In the case of *alloy materials*, the temperature dependent bandgaps of the constituents, E_g^A and E_g^B, are calculated by (3.87). However, for materials where the bandgap changes between direct and indirect, the multiple valley conduction bands are considered. For this purpose, additional model parameters are needed for the higher energy valleys in the respective III-V binary materials (Table 3.12). The resulting bandgaps at 300 K, $E_{g,300}$, are included in Table 3.13. For some of the binary materials, such as InSb, the reported values scatter significantly. The value for the X bandgap at 300 K proposed in [3] is erroneously cited in [532]. Temperature-dependent data for the X valley in InSb is not available, to the authors' knowledge.
The bandgap and the energy offset of an alloy $A_{1-x}B_x$ are calculated by

$$E_{g,X}^{AB} = E_{g,X}^{A} \cdot (1-x) + E_{g,X}^{B} \cdot x + C_{g,X} \cdot (1-x) \cdot x \tag{3.89}$$

$$E_{g,\Gamma}^{AB} = E_{g,\Gamma}^{A} \cdot (1-x) + E_{g,\Gamma}^{B} \cdot x + C_{g,\Gamma} \cdot (1-x) \cdot x \tag{3.90}$$

$$E_{g}^{AB} = \min\left(E_{g,X}^{AB}; E_{g,\Gamma}^{AB}\right) \tag{3.91}$$

The bowing parameters $C_{g,X}$ and $C_{g,\Gamma}$ are summarized in Table 3.14. Additional bowing parameters C_g are given as a reference for the case when a one-valley bandgap fit is used.

Table 3.14. Parameter values for the bandgap energy model for alloy materials

Material	$C_{g,\Gamma}$ [eV]	$C_{g,X}$ [eV]	C_g [eV]	Reported [eV]	References
SiGe	-	-0.4	-0.4	-0.238^X,-0.396^X	[133, 488]
AlGaAs	0.0	-0.055	0.7	0^Γ, -0.143^X,-0.055^X	[87, 406, 444, 532]
InGaAs	-0.475	-	-0.475	-0.4, -0.475, -0.555	[253, 305, 469, 549]
InAlAs	-0.3	-0.713	1.2	-0.689^Γ,-0.24^Γ	[253, 304, 469]
InAsP	-0.32	-	-0.32	-0.101, -0.32	[253, 469]
GaAsP	-0.21	-0.21	0.5	-0.176^Γ,-0.23^Γ	[75, 253]
InGaP	-0.67	-0.17	0.6	-0.786^Γ,-0.6^Γ,-0.18^X	[4, 253, 469]
AlGaN	-1.65	-	-1.2596	3.88 (x=0.2), -1.0	[454, 532, 533]
InGaN	-2.05	-	-2.05	-2.05,-3.0	[500, 533]
InGaN	-0.97	-	-0.97	-0.97	[326]
GaAsSb	-1.2	-	-1.2	-1.2,-1.43	[99, 297, 339, 532]
AlGaSb	-0.47	-0.48	0.4	-0.69^Γ,-0.47^Γ,-0.48^X,-0.368^Γ	[87, 275, 532]
AlAsSb	-0.8	-0.28	-0.28	-0.25^Γ, -0.8^Γ, -0.28^X	[532]
InAsSb	-0.598	-	-0.598	-0.58,-0.59,-0.598,-0.6	[3, 87, 275, 532]
InAlSb	-0.43	-	-0.43	-0.43	[87, 275]
InGaAsP	-	-	-0.13	-0.13	[532]
AlGaInAs	-	-	-0.22	-0.22	[532]

$$E_g^{AB} = E_g^A \cdot (1-x) + E_g^B \cdot x + C_g \cdot (1-x) \cdot x \tag{3.92}$$

For example, such a one-valley bandgap fit is used in the case of the technologically important strained $Si_{1-x}Ge_x$ grown on silicon (see Fig. 3.12). In certain cases depending on the strain, the bandgap can become smaller than the one of pure Ge for $x > 0.6$ [234]. In the unstrained case, however, an X to L gap transition is observed at about $x = 0.85$.

The material composition dependence of the Γ, L, and X bandgaps at 300 K in $Al_xGa_{1-x}As$ is shown in Fig. 3.13. A direct-to-indirect gap transition is observed at about $x = 0.42$. The one-valley bandgap fit which is included for comparison gives good agreement only for $x < 0.28$. Fig. 3.14 shows the temperature dependence of the bandgap in $Al_xGa_{1-x}As$ with aluminum content as a parameter. Fig. 3.15 shows the material composition dependence of the Γ, L, and X bandgaps in $In_xGa_{1-x}As$ at 300 K. As can be seen, this material has a direct bandgap for the entire composition range. Therefore only the energy of the Γ valley is taken into account. However, in the case of strain, the bandgap can differ significantly [267]. For example, a good fit to the strained bandgap values of $In_xGa_{1-x}As$ grown directly on GaAs can be achieved by changing the $E_{g,0}$ of InAs from 0.42 eV to 0.58 eV.

Fig. 3.16 shows the temperature dependence of the bandgap in $In_xGa_{1-x}As$ with indium content as a parameter. In the case of $In_xAl_{1-x}As$, there is a transition from direct to indirect bandgap at about $x = 0.3$, and the use of one-valley

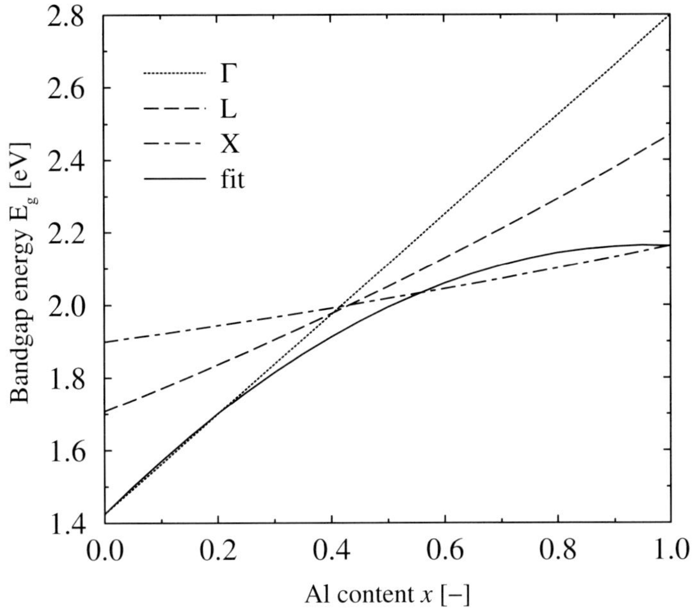

Fig. 3.13. Material composition dependence of the Γ, L, and X bandgaps in $Al_xGa_{1-x}As$ at 300 K.

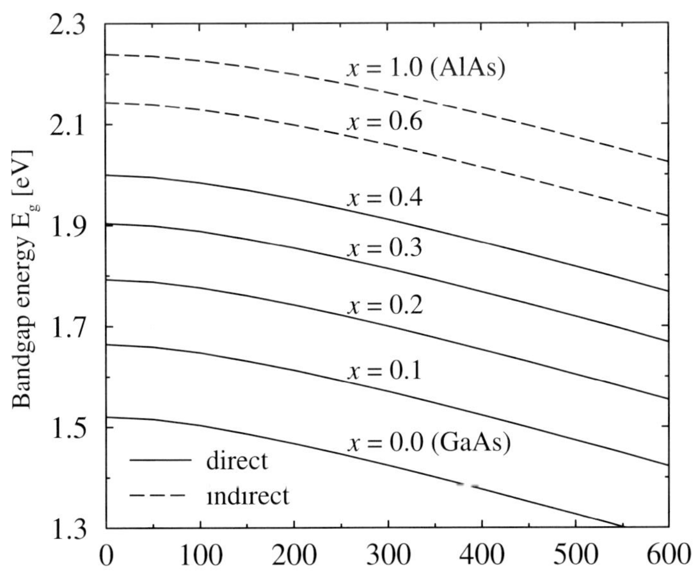

Fig. 3.14. Temperature dependence of the bandgap in $Al_xGa_{1-x}As$ with Al content as a parameter.

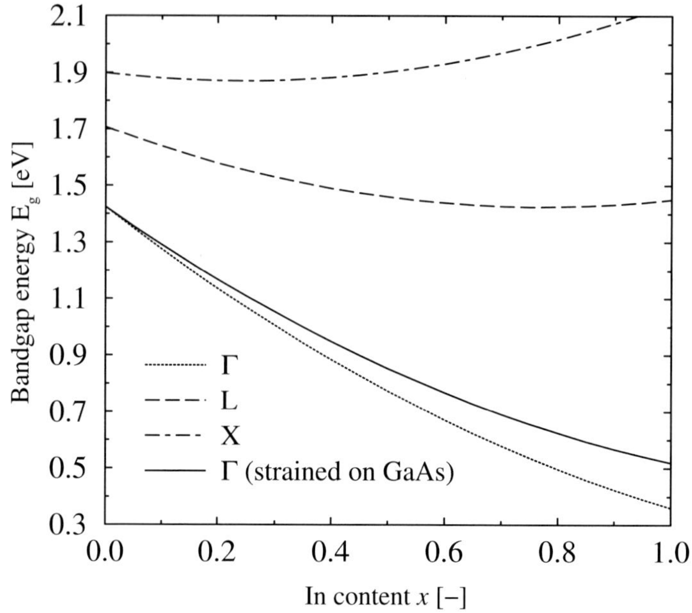

Fig. 3.15. Material composition dependence of the Γ, L, and X bandgaps in $In_xGa_{1-x}As$ at 300 K.

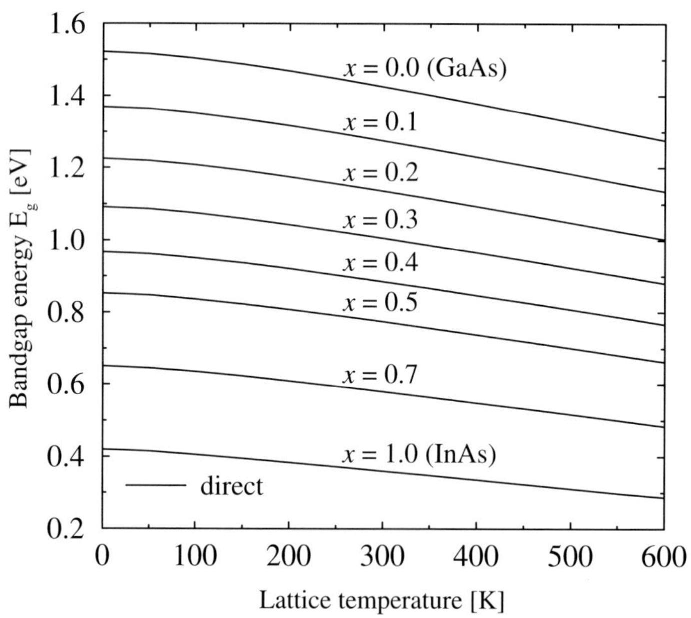

Fig. 3.16. Temperature dependence of the bandgap in $In_xGa_{1-x}As$ with In content as a parameter.

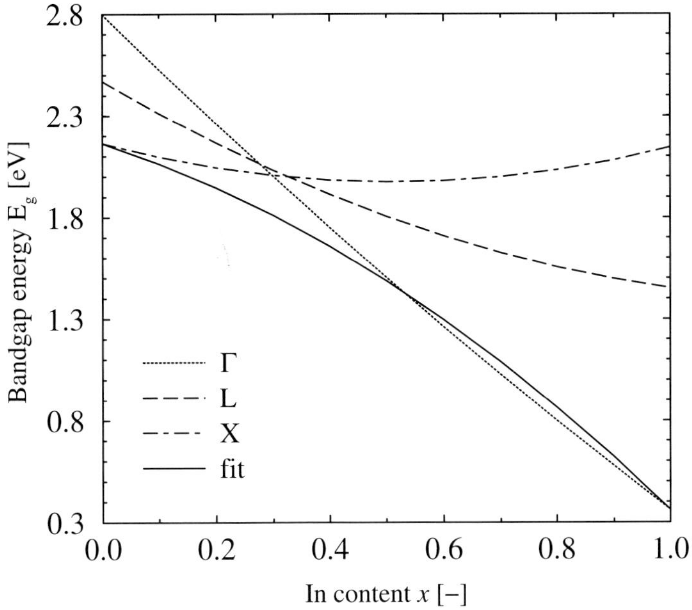

Fig. 3.17. Material composition dependence of the Γ, L, and X bandgaps in $In_xAl_{1-x}As$ at 300 K.

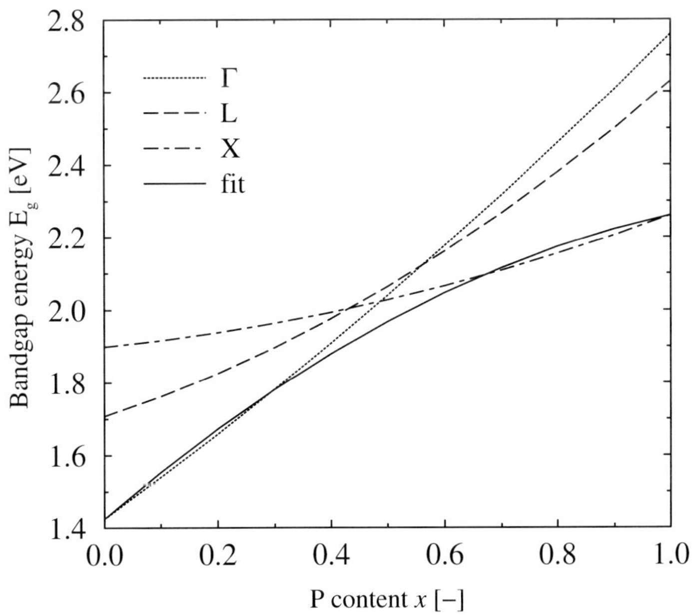

Fig. 3.18. Material composition dependence of the Γ, L, and X bandgaps in $GaAs_{1-x}P_x$ at 300 K.

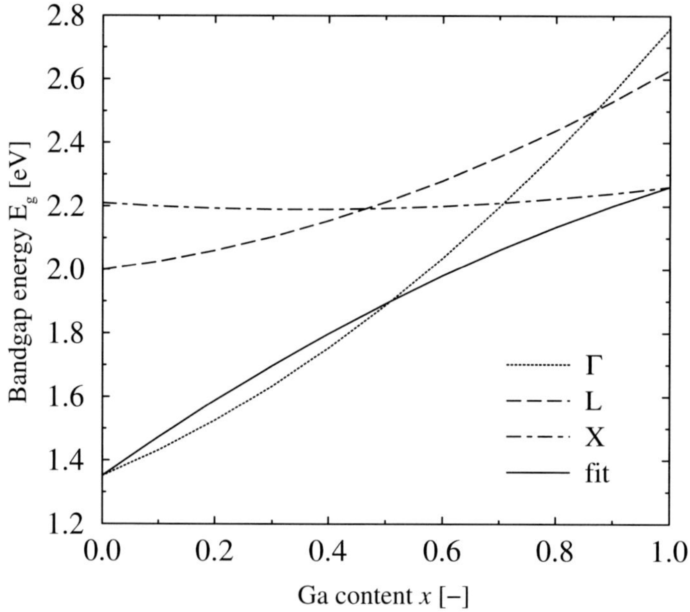

Fig. 3.19. Material composition dependence of the Γ, L, and X bandgaps in $Ga_xIn_{1-x}P$ at 300 K.

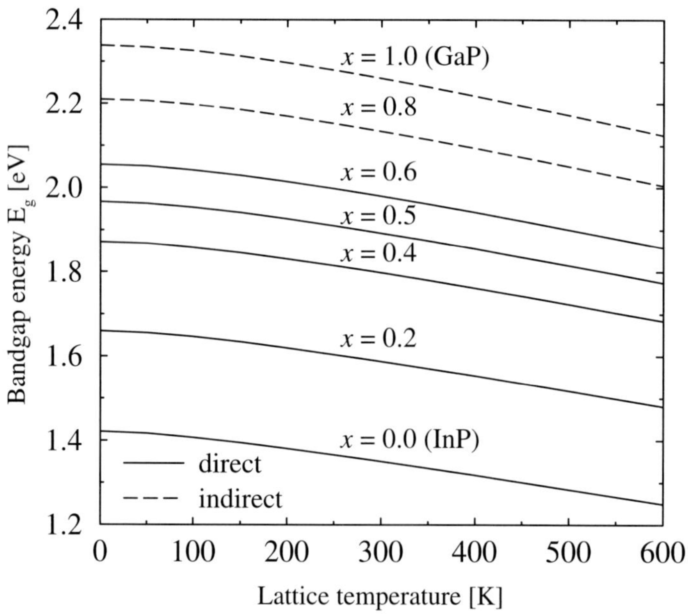

Fig. 3.20. Temperature dependence of the bandgap in $Ga_xIn_{1-x}P$ with Ga content as a parameter.

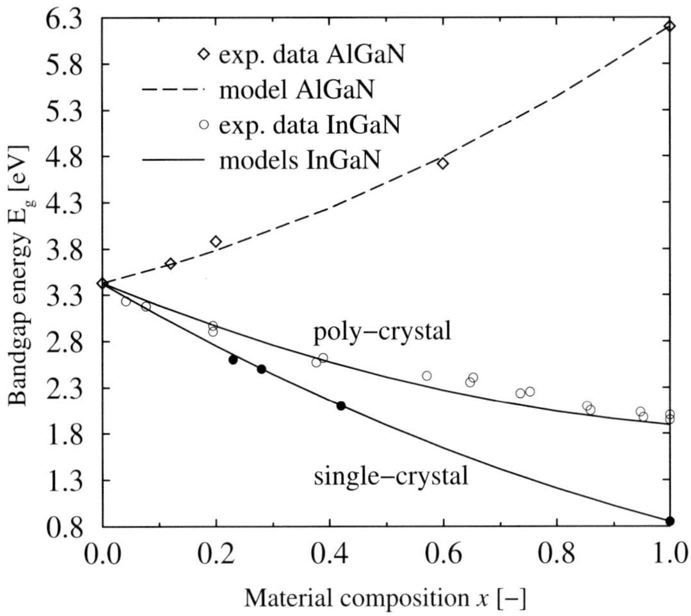

Fig. 3.21. Material composition dependence of the bandgap in $Al_xGa_{1-x}N$ and $In_xGa_{1-x}N$ at 300 K.

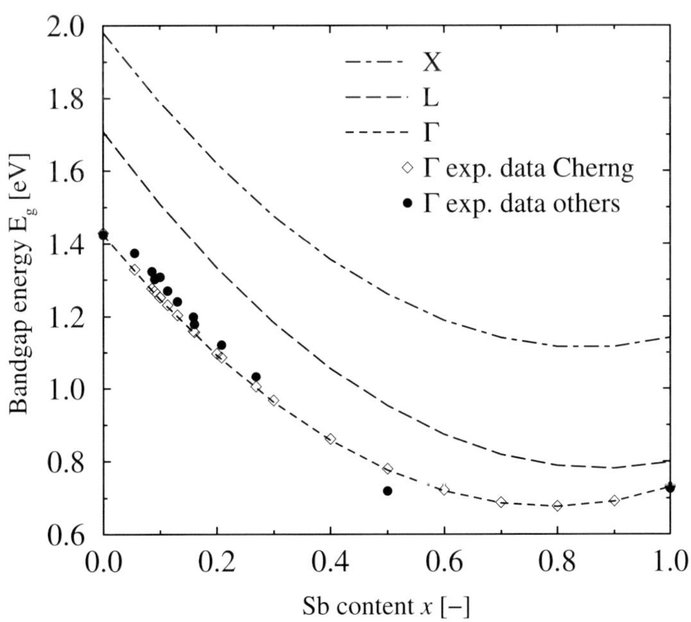

Fig. 3.22. Material composition dependence of the bandgap in $GaAs_{1-x}Sb_x$ at 300 K.

model is nearly impossible. The latter gives comparatively accurate values for the lattice-matched case of $x = 0.52$ and above (see Fig. 3.17). $InAs_xP_{1-x}$ has a direct gap for the complete material composition range, so only the Γ valley energy is calculated. The bowing parameter value suggested in [253] is used. $GaAs_{1-x}P_x$ has a direct-to-indirect gap transition at about $x = 0.5$. As can be seen in Fig. 3.18, a one-valley model can be successfully used for this material. Fig. 3.19 shows the material composition dependence of the Γ, L, and X bandgap energies in $Ga_xIn_{1-x}P$ at 300 K. The direct-to-indirect gap transition is at about $x = 0.7$. Only at the technologically important value of x=0.51, where the $Ga_xIn_{1-x}P$ lattice constant matches that of GaAs, does the one-valley model give a good fit for the bandgap. The temperature dependence of the bandgap in $Ga_xIn_{1-x}P$ is shown in Fig. 3.20 with gallium content as a parameter. The bandgap situation for nitrides is complicated, due to polarization effects and the resulting high carrier densities in the vicinity of the interface. The significant scatter for the bowing parameter for $Al_xGa_{1-x}N$ has been resolved in [533] for $x < 0.16$ and has been attributed to a systematic discrepancy of the two experimental methods applied. From these data, a specific Γ (first) valley value for the bowing parameter has been obtained. Fig. 3.21 shows a consistent description by a one-valley model based on experimental data from [454]. For InGaN, the situation is given in Fig. 3.21 based on the prelimenary experimental data available for poly-crystal material [500] and recent advances for single crystal InN [326]. Fig. 3.22 compares the one-valley model to experimental data from [99, 339, 297, 532] at $T_{\mathrm{L}} = 300$ K for GaAsSb. For the other ternary Sb-based materials, an overview of the bandgaps is given in Table 3.14. The situation is complicated, as some of the materials such as AlGaSb have two crossover (Γ-L and L-X) transitions [288]. The bowing parameters are given for the multi-valley approach including the range for the material composition, and, where available, a one-valley approach is additionally given. For further reading, see the overview papers in [3, 532]

3.3.2 Bandgap Offsets

The problem of the alignment of the bandgaps of two or more different materials is well-discussed in the literature [84, 523, 524, 532] but is not completely solved, since the alignment also depends on the growth conditions. The many approaches suggested differ mainly in whether to use the electron affinity to align the conduction bands, whether to align the midgap energies, or whether to align the valence band edges. In this work the last approach is used. An energy offset, E_{off}, is used to align the band edge energies of different materials. E_{off} is an arbitrary value, which, if it is changed consistently for all materials at the same time, does not change the results of the model. In this work, silicon is chosen as the reference for the Group IV materials. The origin of the energy axis is assumed to be in the middle of the silicon bandgap at 300 K. This corresponds to a fixed value for silicon of $E_{\mathrm{off}} = -E_{\mathrm{g}}/2 = -0.562$ eV. For the III-V material system, GaAs

Table 3.15. Parameter values for the bandgap energy off-sets

Material	E_{off} [eV]	Material	E_{off} [eV]
Si	-0.562	GaN	-1.717
Ge	-0.157	AlN	-2.415
GaAs	-0.712	InN	-0.665
AlAs	-1.008	GaSb	0.012
InAs	-0.286	AlSb	-0.695
InP	-0.724	InSb	0.042
GaP	-1.062		

is chosen as the reference material for arsenides, phosphides, and antimonides, since they are used together in the same device. Therefore the offset for GaAs is $E_{\mathrm{off}} = -E_{\mathrm{g}}/2 = -0.712$ eV. The default offsets for other materials, summarized in Table 3.15, are chosen to give good agreement with reported data from [38, 253, 84, 523, 524]. The midgap of GaN serves as a reference energy for the nitride material system. The energies of the conduction and valence band edges are calculated by

$$E_{\mathrm{V}} = E_{\mathrm{off}} \tag{3.93}$$

$$E_{\mathrm{C}} = E_{\mathrm{V}} + E_{\mathrm{g}}. \tag{3.94}$$

For *alloy materials*, the following expressions for the conduction and valence band energies are often used:

$$E_{\mathrm{C}} = E_{\mathrm{off}}^{\mathrm{A}} + E_{\mathrm{g}}^{\mathrm{A}} + \left(\frac{\Delta E_{\mathrm{C}}}{\Delta E_{\mathrm{g}}}\right) \cdot \left(E_{\mathrm{g}}^{\mathrm{AB}} - E_{\mathrm{g}}^{\mathrm{A}}\right) \tag{3.95}$$

$$E_{\mathrm{V}} = E_{\mathrm{C}} - E_{\mathrm{g}}^{\mathrm{AB}}. \tag{3.96}$$

The change of the bandgap with the material composition is defined by the ratio $\Delta E_{\mathrm{C}}/\Delta E_{\mathrm{g}}$, starting from one of the materials. ΔE_g denotes the total difference of the bandgaps at the heterointerface between the segments. ΔE_{V} and ΔE_{C} denote the difference of the valence and conduction band energies at a heterointerface. For example, it is known for the $\mathrm{Al}_x\mathrm{Ga}_{1-x}\mathrm{As}/\mathrm{GaAs}$ interface that this ratio equals 60%. This means that with increasing x, 60% of the increase of the bandgap ($\Delta E_{\mathrm{g}} = \Delta E_{\mathrm{C}} + \Delta E_{\mathrm{V}}$) is attributed to the conduction band (ΔE_{C}) and 40% to the valence band (ΔE_{V}).

The model is generalized when assuming $\Delta E_{\mathrm{C}}/\Delta E_{\mathrm{g}}$ is constant for the whole composition range, thus one obtains:

$$E_{\mathrm{C}} = E_{\mathrm{off}}^{\mathrm{B}} + E_{\mathrm{g}}^{\mathrm{B}} + \left(\frac{\Delta E_{\mathrm{C}}}{\Delta E_{\mathrm{g}}}\right) \cdot \left(E_{\mathrm{g}}^{\mathrm{AB}} - E_{\mathrm{g}}^{\mathrm{B}}\right). \tag{3.97}$$

Thus from (3.95) and (3.97) the ratio $\Delta E_{\mathrm{C}}/\Delta E_{\mathrm{g}}$ can be expressed as:

$$\Delta E_{\mathrm{C}}/\Delta E_{\mathrm{g}} = 1 + \frac{E_{\mathrm{off}}^{\mathrm{A}} - E_{\mathrm{off}}^{\mathrm{B}}}{E_{\mathrm{g}}^{\mathrm{A}} - E_{\mathrm{g}}^{\mathrm{B}}} \tag{3.98}$$

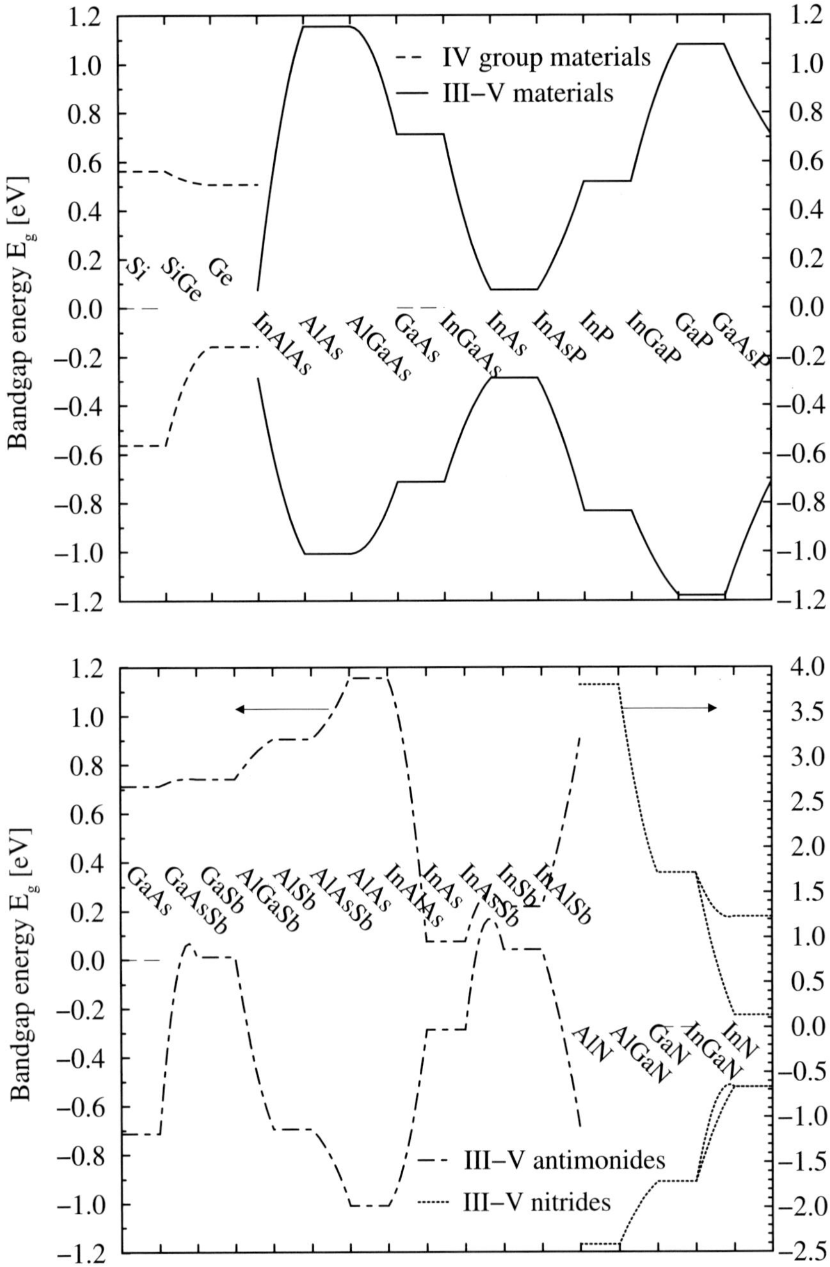

Fig. 3.23. Bandgap energies and bandgap alignment of the semiconductor materials modeled in this work.

Replacing $\Delta E_C/\Delta E_g$ in (3.95) or (3.97), the offset of alloy material is obtained:

$$E_{\mathrm{off}}^{\mathrm{AB}} = \frac{E_{\mathrm{off}}^{\mathrm{A}} \cdot \left(E_{\mathrm{g}}^{\mathrm{AB}} - E_{\mathrm{g}}^{\mathrm{B}}\right) - E_{\mathrm{off}}^{\mathrm{B}} \cdot \left(E_{\mathrm{g}}^{\mathrm{AB}} - E_{\mathrm{g}}^{\mathrm{A}}\right)}{E_{\mathrm{g}}^{\mathrm{A}} - E_{\mathrm{g}}^{\mathrm{B}}} . \tag{3.99}$$

The valence and the conduction band energies are calculated by (3.93) and (3.94) respectively. Using the default model parameters in this work, ratios $\Delta E_C/\Delta E_g$ of 0.12 for SiGe, 0.6 for AlGaAs, 0.5 for InAlAs, 0.6 for InGaAs, InAsP, and GaAsP, and 0.3 for InGaP are obtained, which are in fairly good agreement with experimental data [5, 38, 44, 253, 302]. For the transition GaN/AlN, a valence band discontinuity of $\Delta E_C = 0.7$ eV is assumed [533]. For the transition, a valence band offset of $\Delta E_V = 1.05$ eV is taken, which assumes strained growth. The complete bandgap alignment of all semiconductor materials presented in this work is shown in Fig. 3.23. Special attention is paid to the band offsets at the heterointerfaces, since they are input parameters to thermionic emission or thermionic-field emission models in the case of abrupt heterojunctions (see Subsection 3.1.7) and critically determine the current and energy fluxes.

3.3.3 Bandgap Narrowing

Bandgap narrowing (BGN) is one of the crucial heavy-doping effects to be considered for bipolar devices. With increasing doping concentration in semiconductors, the strict periodicity of the lattice is disturbed by the existence of the impurity atoms and various heavy doping effects occur. These may include impurity band widening, band tailing, electron-electron interaction, and screening effects [321, 392]. According to the rigid band model in [322], all of these effects can be accounted for by a single parameter, an effective bandgap shrinkage as a function of doping concentration. BGN has a pronounced effect on device behavior in HBTs, since several layers, such as the emitter cap, the subcollector, and especially the base, are typically very highly doped. BGN actually enhances the performance of the np$^+$n HBTs unlike the n$^+$pn Si BJTs. A higher doping in the base leads to a larger difference between the bandgaps of the emitter and base. This results in increased emitter injection efficiency [314] and reduced turn-on voltage. This is beneficial, since the power dissipation is thus reduced [515].
Not only the total BGN inside the base of HBTs has implications on device performance, but also the distribution of BGN between the conduction and the valence bands [309]. In abrupt HBTs, BGN can affect the band alignment at the base-emitter heterointerface, which itself influences the tunneling and thermionic-emission current across this junction. Numerical simulation in [309] has shown that different BGN distributions between the conduction and valence bands can result in simulated currents which differ by an order of magnitude. Several publications in the literature have been devoted to modeling of BGN for device simulation. As for all simulation models, the goal of theoretical investigations has been to find an optimal balance between accuracy of the results and simplicity of the model implementation. Commercial device simulators, such as ATLAS

[465], DESSIS [227], and MEDICI [503], use logarithmic fit models for BGN in silicon which are simple to implement but also deliver non-physical results for some input parameters.
Furthermore, the models for silicon are often used for other materials (such as SiGe or III-V compounds [319]), or the BGN effect is even neglected. Therefore, this subsection reviews BGN models suited for simulation of heterostructure devices in comparison to experimental data for different materials.

Bandgap Narrowing Models for Si, Ge, and SiGe. A detailed discussion on BGN in silicon and a comparison of different theories and models can be found in [440]. Fig. 3.24 illustrates the output of the models proposed by Slotboom and de Graaff [472], by Klaassen et al. [258], by del Alamo et al. [119], by Bennet and Wilson [31], by Jain and Roulston [236], and by Schubert [444]. As can be seen from the figure, both experimental data for BGN and the results of the models scatter for this effect.

The Slotboom Model. The model of Slotboom et al. [472] is probably the most popular model used to model BGN in silicon. The model is relatively simple and has three fitting parameters: E_{ref} is a reference energy, C_{ref} denotes a reference concentration, and $\Delta E_{\mathrm{C}}/\Delta E_{\mathrm{g}}$ gives the part of the total BGN ΔE_{g} which is contributed to the conduction band.

$$\Delta E_{\mathrm{g}} = E_{\mathrm{ref}} \cdot \left(\ln\left(\frac{C_{\mathrm{I}}}{C_{\mathrm{ref}}}\right) + \sqrt{\ln^2\left(\frac{C_{\mathrm{I}}}{C_{\mathrm{ref}}}\right) + \frac{1}{2}} \right) \tag{3.100}$$

The shifts of the conduction and valence band edges are calculated by

$$\Delta E_{\mathrm{C}} = \left(\frac{\Delta E_{\mathrm{C}}}{\Delta E_{\mathrm{g}}}\right) \cdot \Delta E_{\mathrm{g}} \tag{3.101}$$

$$\Delta E_{\mathrm{V}} = \Delta E_{\mathrm{g}} - \Delta E_{\mathrm{C}}. \tag{3.102}$$

In a later paper, Klaassen et al. have corrected the values of the original model parameters (see Table 3.16).
In addition, Jain and Roulston have suggested that BGN in p-SiGe with $x < 0.3$ differs insignificantly from BGN in p-Si. Therefore, the usage of the parameters from [472] has been recommended for both materials. This assumption seems to have been very widely accepted for device simulation applications [133, 245].

Table 3.16. Parameter values for the Slotboom bandgap narrowing model

Material	E_{ref} [eV]	C_{ref} [cm^{-3}]	$\Delta E_{\mathrm{C}}/\Delta E_{\mathrm{g}}$	Reference
p-Si	0.009	1.0e17	0.5	[472]
p-Si, n-Si	0.00692	1.3e17	0.5	[258]
n-Si	0.0085	3.5e17	–	[236]
p-Si, p-SiGe	0.009	1.0e17	0.5	[236]
p-GaAs	0.0045	1.7e16	–	[236]

Bandgap Narrowing Models for III-V Materials. Although the base layers of npn III-V HBTs are typically highly doped and thus the BGN effect is quite pronounced, III-V device simulation often underestimates the importance of BGN. However, there are few relatively simple models which can be applied for modeling BGN in various III-V semiconductors.

The Jain and Roulston model. Jain and Roulston have proposed a physics-based model which gives the BGN in several III-V semiconductors depending on their material properties. The model describes not only the magnitude of BGN but also its distribution between the conduction and the valence bands [235].

$$\frac{\Delta E_\mathrm{g}}{R_\mathrm{maj}} = \frac{1.83 \cdot \Lambda}{N_b^{1/3} \cdot r_s} + \frac{0.95}{r_s^{3/4}} + \left(1 + \frac{R_\mathrm{min}}{R_\mathrm{maj}}\right) \cdot \frac{1.57}{N_b^{1/3} \cdot r_s^{3/2}} \tag{3.103}$$

In (3.103) R_maj and R_min denote the majority and the minority Rydberg energies respectively, and r_s is the density parameter. Table 3.17 summarizes the values proposed in [235] for N_b and Λ. N_b denotes the number of valleys for n-type materials, or the valence band multiplicity for p-type materials.

Table 3.17. Parameter values for the Jain and Roulston bandgap narrowing model

Material	carrier type	N_b	Λ
GaAs, InAs, InP, GaSb, InSb	n	1	1.00
GaAs, InAs, InP, GaSb, InSb	p	2	0.75
AlAs, GaP, AlSb	n	6	1.00
AlAs, GaP, AlSb	p	2	0.75

This model has been successfully applied to Si, Ge, p-GaAs, and p-GaSb, and presumably can be also applied (with some modifications) to other III-V compounds [235, 236].

The Schubert Model. In [444] Schubert has proposed a physics-based model for BGN. The basic assumption in [444] is that BGN is a result of five types of many-body interactions (electron-electron, electron-impurity, hole-hole, hole-impurity, and electron-hole). The electron-impurity contribution is assumed to be dominant at high doping concentrations, which leads to a BGN model given by (3.104).

$$\Delta E_\mathrm{g} = -\frac{\mathrm{q}^2 \cdot \beta_\mathrm{F}}{4 \cdot \pi \cdot \varepsilon_0 \cdot \varepsilon_\mathrm{r}} \tag{3.104}$$

$$\beta_\mathrm{F}^2 = \frac{n \cdot \mathrm{q}^2}{\varepsilon_0 \cdot \varepsilon_\mathrm{r} \cdot k_\mathrm{B} \cdot T_\mathrm{L}} \cdot \frac{\mathcal{F}_{-1/2}(\eta)}{\mathcal{F}_{1/2}(\eta)} \tag{3.105}$$

β_F denotes the inverse Thomas-Fermi screening length, and $\mathcal{F}_j(x)$ is the Fermi integral of order j [144].

The Dopant-Dependent BGN Model. Removing the point-charge approximation yields a refined BGN model [380]. Using the physically-based approach from [444], an analytical BGN model has been developed. It is applicable to compound semiconductors [380] and considers the semiconductor material and the dopant species for arbitrary finite temperatures.

Semiconductor Alloys. The validity of the Schubert model extends to alloy semiconductors using the material composition dependence of the effective carrier masses and of the permittivity. As the model is physics-based, it contains no free parameters except the ratio $\Delta E_{\mathrm{C}}/\Delta E_{\mathrm{g}}$ used in (3.101).
The results for boron- and gallium-doped $Si_{1-x}Ge_x$ for different Ge contents using the dopant-dependent BGN model are presented in Fig. 3.25. This is even better illustrated in Fig. 3.26, where the BGN versus material composition in $Si_{1-x}Ge_x$ is compared to the Jain and Roulston model, suggesting increased BGN with rising x. The decrease of the BGN with the increase of the germanium fraction has already been observed experimentally in [300, 327]. The theoretical approach explains this effect by the decreased valence band density of states and the increase of the relative permittivity in the SiGe alloy.
Good agreement of the Schubert and the Jain and Roulston model is obtained with experimental data for p-type GaAs [58, 204, 220, 312, 558], as shown in Fig. 3.27. A fit by a simple power function is also added.
As stated in [314], photoluminescence studies typically deliver larger bandgap shrinkages. Thus, the effective gaps deduced electrically and optically can differ, which explains the scatter of the experimental data. For InP HBT simulation, the BGN is especially important, since high-speed operation requires doping the base to the highest technically possible extent. Therefore, the models extend to ternary compounds, such as $In_{0.53}Ga_{0.47}As$ and the novel base material $GaAs_{0.5}Sb_{0.5}$ (see Fig. 3.29). For p-type GaN and related alloy materials, polarization effects interfere with the bandgap narrowing. Further, active p-doping concentrations in GaN above 10^{18} cm^{-3} can hardly be achieved with Mg doping, due to its large activation energy. Therefore, even for the highest possible p-doping concentrations, BGN is negligible, as confirmed in [301].
Experimental data in n-type III-V materials are available for GaAs in [58, 203, 561], for $Al_{0.3}Ga_{0.7}As$ in [108], for InGaP in [567], and for InP in [76, 464]. Experiments showed higher BGN in n-InP than in n-GaAs. Higher conduction band density of states and lower relative permittivity explain the expected higher values for BGN in AlAs and GaP (see Fig. 3.28) than in InP, GaAs, and InAs. Further, data for n-InGaAs are available in [351].
Note that different BGN models have to be used with Boltzmann or Fermi statistics [106, 309]. The models of Slotboom and Schubert are suited for use with Boltzmann statistics while the model of Jain and Roulston must be used with Fermi statistics or must be adopted for use with Boltzmann statistics [236]. The amount of correction needed depends on N_{C} and N_{V}, which in turn depend on

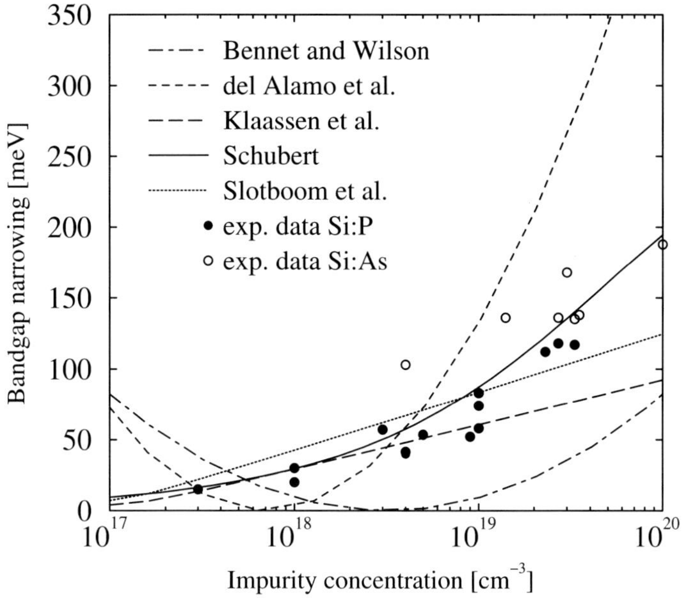

Fig. 3.24. Different bandgap narrowing models for Si in comparison to measurements.

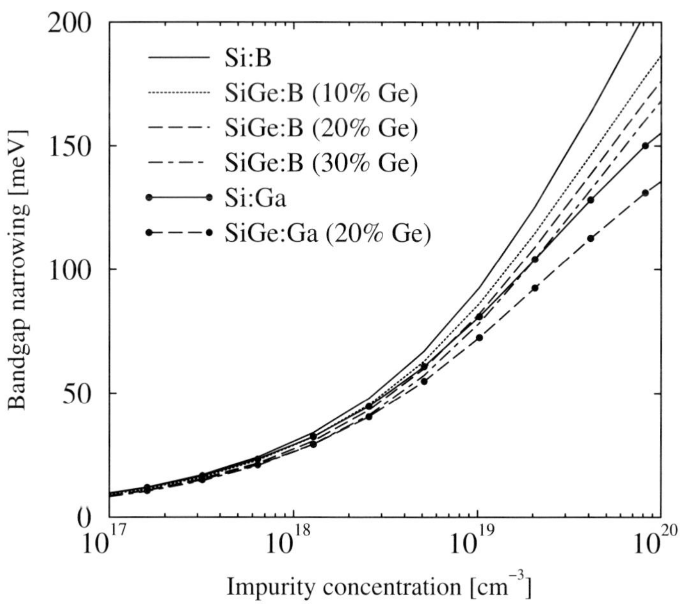

Fig. 3.25. Influence of the dopant material and material composition on bandgap narrowing in p-Si and p-SiGe.

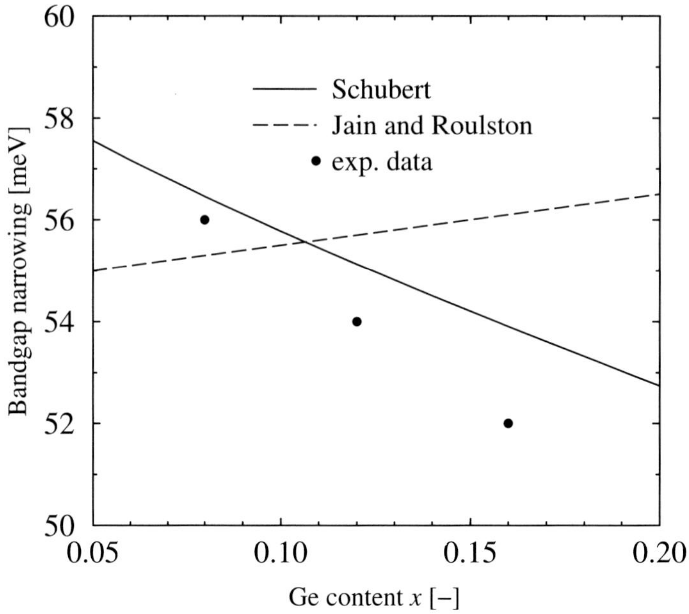

Fig. 3.26. Bandgap narrowing in p-SiGe as a function of Ge content compared to experimental data.

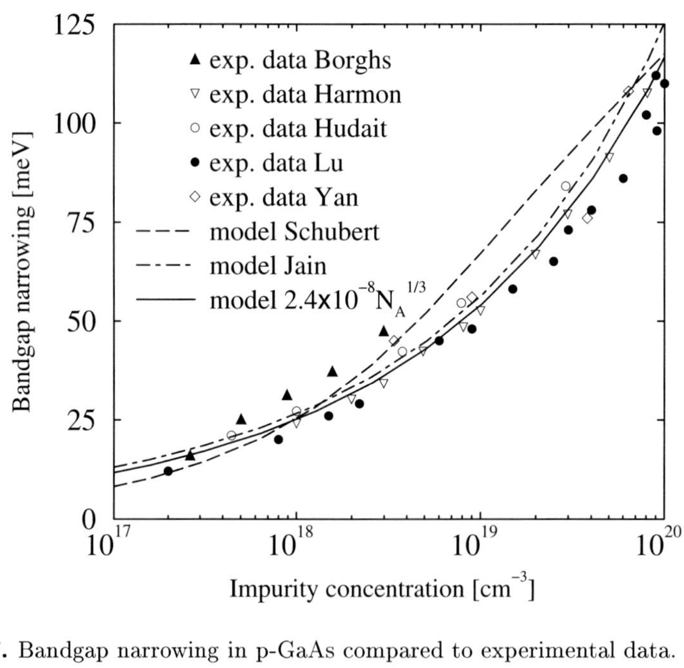

Fig. 3.27. Bandgap narrowing in p-GaAs compared to experimental data.

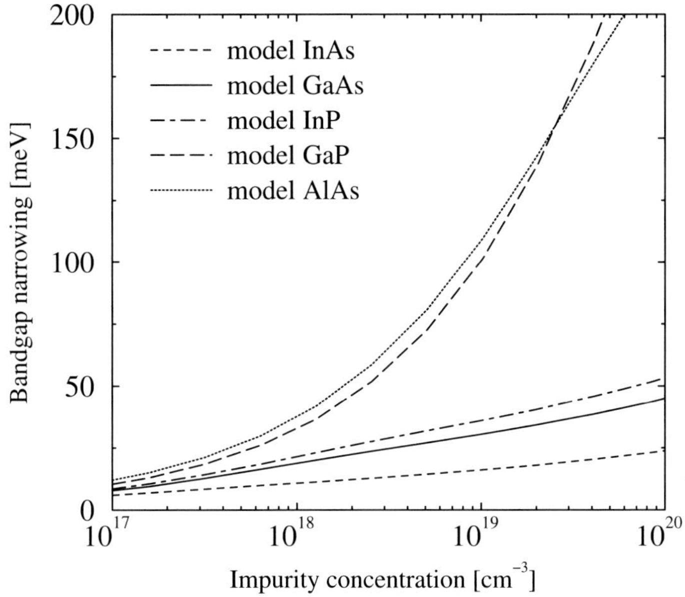

Fig. 3.28. Bandgap narrowing in n-type binary compound semiconductors.

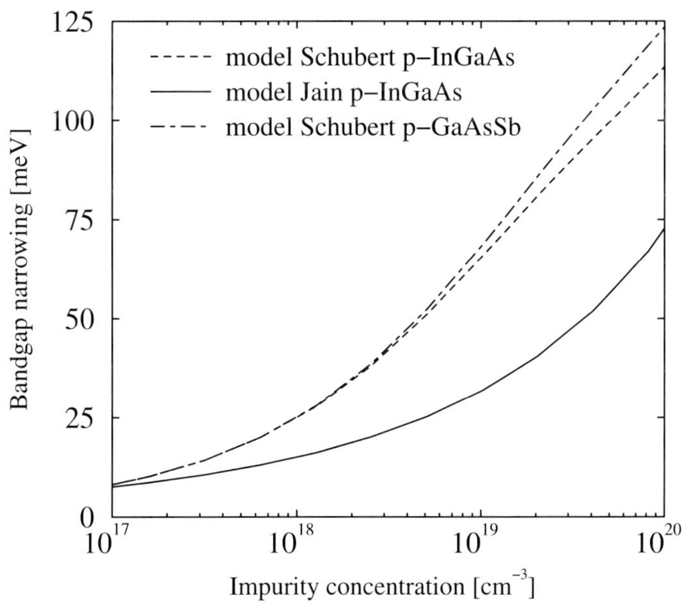

Fig. 3.29. Bandgap narrowing in p-type $In_{0.53}Ga_{0.47}As$ and p-$GaAs_{0.5}Sb_{0.5}$.

the material and the material composition [106]. The problem itself remains for further analysis.

3.3.4 Effective Carrier Mass

In this work, parabolic conduction and valence bands are assumed and the effects of band non-parabolicity are assumed to be negligible. Therefore effective masses for each of the three conduction band minima (Γ, X, and L) can be defined and are used to calculate the density of states (DOS). The DOS in the valence band can be characterized by the effective masses of three sub-bands, all in the center of the Brillouin zone: two degenerate sub-bands of light and heavy holes, and the split-off sub-band. The contribution of the split-off band, which is depressed in energy and is much less populated, is neglected in this work.
The effective DOS masses scaled by the free electron mass give relative carrier masses. In materials with a direct bandgap, the value of the relative mass for electrons from the Γ valley is considered. In the case of materials with indirect bandgaps, namely Si, Ge, AlAs, AlSb, and GaP, the transverse and the longitudinal electron masses (m_{nt} and m_{nl}) from the X valley (or L valley for Ge) are taken into account by (3.106).
The hole masses for all materials are calculated by (3.107), accounting for the heavy and the light hole masses (m_{pl} and m_{ph}) [496].

$$m_{0,\mathrm{n}} = \begin{cases} m_\Gamma & \text{for Ga(As, Sb, N) and In(As, P, Sb, N)} \\ m_\mathrm{X} = \left(m_\mathrm{nt}^2 \cdot m_\mathrm{nl}\right)^{1/3} & \text{for Si, Al(As, Sb, N) and GaP} \\ m_\mathrm{L} = \left(m_\mathrm{nt}^2 \cdot m_\mathrm{nl}\right)^{1/3} & \text{for Ge} \end{cases} \tag{3.106}$$

$$m_{0,\mathrm{p}} = \left(m_\mathrm{pl}^{3/2} + m_\mathrm{ph}^{3/2}\right)^{2/3} \tag{3.107}$$

The temperature dependence of the relative carrier masses is well-studied for silicon (see [188] and references therein). However, such data are scattered or missing for III-V materials, and the effect is often neglected. A linear temperature dependence (3.108) is sufficient to model the relative electron masses for all materials [264]. A quadratic temperature dependence (3.109) is the simplest expression for holes which can model the data for silicon well.

$$m_\mathrm{n} = m_{0,\mathrm{n}} + m_{1,\mathrm{n}} \cdot \left(\frac{T_\mathrm{L}}{300\ \mathrm{K}}\right) \tag{3.108}$$

$$m_\mathrm{p} = m_{0,\mathrm{p}} + m_{1,\mathrm{p}} \cdot \left(\frac{T_\mathrm{L}}{300\ \mathrm{K}}\right) + m_{2,\mathrm{p}} \cdot \left(\frac{T_\mathrm{L}}{300\ \mathrm{K}}\right)^2 \tag{3.109}$$

The model parameters for electrons m_Γ, m_{nt}, m_{nl}, and $m_{1,\mathrm{n}}$ are summarized in Table 3.18. In addition, the calculated values for m_X and m_L are included, after comparison with calculated values reported in the literature. The values for the relative masses of electrons from higher valleys are used for the calculation of the relative electron masses of *alloy materials*.

Table 3.18. Parameter values for modeling the effective electron masses

Mat.	m_{Γ}	m_{L}	m_{X}	m_{nt}	m_{nl}	$m_{1,n}$	References
Si	-	-	0.328	0.19	0.98	0.009	[364, 469, 496]
Ge	-	0.222	0.42	0.082	1.64	0.0068	[147, 364, 469, 496]
GaAs	0.067	-	0.471	0.229	1.987	-0.0036	[237, 354, 496, 537]
AlAs	0.15	-	0.38	0.19	1.5	-0.0036	[253, 305]
InAs	0.023	-	0.64	0.271	3.619	-0.0027	[67, 147, 364, 496]
InP	0.077	-	0.462	0.273	1.321	-0.004	[147, 253, 305, 513]
GaP	0.15	-	0.5	0.25	1.993	-0.0026	[1, 147, 286, 364]
GaN	0.20	-	-	0.2283	0.1846	-	[130, 174, 462, 496]
AlN	0.314(0.48)	-	-	0.3433	0.2938	-	[174, 374, 462]
InN	0.11	-	-	0.1311	0.1190	-	[174, 462]
GaSb	0.0512	-	0.87	0.22	1.51	-0.006	[364, 455, 513, 532]
AlSb	0.0942	-	1.5	0.23	1.64	-0.0052	[364, 455, 513, 532]
InSb	0.0197	-	0.64	-	0.29	-0.0043	[364, 455, 513, 532]

Table 3.19. Parameter values for modeling the effective hole masses

Material	m_{ph}	m_{pl}	$m_{0,p}$	$m_{1,p}$	$m_{2,p}$	References
Si	0.49	0.16	0.55	0.6	-0.1	[364, 469, 496]
Ge	0.29	0.04	0.28	0.1	-	[147, 364, 469, 496]
GaAs	0.49	0.08	0.49	-0.05	-	[237, 354, 496, 537]
AlAs	0.76	0.16	0.81	-	-	[253, 305]
InAs	0.41	0.025	0.41	-0.01	-	[496, 364, 67, 147]
InP	0.6	0.12	0.64	-0.05	-	[147, 253, 305, 513]
GaP	0.79	0.14	0.6	-0.05	-	[1, 147, 286, 364, 455]
GaN	1.4	0.259	1.47 ,0.96[165]	-	-	[297, 462]
AlN	1.42	0.471	1.59	-	-	[174, 297, 462]
InN	1.63	0.27	1.65	-	-	[174, 297]
GaSb	0.4	0.06	0.42	-0.05	-	[297, 323, 455, 496]
AlSb	0.72	0.151	0.77	-0.07	-	[257, 455, 460]
InSb	0.43	0.015	0.43	-0.01	-	[257, 455, 460, 297]

The model parameters m_{ph}, and m_{pl} are summarized in Table 3.19. In addition, the calculated values for $m_{0,p}$ are included and compared to values reported in the literature. The temperature-dependent data is taken from [455] and the model of (3.108) and (3.109) assuming, the heavy hole mass to be temperature independent. For the nitride materials there is still significant uncertainty for the exact band-structure, especially for AlN and InN.

For *alloy materials*, the temperature dependence is introduced first. Then a quadratic interpolation of the masses is used as a function of the material composition of an alloy $A_{1-x}B_x$.

$$m_\nu^{AB} = m_\nu^{A} \cdot (1-x) + m_\nu^{B} \cdot x + C_{m,\nu} \cdot (1-x) \cdot x \tag{3.110}$$

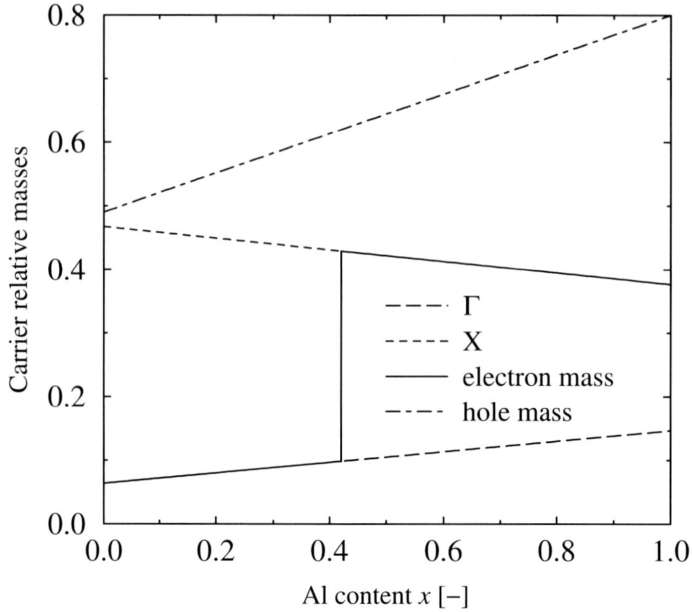

Fig. 3.30. Relative masses of electrons and holes in $Al_xGa_{1-x}As$ as a function of the material composition.

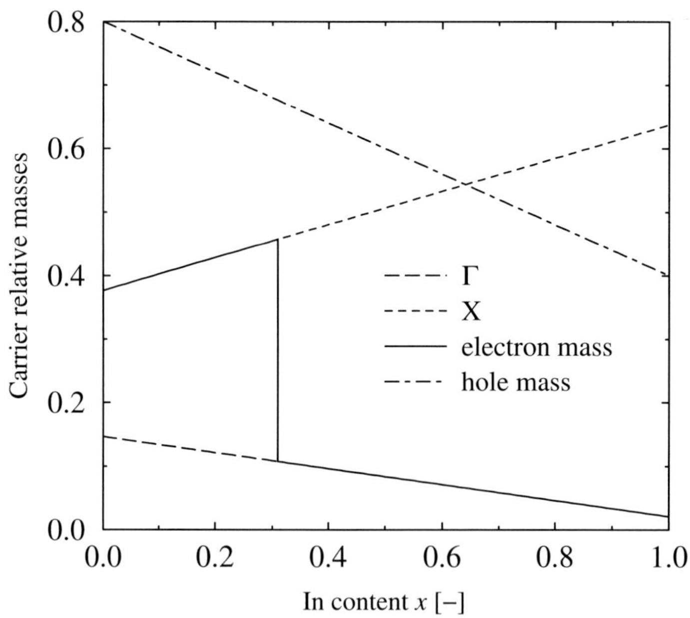

Fig. 3.31. Relative masses of electrons and holes in $In_xAl_{1-x}As$ as a function of the material composition.

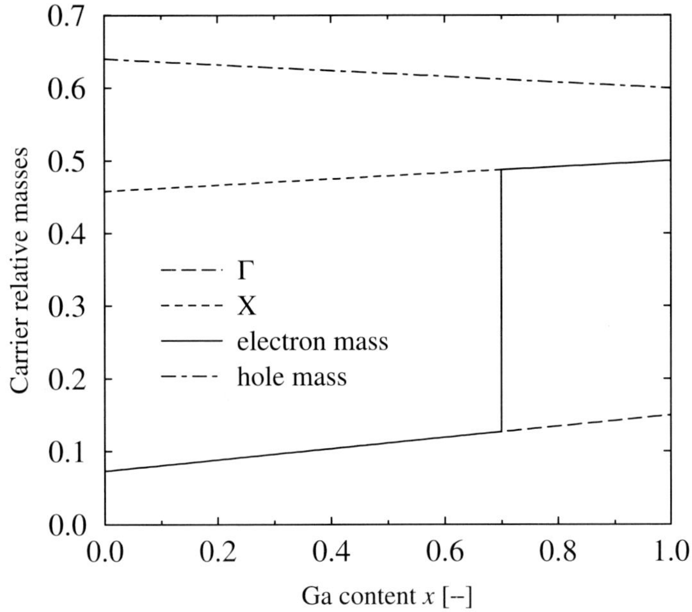

Fig. 3.32. Relative masses of electrons and holes in $In_xGa_{1-x}P$ as a function of the material composition.

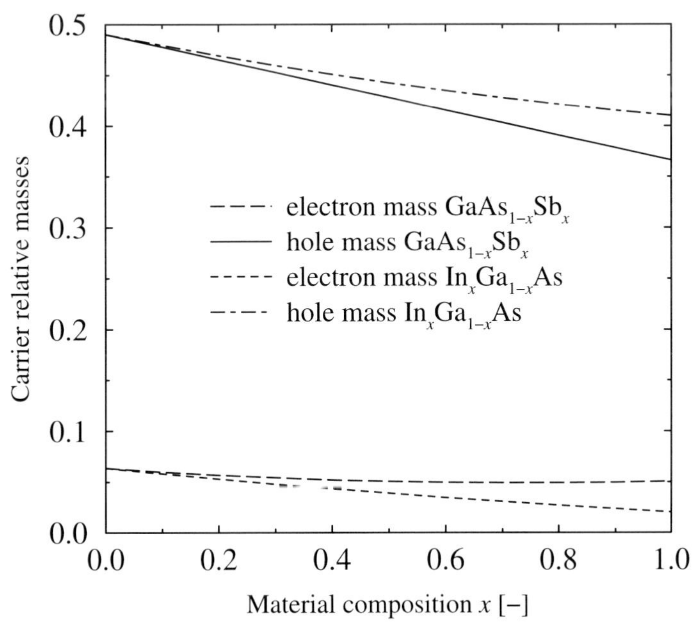

Fig. 3.33. Relative masses of electrons and holes in $In_xGa_{1-x}As$ and $GaAs_{1-x}Sb_x$ as a function of the material composition.

Table 3.20. Bowing parameter values for modeling the effective carrier masses

Material	$C_{\mathrm{m,n}}$	$C_{\mathrm{m,p}}$	Reported Range m_{n}	References
SiGe	-0.183	-0.096	-	[148]
AlGaAs	0.0	0.0	0.07-0.08 (x=0.2)	[5, 444]
InGaAs	-0.012(-0.003)	-0.03	0.038-0.044 (x=0.53)	[127, 226, 298, 304, 444]
InAlAs	0.0	0.0	0.070-0.083 (x=0.52)	[38, 127, 298, 304, 444]
InAsP	0.0	0.0	-	[298]
GaAsP	0.0	0.0	-	[298]
InGaP	0.0	0.0	0.092-0.118(x=0.5)	[35, 70, 353, 567]
AlGaN	0.0048	-	-	[173, 285]
InGaN	-0.1093	-	-	[173]
GaAsSb	-0.026	-	-	[298]
AlGaSb	-0.048	-0.012	-	[288]
InAsSb	-0.03	0.0	-	[298]
InGaSb	-0.009(-0.003)	-	-	[298, 532]
InGaAsP	0	0.03	-	[532]
AlInGaAs	0.016	-	-	[532]

The bowing parameters $C_{\mathrm{m},\nu}$ are summarized in Table 3.20. For materials where the bandgap changes between direct and indirect, the relative electron mass of the respective lowest band is considered. Examples for such materials are AlGaAs (see Fig. 3.30), InAlAs (see Fig. 3.31), InGaP (see Fig. 3.32), GaAsP, and AlGaSb. For other materials, such as InGaAs (see Fig. 3.33), GaAsSb (see Fig. 3.33), InAsP, AlGaN (Fig. 3.34), and InGaN (see Fig. 3.34), only the Γ valley masses are considered. For SiGe, a strained bandgap without X to L transition is assumed and only the X valley masses are used in the interpolation.

3.3.5 Effective Density of States

The effective density of states (DOS) in the conduction and the valence bands are expressed by the following theoretical expressions [496]:

$$N_{\mathrm{C}} = 2 \cdot M_{\mathrm{C}} \cdot \left(\frac{2 \cdot \pi \cdot m_{\mathrm{n}} \cdot k_{\mathrm{B}} \cdot T_{\mathrm{L}}}{h^2} \right)^{3/2} \tag{3.111}$$

$$N_{\mathrm{V}} = 2 \cdot \left(\frac{2 \cdot \pi \cdot m_{\mathrm{p}} \cdot k_{\mathrm{B}} \cdot T_{\mathrm{L}}}{h^2} \right)^{3/2} . \tag{3.112}$$

M_{C} represents the number of equivalent energy minima in the conduction band. Parameter values are given in Table 3.21.

In the model for *alloy materials*, effective carrier masses of the constituents are used in the expressions (3.111) and (3.112). In the case of a transition between a direct and an indirect bandgap in III-V ternary compounds, the valley degeneracy factor M_{C} is modeled by an expression equivalent to the one proposed in [493].

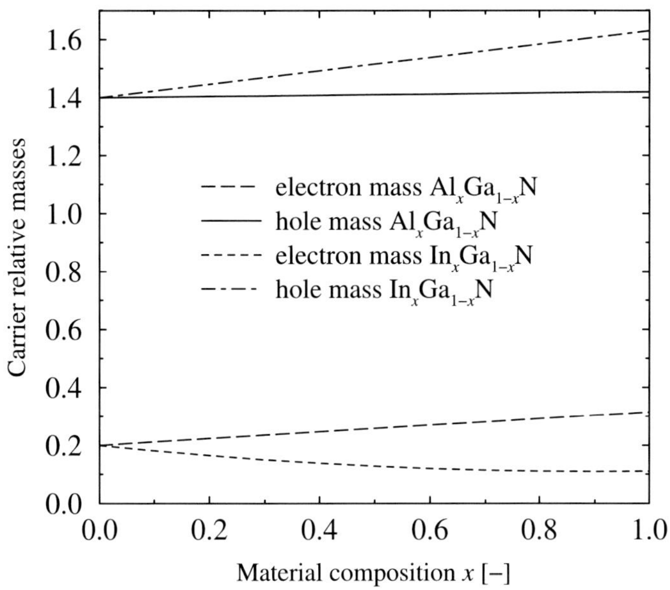

Fig. 3.34. Relative masses of electrons and holes in $Al_xGa_{1-x}N$ and $In_xGa_{1-x}N$ as a function of the material composition.

Table 3.21. Parameter values for energy minima in the DOS model

Materials	Minima	M_C
Si	X	6
Ge	L	4
GaAs, InAs, InP,GaSb, InSb, GaN, AlN, InN	Γ	1
AlAs, GaP, AlSb	X	3

$$M_C = M_C^d \cdot \exp\left(\frac{E_g - E_{g,\Gamma}}{k_B \cdot T_L}\right) + M_C^i \cdot \exp\left(\frac{E_g - E_{g,X}}{k_B \cdot T_L}\right) \tag{3.113}$$

The superscripts d and i denote direct and indirect respectively.
In the case of SiGe, the splitting of the valley degeneracy due to strain is modeled accordingly, as given in [398].

$$M_C = M_C^{Si} \cdot \exp\left(-\frac{\delta E_C}{k_B \cdot T_L} \cdot x\right) + M_C^{Ge} \cdot \left(1 - \exp\left(\frac{\delta E_C}{k_B \cdot T_L} \; x\right)\right) \tag{3.114}$$

In this case, δE_C denotes the energy difference between the valleys shifted down and up in energy. It is set equal to 0.6 eV, as given in [398].

3.4 Carrier Mobility

Electron and hole mobilities, μ_{n} and μ_{p}, are basic carrier transport parameters. They are used as basic inputs for expressing the current in semiconductors.
For industrial engineering applications, the mobility remains the only viable concept for multi-dimensional physical device simulation. Rather elaborate improvements have been made in the modeling to account for high-field effects in small-scale devices, such as FETs. However, physically the concept of mobility is limited when it comes to effects on the atomic scale, high-field effects, or very abrupt gradients of fields. In such cases, very close interaction with Monte Carlo (MC) simulations and with even more elaborate concepts is necessary to control, account for, and calibrate the effects incorporated in the mobility models [72, 247].
Carrier mobilities are influenced by several physical mechanisms, such as scattering by interaction with thermal lattice vibrations, charged or neutral impurities, lattice defects, and surfaces. For *alloy materials*, the mechanism of alloy scattering also has to be accounted for. These important effects must be modeled without generating an endless number of parameters. Thus carrier mobility modeling compiles the current transport properties and their dependence on relevant parameters, such as carrier temperatures, lattice temperature, impurity concentrations, material composition, and driving electric fields.
For silicon, the established computationally-efficient mobility model of MINIMOS 6 [146] can be used. For a more detailed discussion see [449, 451]. This approach is extended to all other semiconductor materials. Other models are proposed where special material effects suggest their introduction, such as a multi-valley concept for the material composition dependence in some *alloy materials*.
Mobility models have to support both the DD and the HD transport models. While the low-field mobility is independent of the transport model used, the high-field mobility is modeled differently. Thus the effects of influence on mobility are grouped into low-field effects (including the impact of lattice scattering, ionized impurities scattering, and surface scattering) and high-field effects.

3.4.1 Lattice Mobility

Temperature Dependence. In solids at finite temperatures the lattice atoms oscillate about their equilibrium sites. The carriers are scattered, and this results in a temperature-dependent mobility μ^{L}. The superscript L stands for lattice scattering. The temperature dependence of the lattice mobility is modeled by a power law for all semiconductors.

$$\mu_{\nu}^{\mathrm{L}} = \mu_{\nu,300}^{\mathrm{L}} \cdot \left(\frac{T_{\mathrm{L}}}{300\ \mathrm{K}} \right)^{\gamma_{0,\nu}} , \qquad \nu = \mathrm{n}, \mathrm{p}. \tag{3.115}$$

ν denotes the carrier type: n for electrons and its concentration, and p for holes and its concentration. The parameter values used in this model for electrons and holes are summarized in Table 3.22.

Table 3.22. Parameter values for the lattice mobility.

Material	Valley	ν	$\mu^{L}_{\nu,300}$ [cm²/Vs]	$\gamma_{0,\nu}$	Range [cm²/Vs]	References
Si	X	n	1430	-2	1500	[364, 469, 496]
		p	460	-2.18(-2)	450-500	[364, 469, 496]
Ge	L	n	3800	-1.66	3900	[148, 364, 469, 496]
		p	1800	-2.33	1900	[148, 364, 469, 496]
GaAs	Γ	n	8500	-2.2	8500-9400	[237, 364, 480, 496]
	X	n	410	-2.2	410	[536]
		p	470	-0.9	400-492 (10^{15})	[136, 237, 364, 480]
AlAs	X	n	410	-2.1	400-1200	[480, 237, 364]
	Γ	n	250	-2.1	250(2000[1])	[536]([180, 304])
		p	130	-2.2	130-420	[444, 237, 364][5]
InAs	Γ	n	32500	-1.7	22600-34000	[237, 480, 496, 513]
	X	n	480	-1.7	570	[180, 423]
		p	510	-2.3	200 - 530	[237, 480, 496, 513]
InP	Γ	n	5300	-1.9	4200-5400	[237, 286, 480, 496]
	X	n	200	-1.9		InGaP
		p	200	-1.2	130-180	[237, 364, 480, 513]
GaP	X	n	150	-1.94	110-300	[286, 364, 496, 513]
	Γ	n	840	-1.94		InGaP
		p	160	-2.0	75-150	[237, 250, 496]
GaSb	Γ	n	5500	-1.5	4000-7600	[97, 109, 489, 519]
		p	1231	-1.5	1200	[110, 548]
AlSb	X	n	202	-1.8 (-1.4-1.6)	200-900	[257, 288]
		p	450	-1.95	400-450	[257, 288]
InSb	Γ	n	77000	-1.66	77000	[297]
		p	774	-1.8	850	[288, 297]
GaN	Γ	n	1478	-1.26 (-1.0-1.7)	1000-1700	[37, 83, 132, 143, 308]
	Γ	n	2000(MC)	-	2000	[368]
	Γ	n	1000	-1	-	[343]
		p	30	-1.5	-	[83, 462]
		p	170	-2	-	[343]
AlN	Γ	n	683	-1.26	215-683	[18, 143]
	Γ	p	14	-	-	[462]
InN	Γ	n	100(exp)	-1.26	-	[373]
	Γ	n	3138(MC)	-1.81	3200,4985	[18, 143, 326]

[1] value only used for InAlAs

Multivalley Modeling Approach for III-V Semiconductors. Following the methodology used in Subsection 3.3.1, different mobilities are applied for the composition of ternary materials from binary mobilities to account for the lowest valley at different material compositions. This allows a more precise modeling, as shown below for the ternary III-V compound semiconductors. The material parameters for the binary materials are compiled from the literature, as given in Table 3.22. The values for AlAs are taken from the broad material base of $Al_xGa_{1-x}As$ extrapolated to $x=1$ in the multivalley modeling. The same is true for AlSb, where the low-field mobility is mostly measured for ternary compounds, e.g. AlGaSb. InSb is well investigated at 77 K, whereas room temperature operation is complicated, due to the extremely small bandgap.

Wide Bandgap Semiconductors. For the low-field mobility of GaN, an evolution of the material quality can be observed. The state-of-the-art can be estimated from the scattering of the low-field values given in Table 3.22. A distinction is made between growth by MBE and by MOCVD, see [132], since MBE growth supplies slightly lower mobility values, while MOCVD growth achieves higher mobility values and is more widely used, due to the higher growth rates achieved. The distinction can be explained by the fact that MBE growth supplies a higher background concentration, which again leads to higher achievable sheet carrier densities. Major contributions to the understanding of the low and high-field properties of wide bandgap semiconductors are summarized in [68]. The high low-field value for undoped material of 2000 cm^2/Vs is due to the improvements in material quality, which yields 1400 cm^2/Vs at 10^{16} cm^2/Vs in this work.
The temperature coefficient for GaN is extracted from experimental data [308] and from MC simulations [37]. A model supporting experimental data for the temperature dependence of GaN can be found in [480]. For AlN, the electron mobility is given as determined in [143]. MC simulations are further available from Anwar et al. in [18], which suggest lower values for pure AlN. Hole data for AlN is very rare, and, though repeatedly questioned, the value given in Table 3.22 represents the only available source.
The data given for the lattice mobility of InN is based on a fit to the doping-dependent MC values in [18] and the model in [143]. There is a tremendous discrepancy between various simulation predictions and the experimental values obtained for InN. The values for the model in [143] are given as well as the fit for the values obtained in [18]. The reason for the discrepancy is the principle constraint of solving In in N during the epitaxial layer growth. The constraints restrict the amount of In brought into a ternary crystal $In_xGa_{1-x}N$, e.g. in [114].

3.4.2 Ionized Impurity Scattering

In semiconductor devices, mobility reduction due to ionized impurity scattering is a dominant effect. The influence of lattice and impurity scattering must be combined to obtain an effective mobility μ^{LI}. The following models are available.

The Mobility Model of MINIMOS 6. To account for mobility reduction due to ionized impurity scattering, the formula of Caughey and Thomas [89] is used in conjunction with temperature dependent coefficients. C_I denotes the concentration of ionized impurities. The model is well applicable for silicon.

$$\mu_\nu^{\mathrm{LI}} = \mu_\nu^{\min} + \frac{\mu_\nu^{\mathrm{L}} - \mu_\nu^{\min}}{1 + \left(\dfrac{C_\mathrm{I}}{C_\nu^{\mathrm{ref}}}\right)^{\alpha_\nu}}, \qquad \nu = \mathrm{n}, \mathrm{p} \tag{3.116}$$

$$\mu_\nu^{\min} = \begin{cases} \mu_{\nu,300}^{\min} \cdot \left(\dfrac{T_\mathrm{L}}{300\ \mathrm{K}}\right)^{\gamma_{1,\nu}} & T \geq 200\ \mathrm{K} \\ \mu_{\nu,300}^{\min} \cdot \left(\dfrac{2}{3}\right)^{\gamma_{1,\nu}} \cdot \left(\dfrac{T_\mathrm{L}}{200\ \mathrm{K}}\right)^{\gamma_{2,\nu}} & T < 200\ \mathrm{K} \end{cases} \tag{3.117}$$

$$C_\nu^{\mathrm{ref}} = C_{\nu,300}^{\mathrm{ref}} \cdot \left(\frac{T_\mathrm{L}}{300\ \mathrm{K}}\right)^{\gamma_{3,\nu}} \tag{3.118}$$

$$\alpha_\nu = \alpha_{\nu,300} \cdot \left(\frac{T_\mathrm{L}}{300\ \mathrm{K}}\right)^{\gamma_{4,\nu}} \tag{3.119}$$

Simpler expressions are applied to other *basic materials*. By setting $\gamma_{1,\nu} = \gamma_{2,\nu}$ and $\gamma_{4,\nu} = 0$, (3.117) and (3.119) reduce to

$$\mu_\nu^{\min} = \mu_{\nu,300}^{\min} \cdot \left(\frac{T_\mathrm{L}}{300\ \mathrm{K}}\right)^{\gamma_{1,\nu}} \tag{3.120}$$

$$\alpha_\nu = \alpha_{\nu,300}. \tag{3.121}$$

The model parameter values are summarized in Table 3.23.

Electron Mobility. The model parameters for silicon are well-investigated [451]. However, as described later in this section, high-doping effects must be taken into account in more elaborate models, as the minority mobility differs significantly from the simpler approach given in (3.116), (3.120), and (3.121). Fig. 3.35 and Fig. 3.36 compare experimental data to results for the electron mobility as a function of doping concentration in various III-V semiconductors. For further reading, a good compilation for low-field mobility in various semiconductor materials is given in [480]. Fig. 3.35 gives the comparison for GaAs and InP. A great variety of data is available for GaAs. Early concise descriptions of the doping dependence in GaAs are found in [274]. The model also agrees well with recent updates of the material properties in [288]. For InP, a set of values is given in Table 3.23 based on our own MC simulations which account for the majority/minority effects at higher doping levels. Additionally, the model values supplied in [480] are given in Fig. 3.35. We found in this approach that, especially for InP, the values are only valid up to a doping concentration of about 10^{19} cm^{-3}. Above these doping levels, the effects of the majority/minority approach must be considered. The comparison of experimental data and the model is given in Fig. 3.36 for AlAs

Table 3.23. Parameter values for the impurity mobility

Material	Valley	ν	$\mu_{\nu,300}^{\min}$ [cm^2/Vs]	$\gamma_{1,\nu}$	$\gamma_{2,\nu}$	$C_{\nu,300}^{\mathrm{ref}}$ [cm^{-3}]	$\gamma_{3,\nu}$	$\alpha_{\nu,300}$	$\gamma_{4,\nu}$	Refs.
Si	X	n	80	-0.45	-0.15	1.12e17	3.2	0.72	0.065	[450]
		p	45	-0.45	-0.15	2.23e17	3.2	0.72	0.065	[450]
Ge	L	n	850	-	-	2.6e17	-	0.56	-	[496]
		p	300	-	-	1.0e17	-	1.0	-	[496]
GaAs	Γ	n	800	-0.9	-0.9	1.0e17	6.2	0.5	-	[274, 288]
	X	n	40	-0.9	-0.9	1.0e17	0.48	0.5	-	AlGaAs
		p	40	-0.9	-0.9	1.0e17	0.5	1.0	-	[288]
AlAs	X	n	10	-	-	5.0e17	-	1	-	[93]
	Γ	n	10	-	-	1.0e17	-	0.5	-	AlGaAs
		p	5	-	-	2.9e17	0.5	1.0	-	[444]
InAs	Γ	n	11700	-0.33	-0.33	4.4e16	3.6	0.5	-	Fig. 3.49
	X	n	40	-0.33	-0.33	4.4e16	0.5	0.5	-	Fig. 3.48
		p	48	-	-	2.55e17	0.5	1.0	-	[180, 513]
InP	Γ	n	1120	2.0	2.0	4e16	3.7	0.6	-	[480]
	X	n	10	-	-	4e16	-	0.6	-	InGaP
		p	24	1.2	1.2	2.5e17	0.47	1.0	-	[274, 288]
GaP	X	n	10	-1.07	-1.07	4.4e18	1.8	0.8	-	[480]
GaP	Γ	n	10	-	-	4.4e18	-	0.8	-	InGaP
		p	27	-	-	5e17	-	1.0	-	[250]
GaSb	Γ	n	680	-	-	3.2e17	-	1.08	-	[288]
		p	154	-	-	3.04e17	-	0.67	-	[548]
		p	85	-	-	4e17	-	0.44	-	[483]
AlSb	X	n	5	-	-	6e18	-	1.21	-	[288]
		p	50	-	-	1e17	-	1	-	[288]
InSb	Γ	n	200	-	-	2e17	-	0.57	-	[288]
		p	48	-	-	2.36e18	-	0.39	-	[288]
GaN	Γ	n	295	-1.02	-	1e17	3.02	0.66	0.81	[143]
	Γ	n	138	-	-	1.89e16	-	0.89	-	Fig. 3.39
	Γ	n	55	-	-	2.0e17	-	1	-	[343]
		p	3	-	-	3.0e17	-	2	-	[343]
AlN	Γ	n	297	-1.82	-	1.0e17	3.78	1.16	0.86	[143]
InN	Γ	n	834	-	-	2.88e17	-	0.76	-	[18]
	Γ	n	774	-6.39	-	1e17	-	8.05	-0.94	[143]

and InAs. The model values for pure AlAs have very little direct experimental foundation; those given in Fig. 3.36 are taken from [93]. The most important experimental support comes from the extrapolation of $Al_xGa_{1-x}As$ data with aluminum contents x close to unity. In combination with the application of the

two-valley approach, the experimental situation can be well-matched, as given in Fig. 3.36.
The experimental data for InAs are mostly extrapolated from $In_xGa_{1-x}As$, especially for higher doping levels. The parameter values given in [480] for the model of InAs are considered too high, as the model is fit to early theoretical data or 77 K values, taken for example from [158]. The experimental confirmation for InAs is given in Fig. 3.49, where the overall modeling approach for $In_xGa_{1-x}As$ delivers the experimental values obtained for heterostructure FETs. Fig. 3.36 further gives the comparison with the experimental data for GaP [250]. For GaSb, the comparison is given to the MC calculations of [109], which are in very good agreement with recent experimental data between $N_D = 2\times10^{15}\mathrm{cm}^{-3}$ and $3.5\times10^{18}\mathrm{cm}^{-3}$ [97, 489, 519], as also given in Fig. 3.36. The values obtained for AlSb and InSb are best-fit values obtained for the experimental data sets cited in Table 3.23.

Hole Mobility. Results delivered by the hole mobility model as a function of doping concentration for group III-V binary semiconductors are shown in Fig. 3.37 and Fig. 3.38 in comparison with measured data. The parameters for the influence of the impurity scattering are given in Table 3.23. The experimental data for GaAs is taken from [288] for the dopants C, Zn, and Be. The hole data for InAs are taken from [180, 297, 513]. Data for InP in Fig. 3.37 are taken from various sources, i.e. [274, 288], and the references therein. The experimental base for the hole mobility of AlAs is obtained from [307] for a doping level of $2.5\times10^{17}\mathrm{cm}^{-3}$ and given in Fig. 3.38. The data obtained for AlAs indirectly from $Al_xGa_{1-x}As$ in [480] based on the compilation of [536] is somewhat ambiguous. For GaSb, the mobility values for holes are taken from the compilation in [548] in agreement with MC calculations of the low-field mobility in [110]. The doping dependence for the high-doping region $\geq 10^{19}$ cm^{-3} is taken from [545]. The model parameters for the hole mobility of AlSb and InSb are obtained from the data compiled in [288].

Wide Bandgap Semiconductors. The doping dependence for wide bandgap semiconductors is often considered less important than in other semiconductors due to the influence of polarization-induced charges. Comparisons of the GaN mobility values for electrons and holes as a function of doping are given in Fig. 3.39 and Fig. 3.40.
The MC electron values obtained for GaN are taken from [143]. Our own experimental data suggest values of 1400 cm^2/Vs for MOCVD grown material with a background concentration of 10^{16} cm^{-3} [261]. The experimental data are taken from the compilation in [164]. The maximum values for extremely low doping levels amount to 2000 cm^2/Vs. This value is in agreement with MC simulations of undoped AlGaN/GaN HEMT structures [368] and thus suitable for their simulation. The modeling situation of low doping levels in III-N materials is complicated by the necessary overcompensation of the background n-doping. A more thorough discussion is given in [405]. For holes in GaN, the situation is

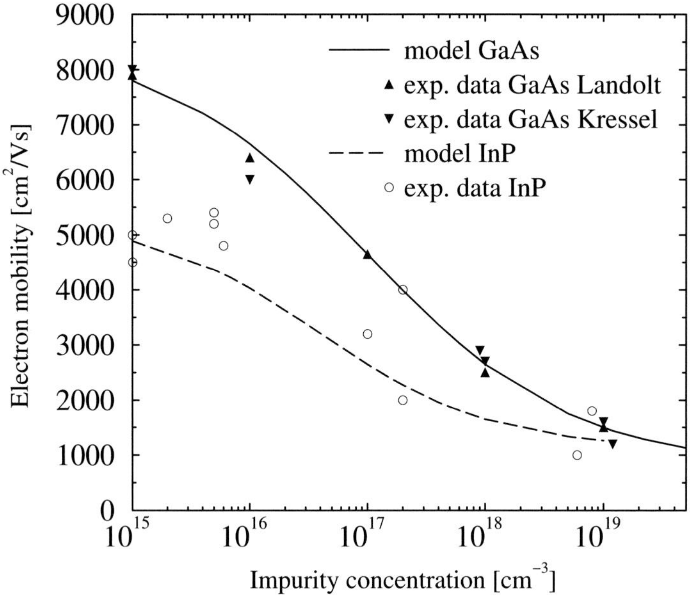

Fig. 3.35. Majority electron mobility as a function of impurity concentration at 300 K: Comparison of the model and experimental data for GaAs and InP.

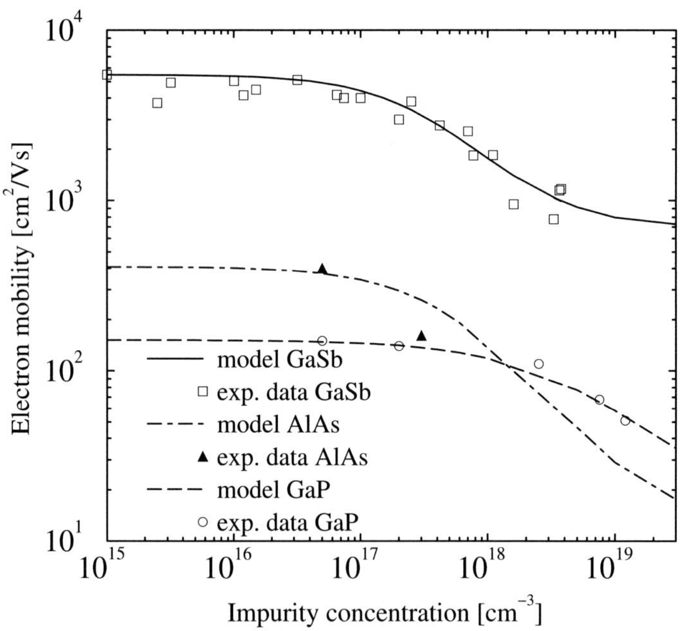

Fig. 3.36. Majority electron mobility as a function of impurity concentration at 300 K: Comparison of the model and experimental data for AlAs, GaP, and GaSb.

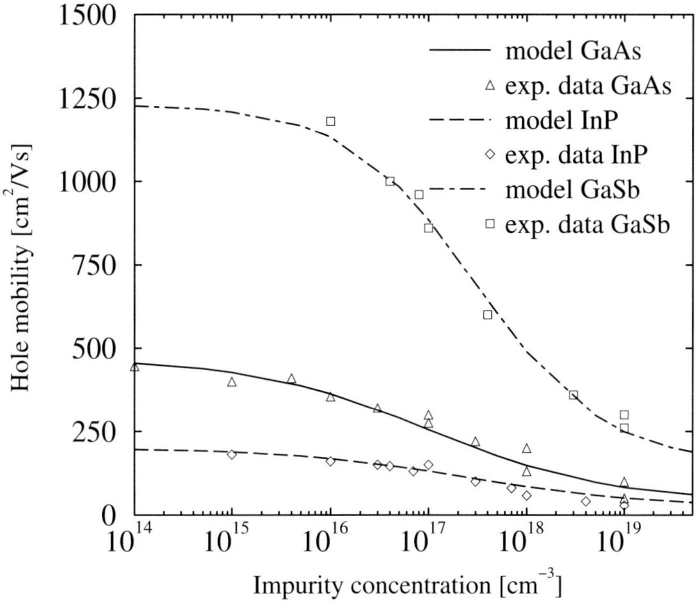

Fig. 3.37. Majority hole mobility as a function of impurity concentration at 300 K: Comparison of the model and experimental data for GaAs, GaSb, and InP.

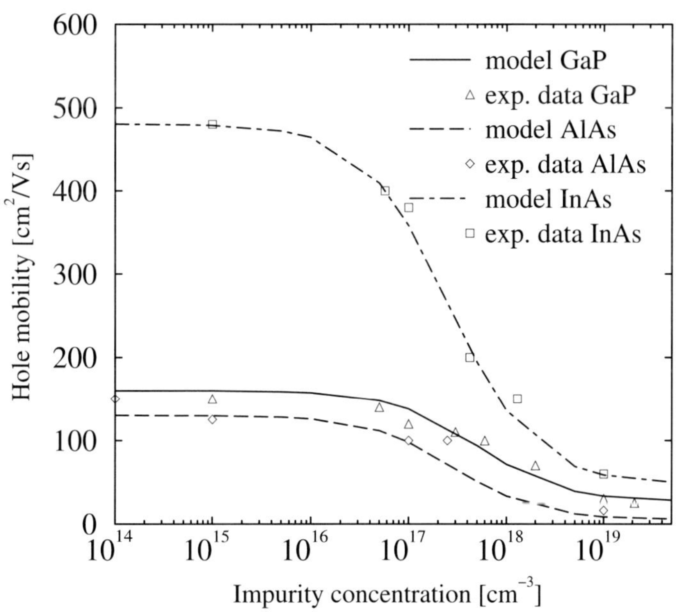

Fig. 3.38. Majority hole mobility as a function of impurity concentration at 300 K: Comparison of the model and experimental data for InAs, AlAs, and GaP.

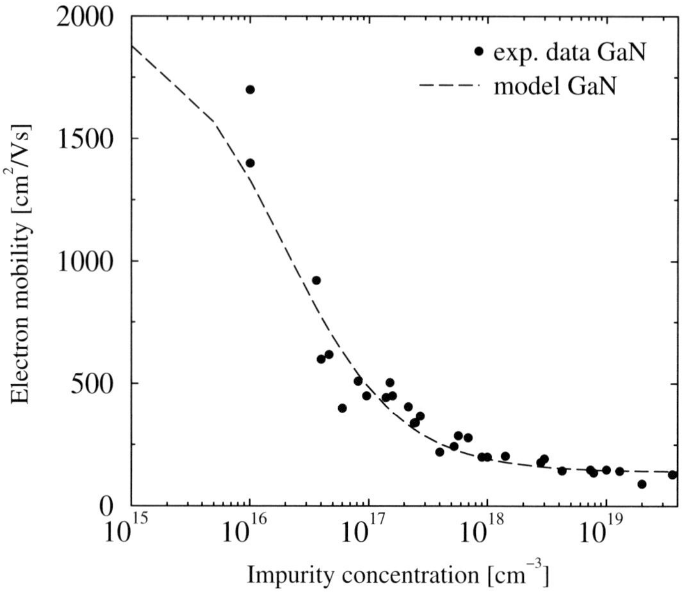

Fig. 3.39. Electron mobility as a function of impurity concentration in GaN at 300 K: Comparison of the model and experimental data.

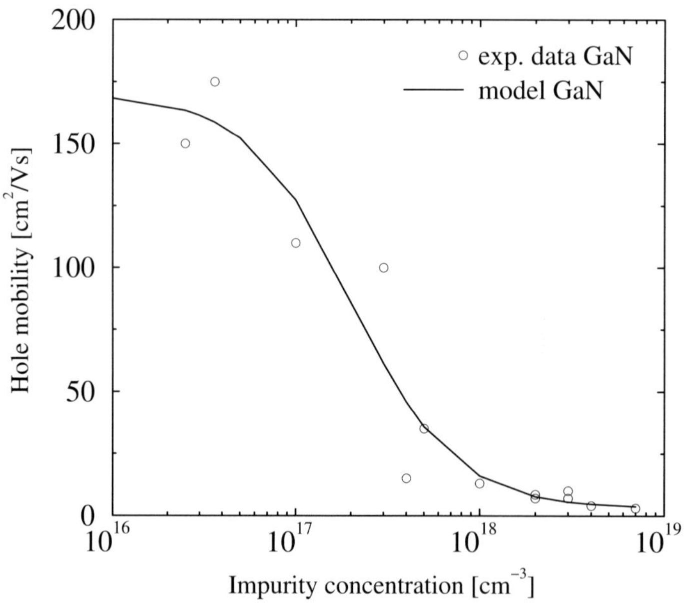

Fig. 3.40. Hole mobility as a function of doping concentration in GaN at 300 K: Comparison of the model and experimental data.

complicated by the compensation mechanism of the different dopants, such as C and Mg. A compilation of experimental data for holes is taken from [164] and given in Fig. 3.40. For InN and $In_xGa_{1-x}N$, the experimental confirmation of the impurity scattering for the MC predictions [18, 143] is relatively poor for most of the material compositions. The doping dependence is fit to the data of [18]. For indium contents of about $x = 7\%$, mobility values suitable for initial InGaN HEMTs can be found in [114]. Further experimental results for InN can be found in the compilation in [288]. Data on the hole transport are not available to the authors' knowledge. Another model to describe both the available impurity and temperature dependence of the mobility in GaN between 50 K and 1000 K is found in [343].

Models for Majority and Minority Electrons. Though numerous theoretical and experimental papers [202, 228, 311] have been published on electron mobility in semiconductors, there are still some issues under discussion, particularly at very high doping concentrations, i.e. $N_\mathrm{D}, N_\mathrm{A} \geq 10^{18}\ \mathrm{cm}^{-3}$. The difference between majority and minority electron mobility is a well-known phenomenon caused by effects such as degeneracy and the different screening behavior of electrons and holes in semiconductors. However, the mobility models usually employed in device modeling do not reflect these effects. The minority mobility at doping levels above $10^{19}\ \mathrm{cm}^{-3}$ exceeds the majority mobility by a factor of more than three at 300 K. The difference gets even stronger at low temperatures. A model which distinguishes between the majority and minority electrons in silicon is given by (3.122)-(3.124). Although initially proposed for the majority electron mobility in silicon, the model offers enough flexibility to describe also the minority electron mobility (see Fig. 3.41).

$$\mu_{\mathrm{n,maj}}^{\mathrm{LI}} = \frac{\mu_\mathrm{n}^\mathrm{L} - \mu_\mathrm{maj}^\mathrm{mid}}{1 + \left(\dfrac{N_\mathrm{D}}{C_\mathrm{maj}^\mathrm{mid}}\right)^{\alpha}} + \frac{\mu_\mathrm{maj}^\mathrm{mid} - \mu_\mathrm{maj}^\mathrm{hi}}{1 + \left(\dfrac{N_\mathrm{D}}{C_\mathrm{maj}^\mathrm{hi}}\right)^{\beta}} + \mu_\mathrm{maj}^\mathrm{hi} \tag{3.122}$$

$$\mu_{\mathrm{n,min}}^{\mathrm{LI}} = \frac{\mu_\mathrm{n}^\mathrm{L} - \mu_\mathrm{min}^\mathrm{mid}}{1 + \left(\dfrac{N_\mathrm{A}}{C_\mathrm{min}^\mathrm{mid}}\right)^{\alpha}} + \frac{\mu_\mathrm{min}^\mathrm{mid} - \mu_\mathrm{min}^\mathrm{hi}}{1 + \left(\dfrac{N_\mathrm{A}}{C_\mathrm{min}^\mathrm{hi}}\right)^{\beta}} + \mu_\mathrm{min}^\mathrm{hi} \tag{3.123}$$

$$\mu_\mathrm{n}^\mathrm{LI} = \left(\frac{1}{\mu_\mathrm{n,maj}^\mathrm{LI}} + \frac{1}{\mu_\mathrm{n,min}^\mathrm{LI}} - \frac{1}{\mu_\mathrm{n}^\mathrm{L}}\right)^{-1} . \tag{3.124}$$

$\mu_\mathrm{n}^\mathrm{L}$ is the mobility for undoped material, μ^hi is the mobility at highest doping. $\mu_\mathrm{maj}^\mathrm{mid}$, $\mu_\mathrm{maj}^\mathrm{hi}$, $\mu_\mathrm{min}^\mathrm{mid}$, $\mu_\mathrm{min}^\mathrm{hi}$, $C_\mathrm{maj}^\mathrm{mid}$, $C_\mathrm{min}^\mathrm{mid}$, $C_\mathrm{maj}^\mathrm{hi}$, $C_\mathrm{min}^\mathrm{hi}$, α, and β are used as fitting parameters. The final low-field electron mobility μ^LI, which accounts for a combination of both acceptor and donor doping, is given by (3.124). Fig. 3.41

Table 3.24. Parameter values for the majority/minority electron mobility at 300 K

Parameter	Si	Ge(on Si)	GaAs	InP	Unit
μ_n^{L}	1430	560	8500	5300	[cm^2/Vs]
μ_{maj}^{mid}	44	80	2460	1120	[cm^2/Vs]
μ_{maj}^{hi}	57	59	130	20	[cm^2/Vs]
μ_{min}^{mid}	141	124	650	1180	[cm^2/Vs]
μ_{min}^{hi}	218	158	1400	1920	[cm^2/Vs]
α	0.65	0.65	0.7	0.6	[–]
β	2.0	2.0	2.0	3.0	[–]
$C_{maj}^{mid} = C_{min}^{mid}$	1.12e17	4.0e17	5e16	4e16	[cm^{-3}]
C_{maj}^{hi}	1.18e20	4.9e18	1.8e19	1.6e19	[cm^{-3}]
C_{min}^{hi}	4.35e19	5.4e19	1.4e19	1.6e19	[cm^{-3}]

Table 3.25. Comparison of model parameters for majority electron mobility in silicon at 300 K and the parameters for the Masetti impurity mobility model

Parameter	Si	Parameter	Masetti Si	Unit
μ_{maj}^{mid}	44	μ_1^{min}	52.5	[cm^2/Vs]
μ_{maj}^{hi}	57	μ_2^{min}	52.5	[cm^2/Vs]
$\mu_{maj}^{mid} - \mu_{maj}^{hi}$	-13	$-\mu_1$	-43.4	[cm^2/Vs]
α	0.65	α	0.68	[–]
β	2.0	$-\beta$	2.0	[–]
C_{maj}^{mid}	1.12e17	C_2	9.68e16	[cm^{-3}]
C_{min}^{mid}	1.18e20	C_1	3.43e20	[cm^{-3}]

demonstrates a good match between the analytical model, MC simulation data, and measurements from [295, 324, 494, 550] for silicon at 300 K.
In general, the model can also be applied for any other material of interest (Fig. 3.42, Fig. 3.43, Fig. 3.44). (3.122) is similar to (3.116), a function with two extreme values (μ_n^{L} as a maximum and μ_n^{min} as a minimum mobility). (3.122) is a mathematical function which can deliver a second maximum or minimum at very high impurity concentrations depending on μ^{mid} and μ^{hi}. Thus, it allows both majority and minority carrier mobilities to be properly modeled.
An optimizer [484] is used for an automated extraction of the model parameters. Most of the existing experimental data on the low-field mobility together with accurate MC simulations for Si, SiGe, and III-V semiconductor compounds [265, 266, 267] are used as input. The temperature dependence of the lattice mobility μ_n^{L} preserves the expression (3.115). The majority electron mobility $\mu_{n,maj}^{LI}$ is modeled as a function of the donor concentration N_D and the lattice temperature. The minority electron mobility $\mu_{n,min}^{LI}$ is modeled as a function of the acceptor concentration N_A and the lattice temperature. The temperature dependence of

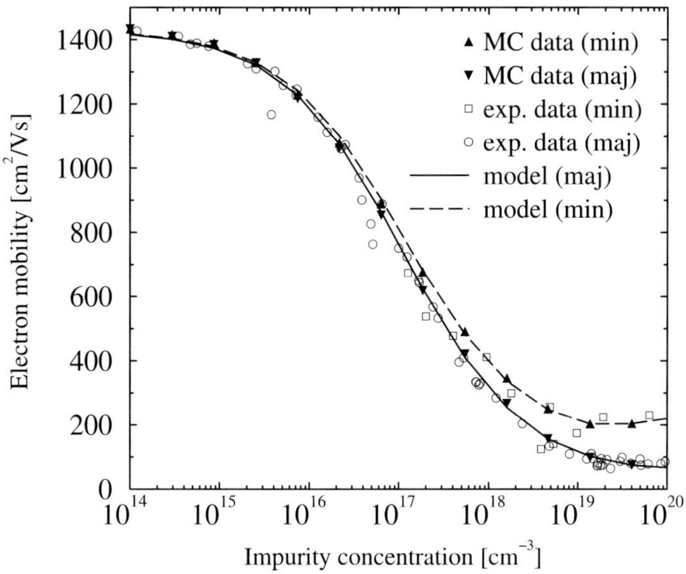

Fig. 3.41. Comparison of the analytical majority/minority model, experimental, and MC simulation data for electron mobility in silicon at 300 K.

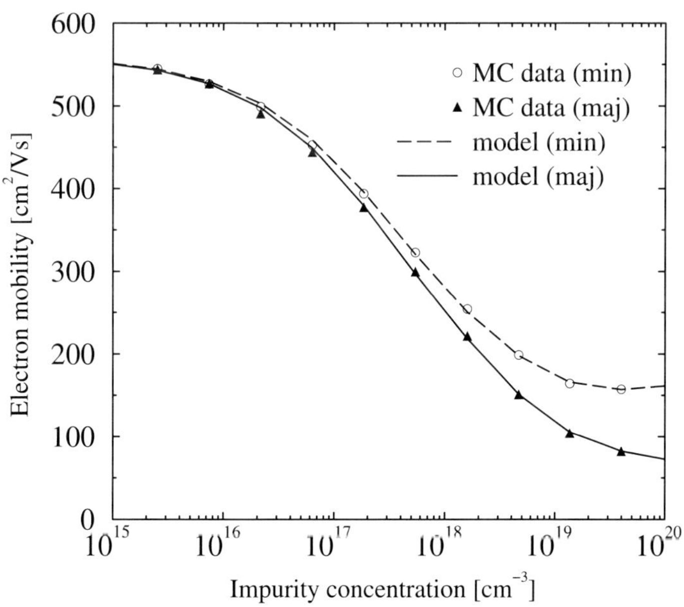

Fig. 3.42. Comparison of the analytical majority/minority model and MC simulation data for electron mobility in Ge (strained on Si) at 300 K.

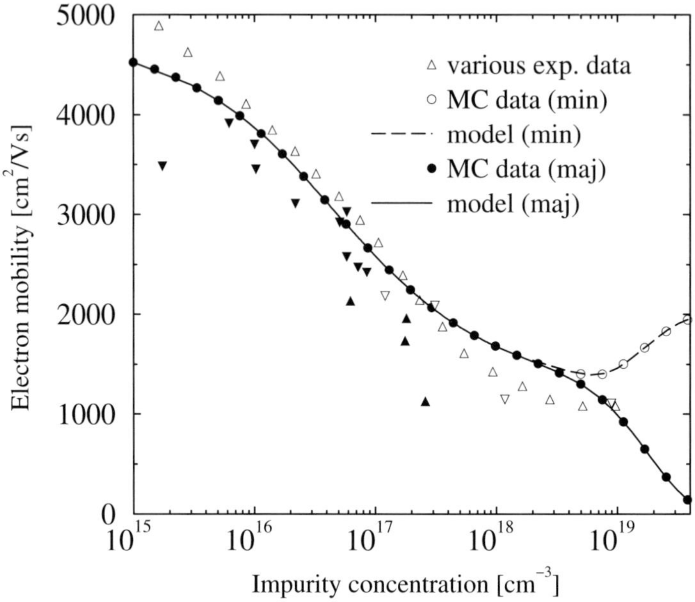

Fig. 3.43. Comparison of the analytical majority/minority model, experimental, and MC simulation data for the electron mobility in InP at 300 K.

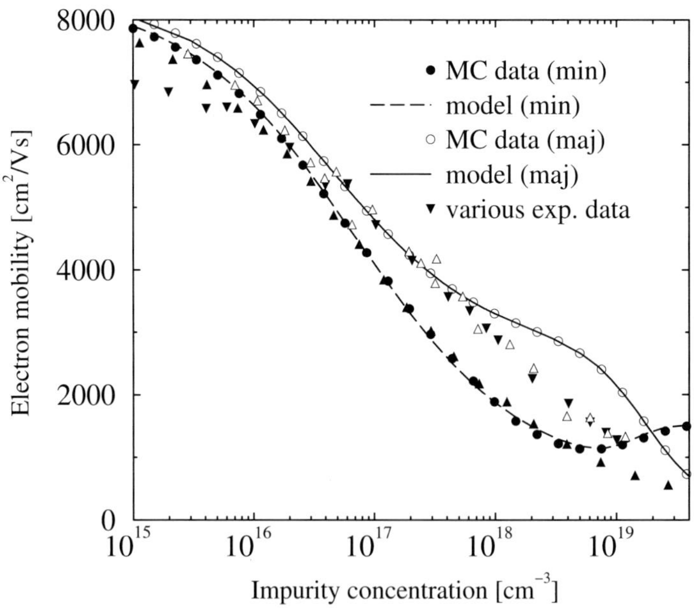

Fig. 3.44. Comparison of the analytical majority/minority model, experimental, and MC simulation data for the electron mobility in GaAs at 300 K.

the parameters $\mu^{\mathrm{mid}}_{\mathrm{maj}}$, $\mu^{\mathrm{hi}}_{\mathrm{maj}}$, $\mu^{\mathrm{mid}}_{\mathrm{min}}$, $\mu^{\mathrm{hi}}_{\mathrm{min}}$, $C^{\mathrm{mid}}_{\mathrm{maj}}$, $C^{\mathrm{mid}}_{\mathrm{min}}$, $C^{\mathrm{hi}}_{\mathrm{maj}}$, $C^{\mathrm{hi}}_{\mathrm{min}}$ are modeled by simple power laws. The model parameters used for Si, Ge (strained on Si), GaAs, and InP at 300 K are summarized in Table 3.24.

Masetti Mobility Model. The Masetti bulk mobility model [324] is a default mobility model in several device simulators, e.g. [227, 503].

$$\mu^{\mathrm{LI}}_{\nu} = \mu^{\mathrm{min}}_{1} \cdot \exp\left(\frac{-C_3}{C_{\mathrm{I}}}\right) + \frac{\mu^{\mathrm{L}}_{300} - \mu^{\mathrm{min}}_{2}}{1 + \left(\dfrac{C_{\mathrm{I}}}{C_2}\right)^{\alpha}} - \frac{\mu_1}{1 + \left(\dfrac{C_1}{C_{\mathrm{I}}}\right)^{\beta}} . \tag{3.125}$$

It can be useful to compare some of the parameters from the majority mobility model (3.122) to the default values of the Masetti model (with $C_3 = 0$) as it is shown in Table 3.25.

3.4.3 Surface Scattering

Mobility Model of MINIMOS 6. Surface scattering is modeled by the following empirical expression [451]

$$\mu^{\mathrm{LIS}}_{\nu} = \frac{\mu^{\mathrm{ref}}_{\nu} + (\mu^{\mathrm{LI}}_{\nu} - \mu^{\mathrm{ref}}_{\nu}) \cdot (1 - F(y))}{1 + F(y) \cdot \left(\dfrac{P_{\nu}}{P^{\mathrm{ref}}_{\nu}}\right)^{\gamma_{6,\nu}}} , \qquad \nu = \mathrm{n}, \mathrm{p} \tag{3.126}$$

$$\mu^{\mathrm{ref}}_{\nu} = \mu^{\mathrm{ref}}_{\nu,300} \cdot \left(\frac{T_{\mathrm{L}}}{300\ \mathrm{K}}\right)^{-\gamma_{5,\nu}} . \tag{3.127}$$

The function $F(y)$, which depends on the surface distance y, describes a smooth transition between the surface and bulk mobility [473, 525]. The parameter y^{ref} describes a critical length.

$$F(y) = \frac{2 \cdot \exp\left(-\left(\dfrac{y}{y^{\mathrm{ref}}}\right)^2\right)}{1 + \exp\left(-2 \cdot \left(\dfrac{y}{y^{\mathrm{ref}}}\right)^2\right)} \tag{3.128}$$

The pressing forces P_{n} and P_{p} in (3.126) are equal to the magnitude of the normal field strength at the interface if the carriers are attracted by the interface, otherwise zero.

Lombardi Mobility Model. Lombardi et al. [306] suggested another surface mobility degradation model. There are two surface scattering contributions due to acoustic phonons, μ_{ac}, and surface roughness, μ_{sr}. They are functions of total doping concentration C_{I}, and the pressing forces, P_{n} and P_{p}. The total mobility is expressed by a Mathiessen rule (3.131).

Table 3.26. Parameter values for surface mobility reduction in silicon: MINIMOS 6 model

ν	$\mu_{\nu,300}^{\mathrm{ref}}$ [cm^2/Vs]	$\gamma_{5,\nu}$	y^{ref} [nm]	P_ν^{ref} [V/m]	$\gamma_{6,\nu}$
n	638	-1.19	10	7.0e7	1.69
p	240	-1.09	10	2.7e7	1.0

Table 3.27. Parameter values for surface mobility reduction in silicon: Lombardi model

ν	B_ν [cm/s]	C_ν [cm$^{5/3}$K/V$^{2/3}$s]	β_ν [-]	D_ν [V/s]
n	4.75e7	1.74e5	0.125	5.82e14
p	9.93e6	8.84e5	0.0317	2.05e14

$$\mu_\nu^{\mathrm{ac}} = \frac{B_\nu}{P_\nu} + \frac{C_\nu \cdot C_{\mathrm{I}}^{\beta_\nu}}{P_\nu^{1/3} \cdot T_{\mathrm{L}}}, \qquad \nu = \mathrm{n}, \mathrm{p} \tag{3.129}$$

$$\mu_\nu^{\mathrm{sr}} = \frac{D_\nu}{{P_\nu}^2} \tag{3.130}$$

$$\frac{1}{\mu_\nu^{\mathrm{LIS}}} = \frac{1}{\mu_\nu^{\mathrm{LI}}} + \frac{1}{\mu_\nu^{\mathrm{ac}}} + \frac{1}{\mu_\nu^{\mathrm{sr}}} \tag{3.131}$$

The parameter values used in this model are summarized in Table 3.27.

3.4.4 Semiconductor Alloys

The model for the material composition dependence of the low-field mobility in *alloy materials* employs the low-field mobilities μ^{LI} of the basic materials (A and B) and combines them by a harmonic mean.

$$\frac{1}{\mu^{\mathrm{AB}}} = \frac{1-x}{\mu^{\mathrm{A}}} + \frac{x}{\mu^{\mathrm{B}}} + \frac{(1-x) \cdot x}{C_\mu} \tag{3.132}$$

C_μ is referred to as a nonlinear or bowing parameter. This modeling approach is useful if the two mobilities to combine are of similar magnitude.

Anisotropic Electron Mobility Model for Silicon-Germanium. Monte Carlo simulation, which accounts for alloy scattering and the splitting of the anisotropic conduction band valleys due to strain [474] in combination with an accurate ionized impurity scattering model [268], yields results for SiGe for the complete range of donor and acceptor concentrations and Ge contents x. The same functional form (3.122) is used to fit the doping dependence of the in-plane mobility component for $x = 0$ and $x = 1$ (Si and strained Ge on Si). The model parameters used at 300 K are summarized in Table 3.24.
The material composition dependence is modeled by (3.132). The bowing parameter C_μ equals 140 cm^2/Vs and 110 cm^2/Vs for doping levels below and above C_{mid} respectively. Fig. 3.45 shows the in-plane minority electron mobility

in $Si_{1-x}Ge_x$ as a function of x at 300 K for different acceptor doping concentrations.
The component of the mobility perpendicular to the surface is then obtained by a multiplication factor given by the ratio of the two mobility components. The good agreement of the model with the measured and the MC simulation data, both for in-plane and perpendicular to the surface directions, is illustrated in Fig. 3.46.

III-V Ternary Compounds. Fig. 3.47 compares measurements, MC simulation data, and the composition-dependent analytical low-field mobility model for $Al_xGa_{1-x}As$. However, for $In_xAl_{1-x}As$, in a one-valley modeling approach, the binary mobilities to combine differ by two orders of magnitude. In this case, (3.132) can supply negative mobility values for certain parameter combinations of bowing parameters, doping, temperature, and material composition. In Fig. 3.48, a one-valley approach is also given, which shows the significant disagreement with the available data at $x= 0.52$. The figures illustrate the improvement of the two-valley modeling concept, since any one-valley bowing approach for electrons will neglect certain compositions where the mobility becomes negative, to allow a precise fit of other compositions. The data are taken from [536]. Fig. 3.48 shows the carrier mobility for $In_xAl_{1-x}As$ as a function of material composition x for a n-type background concentration of 10^{15} cm^{-3}. Although InAlAs is a very important material for both InP-based HEMTs and HBTs, the understanding of the material composition dependence of the properties of InAlAs is still relatively poor, as can be seen from the small amount of experimental data given in Fig. 3.48. Recent data are given in [180]. Table 3.28 supplies the bowing parameters for (3.132).

Table 3.28. Parameter values for mobility model for alloy material

Material	$C_{\mu,n}^{one}$ [cm^2/Vs]	$C_{\mu,n}^{\Gamma}$ [cm^2/Vs]	$C_{\nu,n}^{X}$ [cm^2/Vs]	$C_{\mu,p}$ [cm^2/Vs]	Refs. -
SiGe	140,110	-	-	-100	Fig. 3.45,[148]
AlGaAs	180	-250	1e6	125	[323]
InGaAs	1e6	1e6	-	1e6	Fig. 3.49
InAlAs	1e6[1]	1.5e8	1e6	95	Fig. 3.48,[323]
InGaP	1e6	1e6	1e6	1e6	[480]
GaAsP	1e6	-	-	-	[571]
InAsP	-2.12e4	-2.12e4	-	1e6	[288]
AlGaN	40	40	-	1e6	[143]
GaAsSb	665	779	-	6.62	[334, 323] [43, 545]
$(InP)_{1-x}(In_{0.53}Ga_{0.47}As)_x$	2276	2276	-	-	[4, 502]

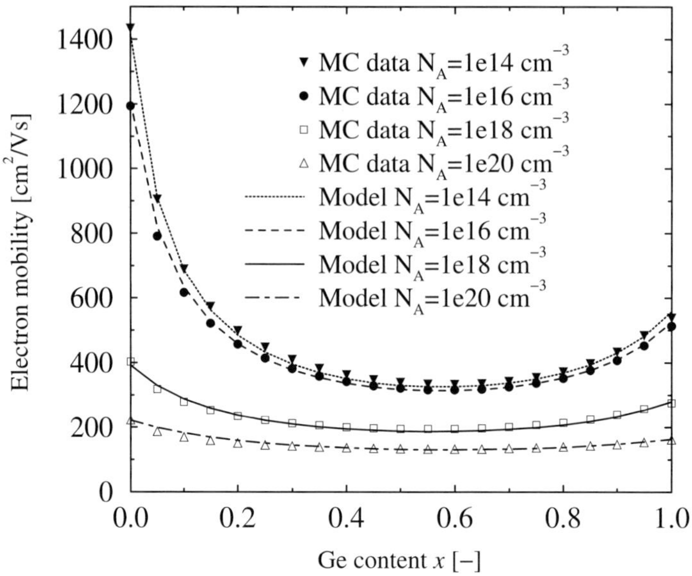

Fig. 3.45. Minority electron mobility in $Si_{1-x}Ge_x$ as a function of material composition for the in-plane direction in comparison with MC simulation data.

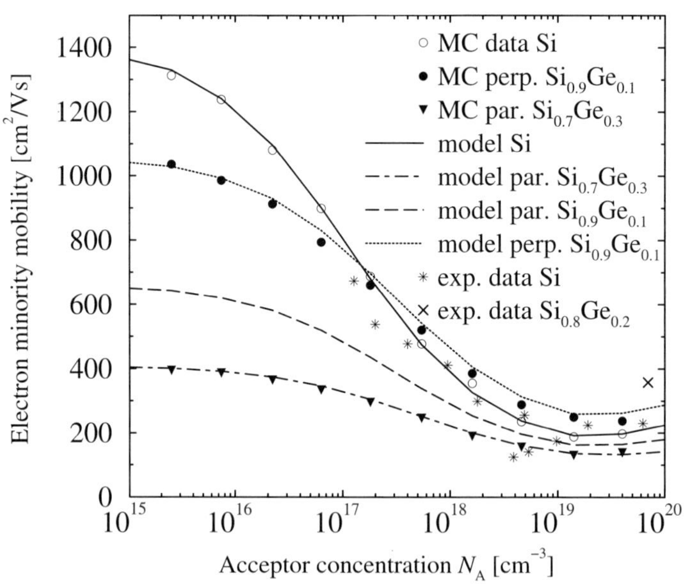

Fig. 3.46. Minority electron mobility in $Si_{1-x}Ge_x$ as a function of material composition x and acceptor concentration N_A in comparison with MC simulation data both for in-plane and perpendicular directions.

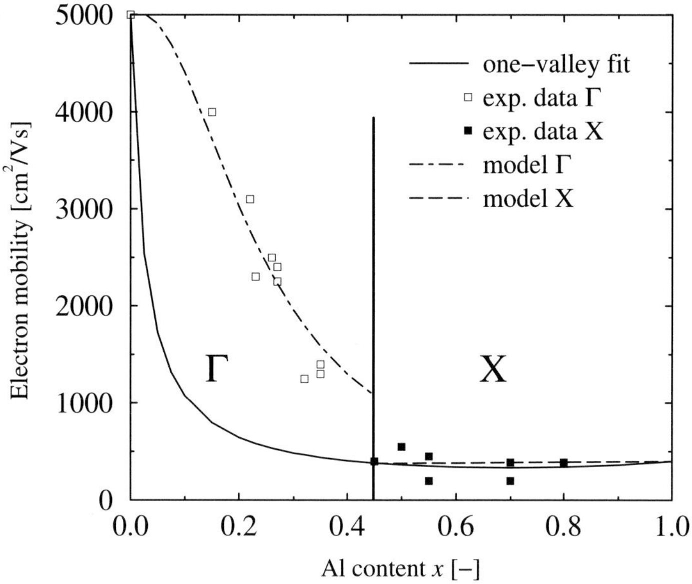

Fig. 3.47. Electron mobility as a function of material composition in comparison with experimental data for $N_D = 5..10 \times 10^{16}$ cm^{-3} in $Al_xGa_{1-x}As$ at 300 K.

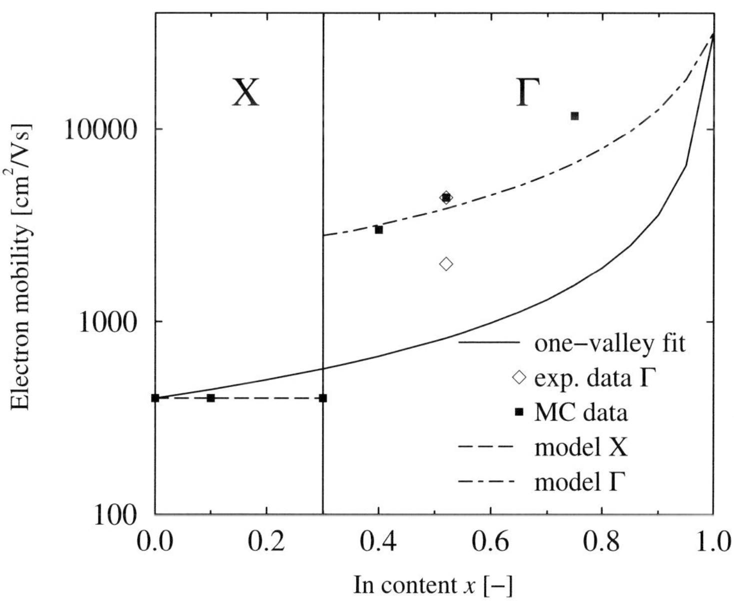

Fig. 3.48. Electron mobility as a function of material composition for $In_xAl_{1-x}As$ in comparison with measurements at 300 K.

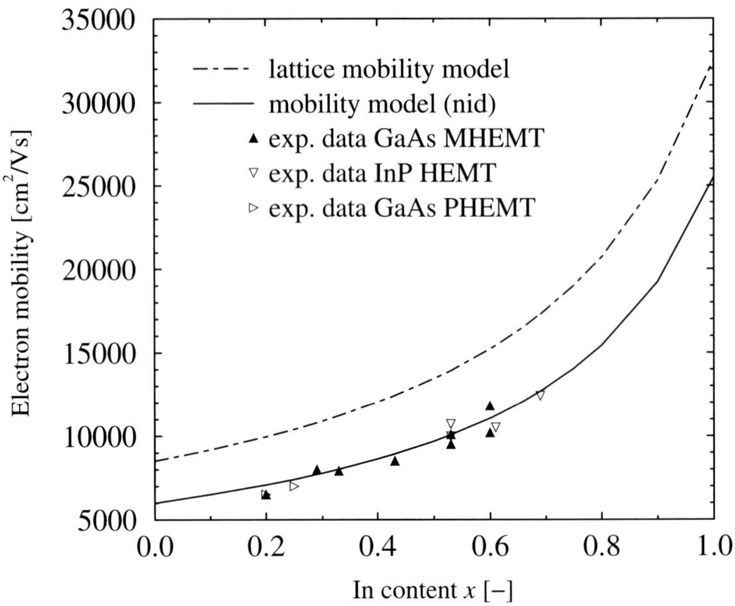

Fig. 3.49. Electron mobility as a function of material composition for $In_xGa_{1-x}As$ at 300 K.

Fig. 3.49 shows the material dependence of the mobility in $In_xGa_{1-x}As$. The simple quadratic bowing approach applies for the low-field mobility shown for the whole range of material compositions. The experimental data given for comparison are Hall data taken from HEMTs of the AlGaAs/InGaAs and the InAlAs/InGaAs material systems. The data for the pseudomorphic HEMTs are taken from [125] and [131]. For the metamorphic HEMTs grown on GaAs, the data are taken from [547] and the compilation of [201]. The data for InP-based devices are taken from [553]. If a double channel concept is applied, a weighted average is determined for x. It can also be seen that the metamorphic devices reach mobility values similar to those of the corresponding InP devices. The bowing factors for holes in AlGaAs and InAlAs are obtained from least square fits of the MC simulations in [323].

A good compilation of the experimental low-field mobility for various growth methods and doping concentrations in $Ga_{0.51}In_{0.49}P$ is given in [480]. MC simulation data for $Ga_{0.51}In_{0.49}P$ is given in [70]. Due to the transition of $Ga_xIn_{1-x}P$ from a direct to an indirect semiconductor at $x = 0.7$, the material dependence of the mobility cannot be modeled in a one-valley approach. The offset between measured data and model is considerable, especially for $x = 0.51$. As given in Table 3.22, a two-valley approach is used which allows for a precise modeling of the mobility for $x \leq 0.7$ (Γ-valley) and $x > 0.7$ (X valley). For InAsP, a precise

modeling of the experimental data is possible over the whole composition range with the bowing parameter supplied in Table 3.28. However, precise modeling of the data in [571] for GaAsP is not possible with (3.132), due to the strong nonlinearity, even when using the two-valley approach.
Fig. 3.50 gives the carrier mobility as a function of material composition x for the most important GaN-based compound, $Al_xGa_{1-x}N$. The composition-dependent mobility of AlGaN is of major importance to AlGaN/GaN HEMTs, since increased aluminum content x promises higher sheet carrier concentrations in the GaN HEMT channel layers, while the transport properties of AlGaN deteriorate, and thus the contributions to the access resistances increase. Fig. 3.50 compares the one-valley modeling approach to experimental and MC simulation data. The experimental database for AlGaN is relatively poor. The good transport properties of GaN can be observed in Fig. 3.50 and are confirmed by MC calculations [143] for similar background doping concentration. The MC simulation data in [18] cannot be fit by using (3.132). Characteristic for the data in [18] is the relatively small influence of the alloy scattering, as can be seen in Fig. 3.50.
On the contrary, this work assumes the strong alloy scattering of the MC simulations in [143]. The bowing parameter C_μ is extracted from this data for a doping concentration of $N_D = 10^{16}$ cm^{-3}. For AlN, experimental data are available which are given in Fig. 3.50.
Fig. 3.51 gives the electron mobility of $GaAs_{1-x}Sb_x$ as a function of material composition at a doping level of 5×10^{16}cm^{-3}. The data is taken from the well-investigated binaries GaAs and GaSb and from [43, 334]. A bowing factor can be extracted from this data, as given in Fig. 3.51. Fig. 3.52 gives the hole mobility as a function of material composition x in $GaAs_{1-x}Sb_x$ for $10^{19} \leq N_A \leq 10^{20}$ cm^{-3}, as given in [545] and earlier in [492]. From this modeling, a strong bowing factor is extracted, which exceeds the bowing in [323] suggested from MC calculations and is caused by the strong alloy scattering. The bowing for the quaternary InGaAsP is determined from the data as given in Table 3.28.

Impurity Scattering in Semiconductor Ternary Alloys. In the modeling concept applied, the ternaries are derived from the binary materials and the doping dependence is calculated first. However, this paragraph gives a comparison of modeled and measured mobility values as a function of impurity concentration for important ternary semiconductors for the verification of the modeling. A comparison of MC simulation data, measurements, and the analytical model for $In_{0.48}Al_{0.52}As$ as a function of n-doping is shown in Fig. 3.53. The data are taken from the compilation of Littlejohn et al. in [304] and selectively from Goto et al. in [180]. By using a special value for Γ AlAs, given in Table 3.22 and suitable for InAlAs only, very good agreement is achieved, as shown in Fig. 3.53. The problem of possible negative mobility values in (3.132) can be overcome by the bowing parameter in Table 3.28 without having to apply a second material composition concept for the ternary alloys in the simulator, as suggested for example by Sotoodeh et al. in [480]. Fig. 3.54 gives a similar comparison for $In_{0.53}Ga_{0.47}As$.

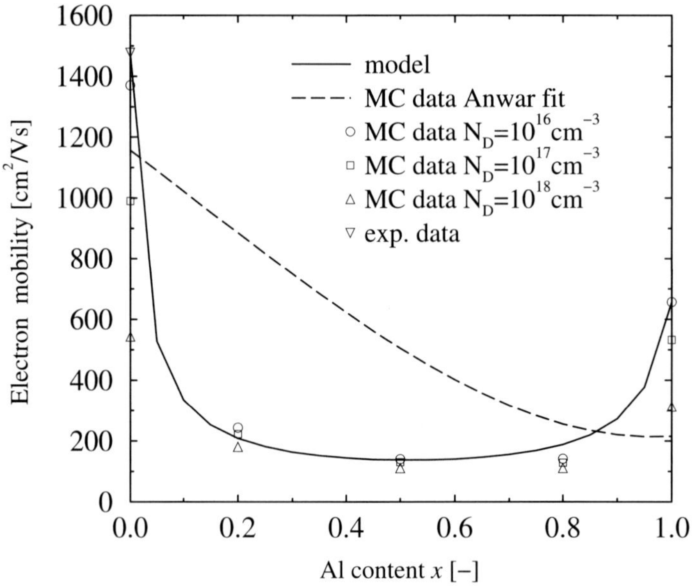

Fig. 3.50. Electron mobility as a function of material composition for $Al_xGa_{1-x}N$ at 300 K.

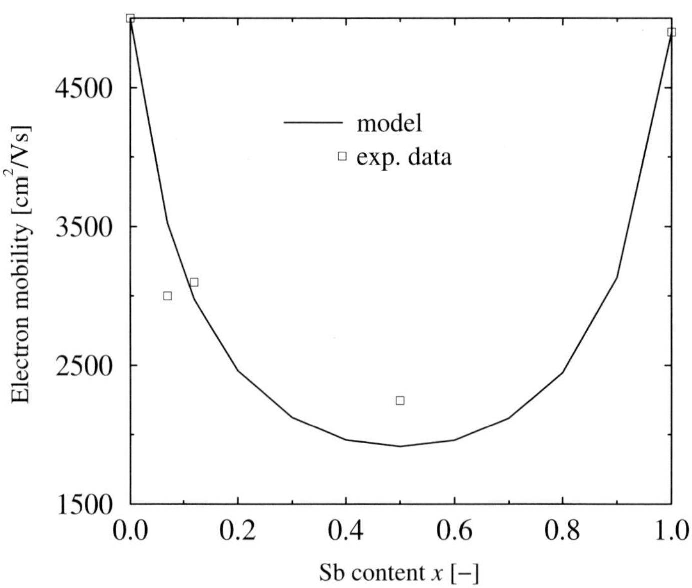

Fig. 3.51. Electron mobility as a function of material composition for $GaAs_{1-x}Sb_x$ with doping levels of $5\times10^{16}cm^{-3}$ at 300 K.

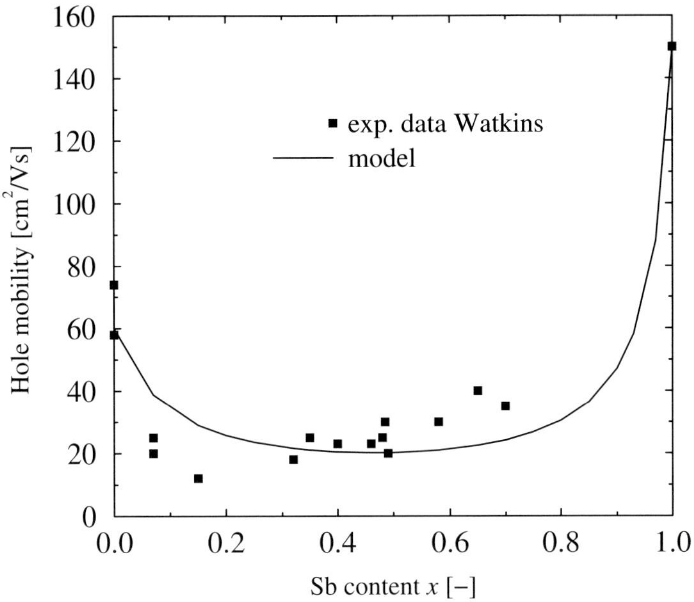

Fig. 3.52. Comparison of the majority mobility model and experimental data in p-$GaAs_{1-x}Sb_x$ as a function of material composition at doping levels $\geq 10^{19}cm^{-3}$ at 300 K.

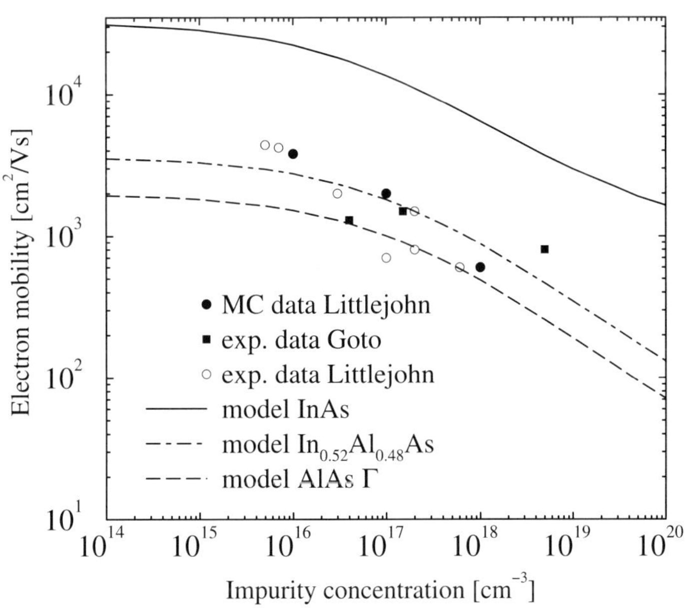

Fig. 3.53. Comparison of the MC simulation, measurement data, and the analytical model for the mobility of $In_{0.52}Al_{0.48}As$ as a function of impurity concentration at 300 K.

The figure stresses the importance of the residual background concentration for the low-field properties [355]. To allow for the simulation of metamorphic and pseudomorphic HEMTs on InP substrate, MC simulations have been performed for $x \neq 0.52$ to support the available MC investigations in [209] and data compiled in [304].
The mobility modeling is also given for the p-material for devices with highly doped ternary materials, such as p-doped $In_{0.53}Ga_{0.47}As$ in InP-based HBTs. Fig. 3.55 gives the hole mobility in p-$In_{0.53}Ga_{0.47}As$ at 300 K as a function of acceptor doping concentration up to $3{\times}10^19$ cm^{-3}. The data are for MOCVD-grown material [96].

3.4.5 High-Field Mobility Models for DD Equations

Silicon, Germanium, and Silicon-Germanium. The mobility reduction in the high-field domain is modeled for silicon by

$$\mu_\nu^{\mathrm{LISF}} = \frac{2 \cdot \mu_\nu^{\mathrm{LIS}}}{1 + \left(1 + \left(\dfrac{2 \cdot \mu_\nu^{\mathrm{LIS}} \cdot F_\nu}{v_{\mathrm{sat},\nu}}\right)^{\beta_\nu}\right)^{1/\beta_\nu}}, \qquad \nu = \mathrm{n}, \mathrm{p} \tag{3.133}$$

$$F_\mathrm{n} = \left| \operatorname{grad} \psi - \frac{1}{n} \cdot \operatorname{grad} \left(\frac{k_\mathrm{B} \cdot T_\mathrm{L}}{\mathrm{q}} \cdot n \right) \right| \tag{3.134}$$

$$F_\mathrm{p} = \left| \operatorname{grad} \psi + \frac{1}{p} \cdot \operatorname{grad} \left(\frac{k_\mathrm{B} \cdot T_\mathrm{L}}{\mathrm{q}} \cdot p \right) \right| \tag{3.135}$$

In (3.133) F_ν represents the driving forces for carrier type ν = n,p. The saturation velocities $v_{\mathrm{sat},\nu}$ for carrier ν are calculated in a separate model in (3.144). The default parameter values used in (3.133) are summarized in Table 3.29.

Table 3.29. Parameter values for DD high-field mobility model

Material	β_n [-]	β_p [-]	References
Si	2.0	1.0	[448]
Si	1.09	1.203	[227]
Ge	2.0	1.0	[496]
GaAs	2.0	1.0	[496]
InP	1.25	1.0	[67]
GaP	1.25	1.0	[20, 241]
GaSb	2.0	1.0	[109, 110]
GaN	2.0	1.0	[143]

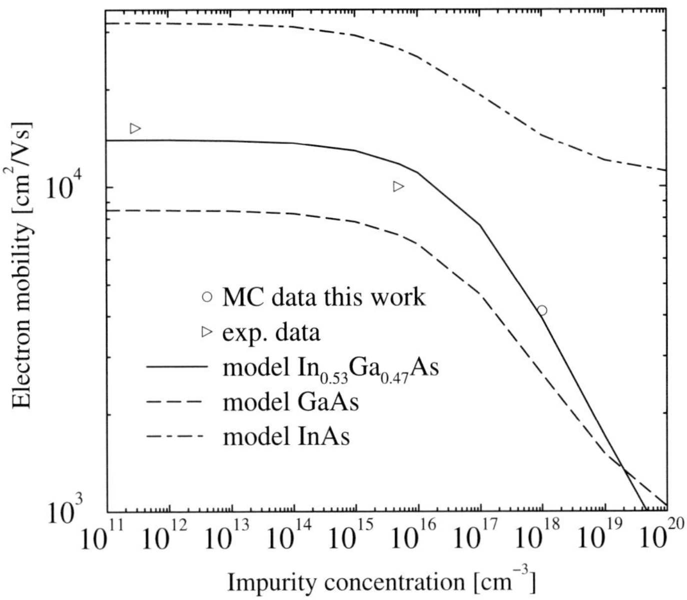

Fig. 3.54. Comparison of the MC simulation, measurement data, and the majority mobility model for n-$In_{0.53}Ga_{0.47}As$ as a function of doping concentration at 300 K.

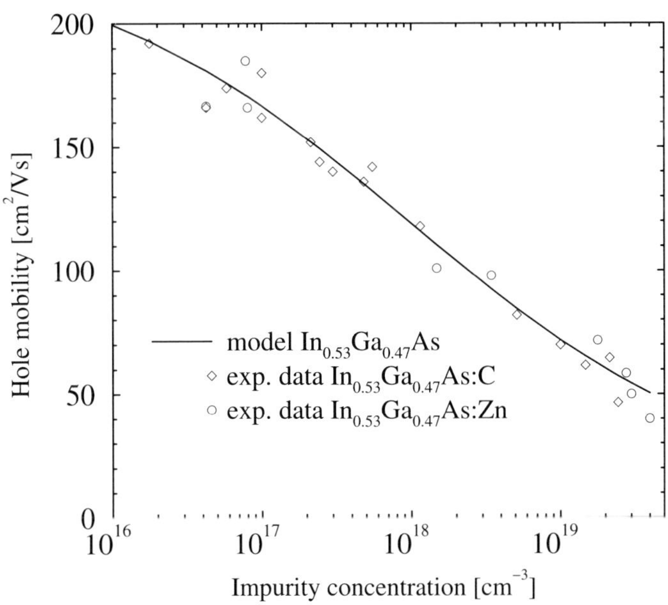

Fig. 3.55. Comparison of the majority mobility model and experimental data in p-$In_{0.53}Ga_{0.47}As$ at 300 K.

III-V Semiconductors. Several model options are available for III-V materials; one gives a monotonic velocity versus field curve (3.136), while another, (3.137), includes a domain with negative differential mobility [273].

$$\mu_\nu^{\mathrm{LIF}}(F_\nu) = \frac{\mu_\nu^{\mathrm{LI}}}{\left(1 + \left(\frac{\mu_\nu^{\mathrm{LI}} \cdot F_\nu}{v_{\mathrm{sat},\nu}}\right)^{\beta_\nu}\right)^{1/\beta_\nu}}, \qquad \nu = \mathrm{n}, \mathrm{p} \tag{3.136}$$

$$\mu_\mathrm{n}^{\mathrm{LIF}}(F_\mathrm{n}) = \frac{\mu_\mathrm{n}^{\mathrm{LI}} + v_{\mathrm{sat,n}} \cdot \frac{F_\mathrm{n}^{\beta_\mathrm{n}-1}}{F_0^{\beta_\mathrm{n}}}}{1 + \frac{F_\mathrm{n}^{\beta_\mathrm{n}}}{F_0^{\beta_\mathrm{n}}}} \tag{3.137}$$

Similar to (3.133), F_n represents the driving force and $v_{\mathrm{sat,n}}$ the saturation velocity for electrons according to (3.134) and (3.144) respectively. The default parameter values used in (3.133) and (3.136) are summarized in Table 3.29 and Table 3.30. (3.137) can be used especially for the mobility in III-V materials. It includes the additional model parameter F_0, which denotes the critical driving force approximately at which the overshoot in the velocity-field characteristics appear. The default parameter values are summarized in Table 3.30. In the model of (3.137), F_0 is considered a free parameter, contrary to (3.136), where F_0 is determined by saturation velocity and by $\mu_\mathrm{n}^{\mathrm{LI}}$. For GaAs, InP, and GaSb, F_0 describes the field where negative differential mobility typically occurs. The values obtained for AlAs are taken from the investigations of transport of its most important compounds, AlGaAs [192] and InAlAs [304]. Parameter values for (3.137) are given in Table 3.30. The comparison of the different F_0 shows higher values for InP than for GaAs. For GaP, no occurrence of static overshoot effects is visible in [20, 241], so the model (3.133) should be used. For InSb, no data are available for the velocity-field characteristics except at $T_\mathrm{L} = 77$ K, so only these values can be given. In comparison to (3.136), the F_0 determined for GaN in (3.137) exceeds the value of $v_{\mathrm{sat}}/\mu^{\mathrm{LI}}$ by an order of magnitude. The data for the wide-bandgap materials are based on [143]. A more detailed discussion of mobility models for GaAs-based semiconductors can be found in [217, 555].

Wide Bandgap Semiconductors. In nitrides, negative differential mobility occurs at magnitudes of the fields which are considered the extreme high-field region for arsenides, phosphides, or antimonides. The model given in (3.137) is an estimate of the underlying high-field physics and is used only to match the very basic physics of GaN with the overshoot occurring at much higher fields than those known for GaAs. Several other mobility models have been proposed to model the high-field properties of GaN-based materials, which are still subject to major research [68]. A model has been proposed for GaN, AlN, and InN in [143] increasing the number of parameters involved relative to (3.137).

Table 3.30. Parameter values for DD III-V material high-field mobility model

Material	β_n	F_0 [V/cm]	Refs.
GaAs	4.0	4e3	[448]
AlAs	4.0	6e3	[71, 304]
InAs	4.0	2.2e3	[68]
InP	4.0	1e4	[67]
GaN	4.0	2.2e5	[143]
AlN	4.0	4.47e5	[143]
InN	4.0	5.24e4	[143]
GaSb	4.0	7.4e2	[109]
InSb	4.0	5e2 (77 K)	[21, 361]

Table 3.31. Parameter values for DD wide-bandgap high-field mobility model

Material	v_{sat} [10^7cm/s]	F_0 [V/cm]	β_{n1}	β_{n2}	a
GaN	2.0	2.21e5	7.2	0.79	6.20
AlN	2.16	4.47e5	17.37	0.85	8.73
InN	1.36	5.24e4	3.85	0.61	2.26

$$\mu_n^{LIF}(F_n) = \frac{\mu_n^{LI} + v_{sat,n} \cdot \dfrac{F_n^{\beta_{n1}-1}}{F_0^{\beta_{n1}}}}{1 + \dfrac{F_n^{\beta_{n1}}}{F_0^{\beta_{n1}}} + a \cdot \dfrac{F_n^{\beta_{n2}}}{F_0^{\beta_{n2}}}} \tag{3.138}$$

The parameter values in (3.138) for GaN, AlN, and InN in are given in Table 3.31. The number of investigations is very limited for the hole transport. Some low-field investigations for p-GaN are given in [165].

3.4.6 High-Field Mobility Models for HD Equations

Silicon, Germanium, and Silicon-Germanium. The deviation from the ohmic low-field mobility is modeled as a function of the carrier temperature T_ν after Hänsch et al. [199].

$$\mu_\nu^{LIST} = \frac{\mu_\nu^{LIS}}{1 + \alpha_\nu \cdot (T_\nu - T_L)}, \qquad \nu = n, p \tag{3.139}$$

$$\alpha_\nu = \frac{3 \cdot k_B \cdot \mu_\nu^{LIS}}{2 \cdot q \cdot \tau_\epsilon \cdot (v_{sat,\nu})^2} \tag{3.140}$$

$\tau_{\epsilon,\nu}$ denote the energy relaxation times and $v_{sat,\nu}$ are the saturation velocities of the carrier type ν calculated in (3.147) and in (3.144). For calculation of the high-field mobilities, no additional parameters need to be specified for the model. The interpolations between the *basic materials* are carried out in the models for the saturation velocities and for the energy relaxation times.

Alternative HD High-Field Mobility Model. A second variation of the high-field mobility is supplied to clarify the interaction of the model and model parameters. (3.141) represents a modification of the Hänsch model, used, for example, in [337]. The formula reads

$$\mu^{\mathrm{LIST}}(T_\nu) = \frac{\mu_\nu^{\mathrm{LI}}}{\left(1 + \alpha_\nu \cdot (T_\nu - T_{\mathrm{L}})^{\frac{1}{\beta}}\right)^{\beta}}, \qquad \nu = \mathrm{n}, \mathrm{p}. \tag{3.141}$$

In this approach an additional parameter $1 \leq \beta \leq 2$ is introduced as demonstrated in HEMT simulations in [64]. It was used in [64] with constant energy relaxation times. The parameter β allows the static overshoot effects in the channel obtained by MC simulation to be matched. The interaction of the high-field parameters obtained within one model needs to be stressed at this point. The acquisition of high-field parameters when using the DD, the HD approach in (3.141), and a HD approach using (3.139) and (3.140) with non-constant energy relaxation times is independent. The results, however, need to be consistent. Physical effects such as static velocity overshoot can be introduced with the different high-field mobility concepts. The parameters cannot be mixed, once calibrated for a particular approach. Furthermore, each approach supplies a different amount of information. For example, the V_{DS} bias dependence of the channel velocities cannot be fit using the DD approach. However, using v_{sat} and β in (3.136) as free parameters, agreement of the channel velocities can be obtained for one bias, as was shown by Bude et al. in [72].

Multi-Valley Model for III-V Semiconductors. For III-V semiconductor materials, a multi-valley mobility model can be used. The term multi-valley is used here in a different sense than before. In the following model a weighted mean is calculated from the low-field mobilities of the Γ and L valley electrons [122].

$$\mu_{\mathrm{n}}^{\mathrm{LIT}} = \frac{\mu_{\mathrm{n},\Gamma}^{\mathrm{LI}} + \mu_{\mathrm{n,L}}^{\mathrm{LI}} \cdot P_{\mathrm{L}}}{1 + P_{\mathrm{L}}} \tag{3.142}$$

$$P_{\mathrm{L}} = 4 \cdot \left(\frac{m_{\mathrm{L}}}{m_\Gamma}\right)^{3/2} \cdot \exp\left(-\frac{E_{C,\mathrm{L}} - E_{C,\Gamma}}{k_{\mathrm{B}} \cdot T_{\mathrm{n}}}\right) \tag{3.143}$$

P_{L} denotes the ratio of the Γ and L valley populations [496] and is constant. The valley mobilities $\mu_{\mathrm{n},\Gamma}^{\mathrm{LI}}$ and $\mu_{\mathrm{n,L}}^{\mathrm{LI}}$ account for impurity scattering after (3.116) and are constant with respect to the lattice temperature. The available data for wurtzite GaN are taken from [12].

HD Mobility Models for GaN and other Wide Bandgap Devices. As GaN-based wide bandgap devices are genuine high-voltage devices, high-field

Table 3.32. Parameter values for the two-valley HD mobility model

Material	$\mu_{n,\Gamma}^{L}$ [cm²/Vs]	$\mu_{n,L}^{L}$ [cm²/Vs]	μ_{n}^{min} [cm²/Vs]	C_{n}^{ref} [cm⁻³]	$\tau_{n,\Gamma}$ [ps]	$\tau_{n,L}$ [ps]	$v_{sat,n}^{\Gamma}$ [cm/s]	$v_{sat,n}^{L}$ [cm/s]
GaAs	8000	400	800	1e17	1.2	0.6	2.5e7	0.9e7
GaN	2000	280	295	1e17	0.15	0.15	2.9e7	2e7

effects require particular attention. This includes the analysis of the high-field domain by MC and other methods and the formulation of analytical high-field models, similar to the DD approaches given in Subsection 3.4.5. A basic methodology to calibrate the model parameters from material analysis and simulation has been discussed in [68]. Hydrodynamic simulations of GaN devices have so far been reported in [12] based on the mobility models reported in the previous subsections.

3.4.7 Saturation Velocity

For all materials, the bulk saturation velocity used in the high-field mobility is modeled as a function of lattice temperature T_L with the following two-parameter model:

$$v_{sat,\nu}(T_L) = \frac{v_{sat,\nu,300}}{(1 - A_\nu) + A_\nu \cdot \left(\frac{T_L}{300\ \mathrm{K}}\right)}. \tag{3.144}$$

The model (3.144) allows a precise fit of the experimental data, while preserving computational efficiency. The model assumes that the saturation velocity is independent of the doping concentration, which is in good agreement with various publications [448, 412]. The model parameter values are summarized in Table 3.33.

For silicon and germanium, the data given in Table 3.33 refers to well-known sources for the saturation velocity in unstrained materials. They are used to model relaxed and strained SiGe as a function of material composition using different bowing factors given in Table 3.34.

The saturation velocity for arsenides, phosphides, and antimonides is extracted for field values of $\approx$ 100 kV/cm. It is considered a material property here and should not be mixed with effective velocities and related terms. It should be noted that the X values for GaAs are used only for the material compositions $x > 0.42$ in $Al_xGa_{1-x}As$ to obtain a good fit and have no meaning for other material compositions. The data for GaSb are taken from [109, 110]. There are no data for AlSb on the high-field transport available to the authors' knowledge. There is a significant scatter for the saturation velocity in GaN and AlN. One reason for these discrepancies is the unusually high value for the peak drift velocity of the static overshoot, so that a saturated region is reached only at extremely high

Table 3.33. Model parameters for the bulk saturation velocity

Material (Electric field)	Valley	$v_{sat,n,300}$ [10^7cm/s]	A_n	$v_{sat,p,300}$ [10^7cm/s]	A_p	References
Si	X	1.0	0.26	0.704	0.63	[82, 147, 140] [230, 556]
Ge	L	0.7	0.55	0.63	0.61	[232, 417]
GaAs	Γ	0.72	0.56	0.9	0.41	[413]
	X	2.73	-	-	-	-
AlAs	X	0.85	0.55	0.8	0.3	[413]
	Γ	0.39	-	-	-	Fig. 3.58
InAs	Γ	0.9	0.57	0.5	0.3	[413]
	X	0.9	-	-	-	Fig. 3.58
InP	Γ	0.67	0.68	0.7	0.3	[413]
	X	0.67	0.68	0.5	0.3	-
GaP	X	0.88 (1.25)	0.30	0.5	0.3	[20, 70, 147, 241]
	Γ	0.88	0.30	0.5	0.3	-
GaN (E=5e5 V/cm)	Γ	1.34	0.15	-	-	[18, 37]
GaN	Γ	1.9	-	-	-	[143]
AlN (E=1e6 V/cm)	Γ	1.6	0.135	-	-	[18, 374]
AlN	Γ	2.16	-	-	-	[143]
InN	Γ	1.32	-	-	-	[18, 373]
	Γ	1.36	-	-	-	[143, 566]
GaSb	Γ	0.9	0.0165	1	-	[109, 110, 299, 430]
InSb (T_L= 77 K)	Γ	5.0	-	-	-	[21, 361]

electric fields > 500 kV/cm. A second reason is the lack of information on the band-structure of III-nitrides, especially for higher bands. Fig. 3.56 shows the temperature dependence of the saturation velocity for GaN and AlN in comparison to MC simulation data from [18, 83]. Significant experimental data for the drift mobility are missing for the ternary compounds to properly account for the alloy scattering [143].

Composition Dependence. The composition dependence of the saturation velocity is modeled according to (3.145). The model values are found in Table 3.34.

$$v_{sat,\nu}^{AB} = (1-x)\cdot v_{sat,\nu}^{A} + x\cdot v_{sat,\nu}^{B} + x\cdot(1-x)\cdot C_{vsat}. \quad (3.145)$$

Table 3.34 supplies values for the saturation velocity for the modeling of $Si_{1-x}Ge_x$ supplied in the literature. For $x \leq 0.3$, Bufler et. al supplied saturation velocities for strained and relaxed in-plane SiGe [74] for electrons.

Fig. 3.57, Fig. 3.58, and Fig. 3.59 illustrate the saturation velocity as a function of material composition for $Al_xGa_{1-x}As$, $In_xAl_{1-x}As$, and $In_xGa_{1-x}As$.

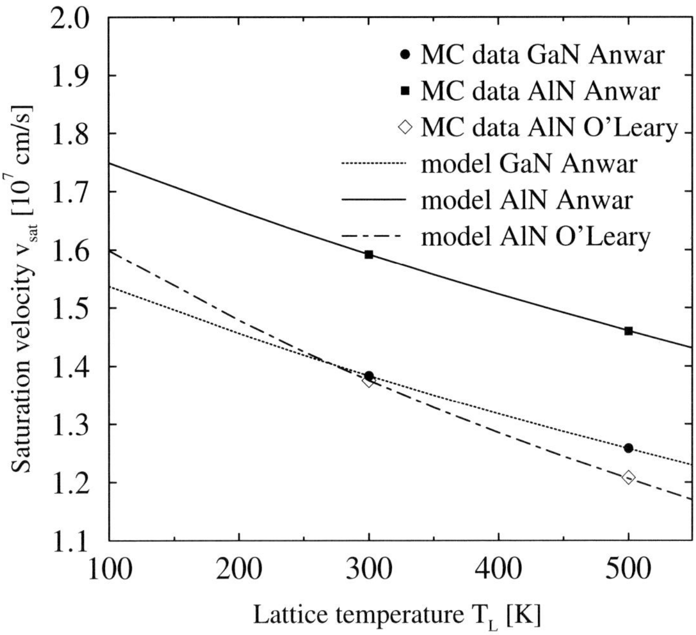

Fig. 3.56. Saturation velocity as a function of temperature for GaN and AlN.

A multi-valley concept is used to account for the crossover transition in AlGaAs and to reproduce our MC simulation data. The one-valley approach models the available data relatively well for $In_xAl_{1-x}As$, as can be seen in the comparison to the Γ-model. Experimental data for $x \leq 0.3$ are not available.
The data for $In_{1-x}Ga_xP$ are obtained from data in [70]. The bowing factors for $Al_xGa_{1-x}N$ is obtained from the MC simulation data of [18], as shown in Fig. 3.60. The hole data for $In_{0.53}Ga_{0.47}As$ is taken from MC simulations in [66] and measurements in [216]. Fig. 3.61 shows the temperature dependence of the electron saturation velocity in $In_{0.53}Ga_{0.47}As$. Fig. 3.60 gives the comparison of the MC simulations available for AlGaN and InGaN. A significant scatter is observed for the values for GaN and AlN and for the assumptions of the bowing. The values of Anwar et al. show lower values for v_{sat} for both GaN and AlN and a positive bowing for both AlGaN and InGaN. Farahmand et al. supply strong alloy scattering and negative bowing, which is a more realistic assumption. Both publications find a value of $\approx 1.3 \times 10^7$ cm^{-3} for the saturation velocity in InN. For GaAsSb, the modeling and data for electron and holes are given in Fig. 3.62. Due to the miscibility gap of GaSb and GaAs for MOCVD growth of GaAsSb, the amount of experimental data as a base for MC simulations is limited for the alloy material. The experimental and MC simulation data in Fig. 3.62 are taken from [109, 110, 323, 430]. Fig. 3.62 gives the saturation velocity for electrons

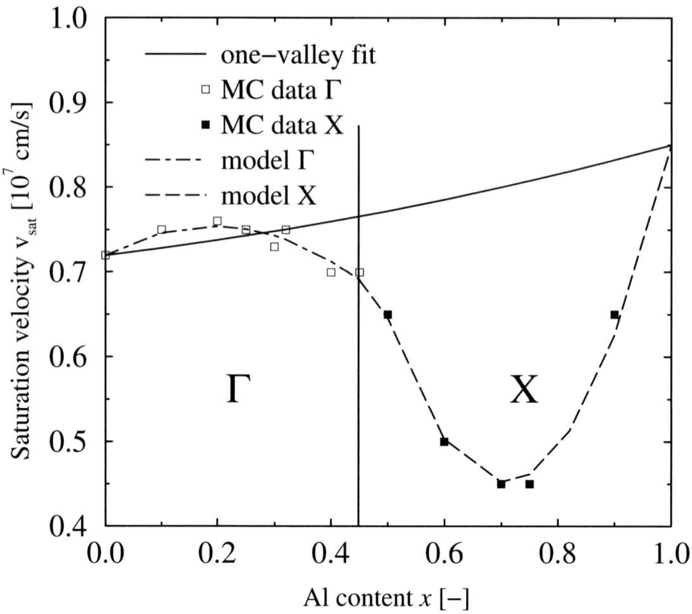

Fig. 3.57. Saturation velocity as a function of material composition at 300 K for $Al_xGa_{1-x}As$.

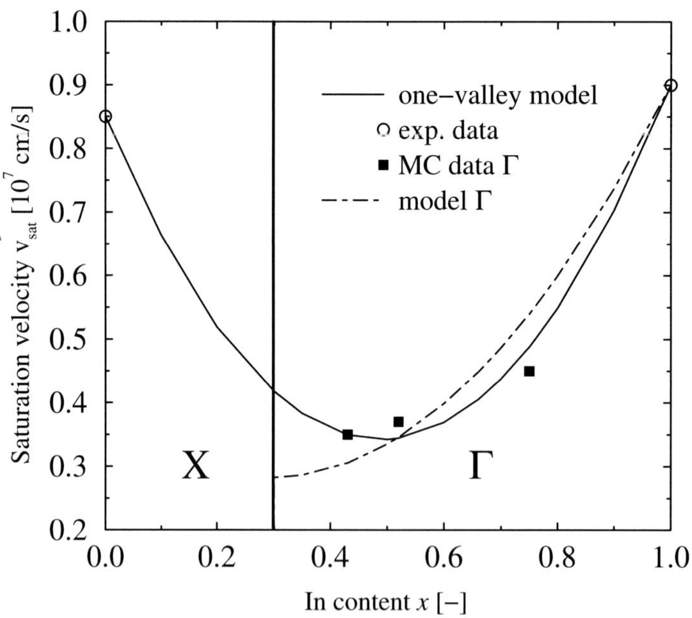

Fig. 3.58. Saturation velocity as a function of material composition at 300 K for $In_xAl_{1-x}As$.

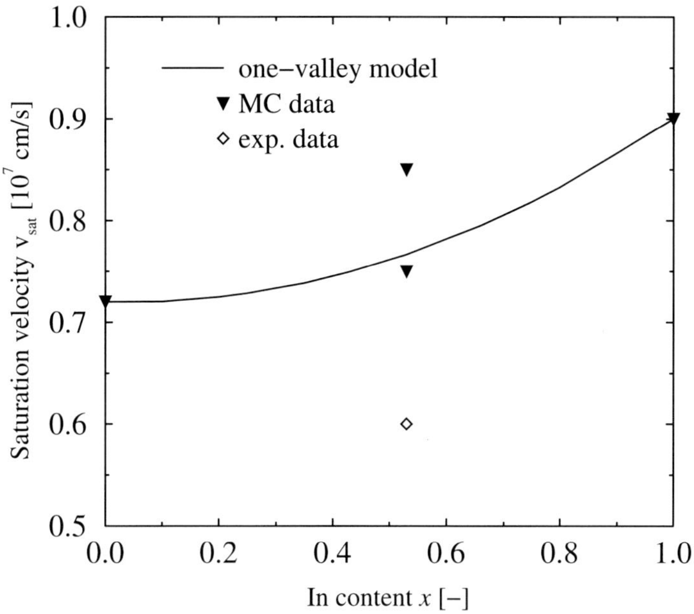

Fig. 3.59. Saturation velocity as a function of material composition at 300 K for $In_xGa_{1-x}As$.

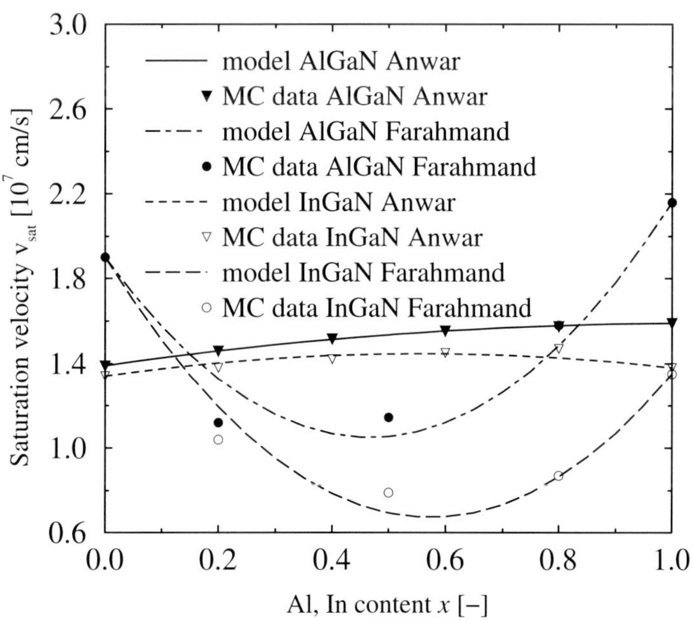

Fig. 3.60. Saturation velocity as a function of material composition at 300 K for $Al_xGa_{1-x}N$ and $In_xGa_{1-x}N$.

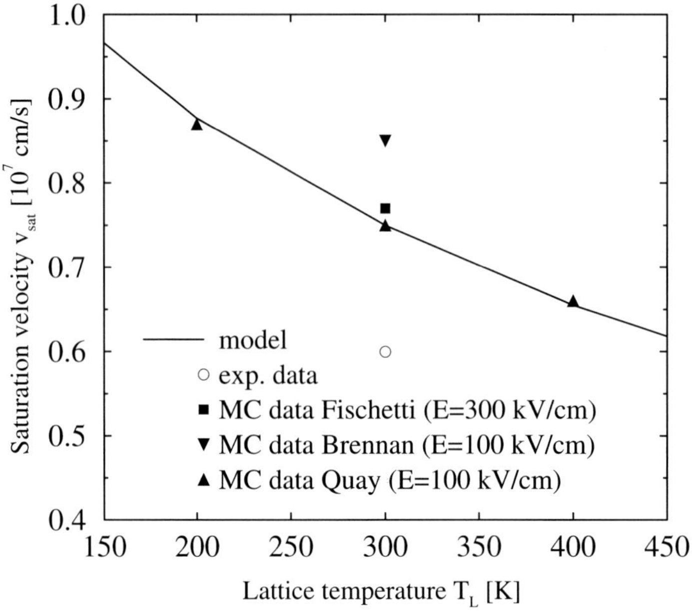

Fig. 3.61. Saturation velocity as a function of temperature for $In_{0.53}Ga_{0.47}As$.

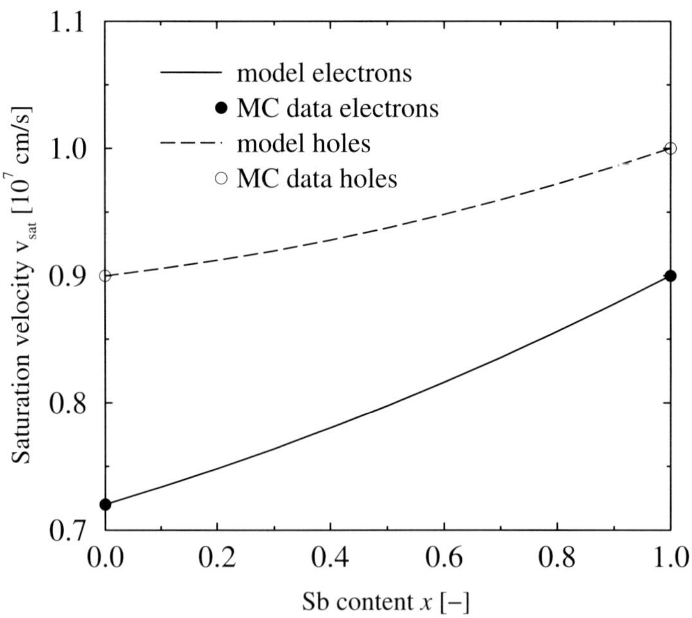

Fig. 3.62. Saturation velocity as a function of material composition at 300 K for $GaAs_{1-x}Sb_x$.

Table 3.34. Parameter values for the saturation velocity model in *alloy materials*

Material	$C^{one}_{vsat,n}$ [10^7cm/s]	$C^{\Gamma}_{vsat,n}$ [10^7cm/s]	$C^{X}_{vsat,n}$ [10^7cm/s]	$C_{vsat,p}$ [10^7cm/s]	References
SiGe (unstrained)	-2.28	-	-	-	[74]
SiGe (strained)	-2.1	-	-	-	[74]
AlGaAs	-0.0512	0.94	-4.58	-	[413]
InGaAs	-0.196	-0.196	-	-0.35	[66, 216, 304]
InAlAs	-2.13	-1.24	-	-	[304]
InGaP	-0.51	-0.51	-	0.63	[70]
AlGaN	0.182	0.182	-	-	[18]
AlGaN	-3.85	-3.85	-	-	[143]
InGaN	0.34	0.34	-	-	[18]
InGaN	-3.72	-3.72	-	-	[143]
GaAsSb	-0.05	-0.05	-	-0.05	[323]
$(InP)_{1-x}(In_{0.53}Ga_{0.47}As)_x$	-0.17	-	-	-	[4, 163]

with a bowing assumed for GaAsSb similar to that of AlGaAs, as is found for the mobility bowing of holes in [323]. For the quaternary $(InP)_{1-x}(In_{0.53}Ga_{0.47}As)_x$ lattice-matched to InP, the data for the electron bowing factor is taken from [4]. No data on the high-field hole transport are available for this material, to the authors' knowledge.

3.5 Energy and Momentum Relaxation

In this section, models for the energy relaxation times of electrons and holes in SiGe and III-V materials are presented. The energy relaxation times are used in the relaxation terms of the energy balance equations (3.8) and (3.9) and, therefore, in the hydrodynamic mobility model (3.139). Additionally, if self-heating is included, the energy relaxation times are used in the relaxation terms of the lattice heat flow equation (3.21).
The extraction of energy-dependent energy relaxation times is based on several assumptions and cannot be performed in a fully universal manner. It is first of all a bulk semiconductor approach. The energy transport models are approximations (see Section 3.1.4) for the sake of computational efficiency. Using them for device simulation requires verification by MC simulation, especially for sharp field gradients occurring at the drain end of submicron FET gates. This is especially true as closure relations and other aspects [187] introduce shortcomings that can be blamed neither on the mobility model nor on the approximations to the energy relaxation times.
A constant energy relaxation time ($\tau_{\epsilon,n}$), or a quadratic dependence on the electron temperature [6, 64], are usually assumed. Another approach to model the

energy dependence of $\tau_{\epsilon,n}$ include the use of table functions [358]. A precise simulation needs to include the dependence of $\tau_{\epsilon,n}$ on the lattice and carrier temperatures. An empirical model for the electron energy relaxation time has been suggested in [176], which is used here for a bulk approach. It is based on Monte Carlo simulation results [147], and is applicable to all relevant semiconductors with diamond and zinc-blende structures. The energy relaxation times are expressed as functions of the carrier and lattice temperatures and, in the case of semiconductor alloys, as a function of the material composition. The influence of doping concentration is not taken into account in this approach.

3.5.1 Modeling Methodology

Depending on the semiconductor material under investigation, different results are available from MC simulations. Two methods, named direct and indirect, are used to obtain $\tau_{\epsilon,n}$.

The Direct Method. For GaAs, the dependence of the electron energy relaxation time and the average electron energy w_n on the electric field are available in [147]. The average energy is approximated by the thermal energy, with the kinetic term being neglected, as typical for energy balance approaches. Using this approximation (3.16) together with the interpolation of the MC simulation results for different electric fields, $\tau_{\epsilon,n}$ can be obtained directly as a function of the electron temperature at different lattice temperatures and lattice directions.

The Indirect Method. The dependence of $\tau_{\epsilon,n}$ on the electric field is not available for binary and ternary III-V compounds. In this case $\tau_{\epsilon,n}$ is calculated in an indirect way, using the dependence of the electron velocity on the electric field from [147]. The local energy balance equation [496] is obtained by neglecting the energy flux

$$\tau_{\epsilon,n} = \frac{3 \cdot k_B}{2 \cdot q} \cdot \frac{T_n - T_L}{v_n \cdot E} \tag{3.146}$$

where q is the electron charge, T_L the lattice temperature, and E is the electric field. Using (3.16) and the average electron energy and the electron velocity as a function of the electric field, $\tau_{\epsilon,n}$ can be extracted indirectly.
Fig. 3.63 shows $\tau_{\epsilon,n}$ for GaAs as a function of the electron temperature at $T_L = 300$ K, as it results from both the direct and indirect methods. The correct values are extracted by the direct method, since it is based on fewer approximations. It turns out that $\tau_{\epsilon,n}$ is overestimated by using (3.146) in the indirect method. Nevertheless the saturation value $\tau_{\epsilon,sat}$ of $\tau_{\epsilon,n}$ at high electron temperatures and the location of the peak $T_{n,peak}$ are independent of the methodology used. These criteria are used for correct estimation of $\tau_{\epsilon,n}$ in materials where only the indirect method can be applied.

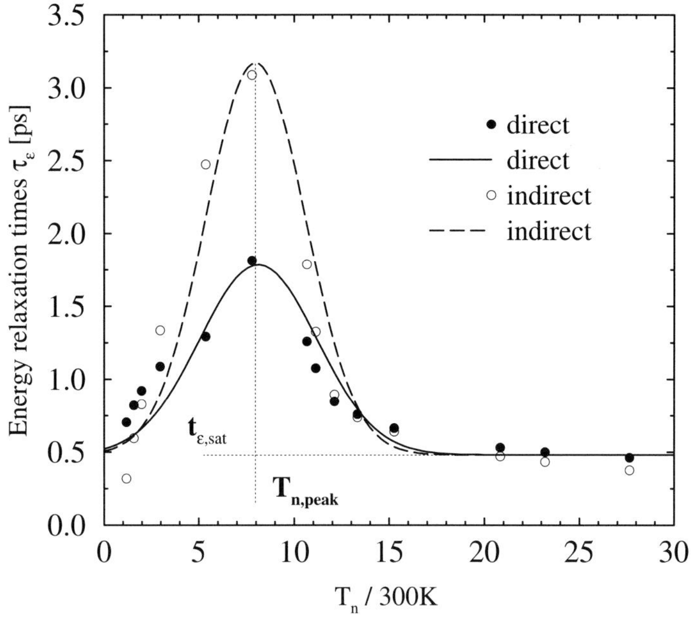

Fig. 3.63. Energy relaxation time as a function of electron temperature extracted by the direct and the indirect method for GaAs.

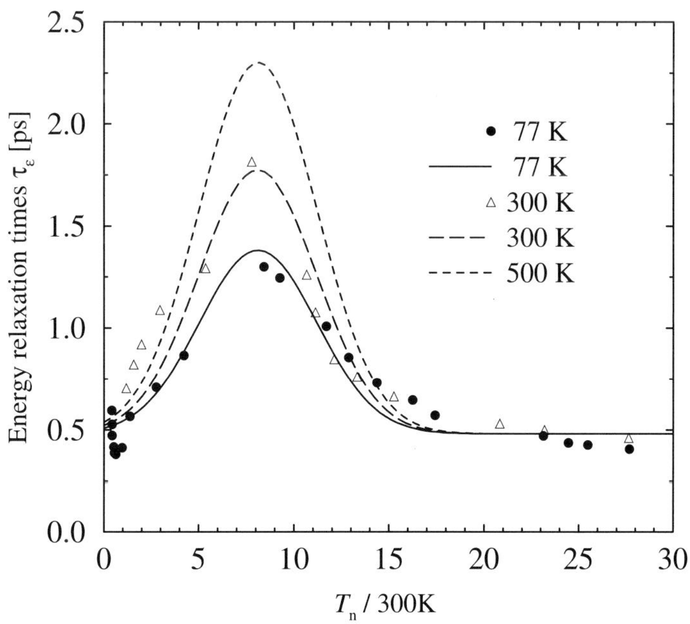

Fig. 3.64. Energy relaxation time as a function of electron temperature compared to MC data for GaAs at several lattice temperatures.

3.5.2 Relaxation Time Model

The following expression is used to model the electron relaxation time as a function of the carrier and lattice temperatures:

$$\tau_{\epsilon,\mathrm{n}} = \tau_{\epsilon,0} + \tau_{\epsilon,1} \cdot \exp\left(C_1 \cdot \left(\frac{T_\mathrm{n}}{300\ \mathrm{K}} + C_0\right)^2 + \right.$$
$$\left. + C_2 \cdot \left(\frac{T_\mathrm{n}}{300\ \mathrm{K}} + C_0\right) + C_3 \cdot \left(\frac{T_\mathrm{L}}{300\ \mathrm{K}}\right)\right) . \tag{3.147}$$

The flexibility of this function allows its easy adaption to all materials under consideration. Table 3.35 summarizes the model parameters of (3.147) for *basic materials*. A constant energy relaxation time $\tau_{\epsilon,\mathrm{p}}$ is assumed for holes in all materials.
For SiGe and III-V semiconductor alloys, the dependence of $\tau_{\epsilon,\mathrm{n}}$ on the material composition x is included. $\tau_{\epsilon,0}$ and C_0 are modeled as quadratic functions of x. The parameter values are summarized in Table 3.36.

Table 3.35. Parameter values for the energy relaxation time model for *basic materials*

Material	$\tau_{\epsilon,0}$ [ps]	$\tau_{\epsilon,1}$ [ps]	C_0	C_1	C_2	C_3	$\tau_{\epsilon,\mathrm{p}}$ [ps]
Si	0.3	-	-	-	-	-	0.1
Ge (on Si)	1.23	-	-	-	-	-	0.3
GaAs	0.48	0.025	0	-0.053	0.853	0.5	0.1
AlAs	0.17	0.025	61	-0.053	0.853	0.5	0.1
InAs	0.08	0.025	3	-0.053	0.853	0.5	0.1
InP	0.5	0.21	-18	-0.04	0	0	0.1
GaP	0.04	0.21	10	-0.04	0	0	0.1

Elementary and Binary Semiconductors. The direct method is used for GaAs, and the indirect method is used for AlAs and InAs, depending on the MC data available.
The results for GaAs are shown in Fig. 3.64. $\tau_{\epsilon,\mathrm{n}}$ tends to saturate at high electron temperatures and becomes independent of the lattice temperature. For low and intermediate electron temperatures, the behavior can be attributed to the transition of electrons from the Γ to the L valleys. The electron temperature at which $\tau_{\epsilon,\mathrm{n}}$ reaches the peak value is independent of the lattice temperature. The associated average energy, 0.31 eV, is close to the energy difference between the two valleys, 0.27 eV.

Semiconductor Alloys. The behavior of the energy relaxation times $\tau_{\epsilon,\mathrm{n}}$ in $\mathrm{In}_x\mathrm{Ga}_{1-x}\mathrm{As}$ and $\mathrm{Al}_x\mathrm{Ga}_{1-x}\mathrm{As}$, similar to that in GaAs, can be seen in Fig. 3.65 and Fig. 3.66. Thus $\tau_{\epsilon,\mathrm{n}}$ of *alloy materials* is modeled by preserving the standard

Table 3.36. Parameter values for energy relaxation times in *alloy materials*

Material	τ_C [ps]	$\tau_{\epsilon,1}$ [ps]	C	C_1	C_2	C_3
SiGe	-0.5	-	-	-	-	-
AlGaAs	-0.35	0.025	-61	-0.053	0.853	0.5
InGaAs	1.8	0.025	-34	-0.053	0.853	0.5
InGaP	-0.4	0.21	-5.2	-0.04	0	0

deviation and the amplitude of the function obtained for GaAs with the direct method and by adjusting the position with the material composition dependence of $\tau_{\epsilon,0}$ ($\tau_{\epsilon,sat}$) and C_0 ($T_{\mathrm{n,peak}}$). For $\tau_{\epsilon,0}$ and C_0 of *alloy materials* ($\mathrm{A}_{1-x}\mathrm{B}_x$), the model (3.149) suggests a quadratic interpolation between the values from Table 3.35 for the *basic materials* (A and B) as a function of the material composition x. τ_C and C are referred to as bowing parameters. The parameter values used in this model are summarized in Table 3.36.

$$\tau_{\epsilon,0}^{\mathrm{AB}} = \tau_{\epsilon,0}^{\mathrm{A}} \cdot (1-x) + \tau_{\epsilon,0}^{\mathrm{B}} \cdot x + \tau_C \cdot (1-x) \cdot x \tag{3.148}$$

$$C_0^{\mathrm{AB}} = C_0^{\mathrm{A}} \cdot (1-x) + C_0^{\mathrm{B}} \cdot x + C \cdot (1-x) \cdot x. \tag{3.149}$$

The indirect method is applicable for all semiconductor alloys. The lattice temperature dependence of $\tau_{\epsilon,\mathrm{n}}$ of GaAs is preserved for two similar semiconductor alloys considered, $\mathrm{Al}_x\mathrm{Ga}_{1-x}\mathrm{As}$ and $\mathrm{In}_x\mathrm{Ga}_{1-x}\mathrm{As}$. This approximation is more accurate for low material compositions, which are more frequently used ($x < 0.3$). The results of the model for $\mathrm{Al}_x\mathrm{Ga}_{1-x}\mathrm{As}$ at 300 K for different material compositions x are shown in Fig. 3.65. Note the shift of the electron temperature at which $\tau_{\epsilon,\mathrm{n}}$ reaches its maximum to lower values with the increase of x. For high values ($x > 0.4$), no peak value of $\tau_{\epsilon,\mathrm{n}}$ is observed. This behavior can be attributed to the composition dependence of the Γ, L, and X valley minima. When the Al content changes from 0 to 0.3, the energy difference between the Γ and L valleys varies between 0.27 and 0.1 eV. The corresponding change of the electron energy associated with the peak of $\tau_{\epsilon,\mathrm{n}}$ varies between 0.31 and 0.1 eV. Furthermore, for Al contents $x > 0.4$, the X valleys are the lowest in energy and the bandgap becomes indirect. This explains the absence of a peak of $\tau_{\epsilon,\mathrm{n}}$ for $x > 0.4$.

Similar results are obtained for $\mathrm{In}_x\mathrm{Ga}_{1-x}\mathrm{As}$, as given in Fig. 3.66. There is a shift of the maximum $\tau_{\epsilon,\mathrm{n}}$ to higher values with increasing indium composition up to x=0.53. This can be explained by the electron population change due to Γ-L transitions. A quick shift to lower values is observed for InAs, which cannot be explained by the dependence of the energy valleys on x. The MC simulation results show that at very high indium contents the average electron energy decreases and the saturation drift velocity increases very much, but no clear results are available for this case.

The results obtained for strained SiGe are based on our own MC simulations in the complete range of material compositions $0 \leq x \leq 1$, i.e. between silicon and strained germanium grown on a silicon substrate. Fig. 3.67 shows the electron energy relaxation times as a function of electron temperature for different material compositions in $\mathrm{Si}_{1-x}\mathrm{Ge}_x$ at 300 K (symbols). MC data from Bufler [73] for $0 \leq x \leq 0.3$ are included as a reference (solid lines). Note that $\tau_{\epsilon,\mathrm{n}}$ in Si and SiGe exhibits an almost constant behavior over a wide range of electron temperatures. At very low electron temperatures, higher $\tau_{\epsilon,\mathrm{n}}$ are observed. However, when the electron temperature is close to the lattice temperature, the term $(T_\mathrm{n} - T_\mathrm{L})/\tau_{\epsilon,\mathrm{n}}$ appearing in the energy balance tends to be zero [64], thus the influence of $\tau_{\epsilon,\mathrm{n}}$ is negligible, and therefore its increase can be neglected. The sharp initial drop is attributed to the increase of optical and inter-valley scattering as the electrons are heated by the field [232].
The model parameters for the hole energy relaxation times in strained SiGe are based on MC simulation data in [73] for $0 \leq x \leq 0.4$. A linear dependence on the material composition is assumed for all materials.

Application of Energy Relaxation Times.

HBT Simulation. Using non-constant electron energy relaxation times in the mobility model (3.139) allows better modeling of the velocity overshoot in HBTs. The results have to be verified by MC device simulations for identical potential distributions, see [247]. Contrary to submicron FETs, quantization effects do not play a dominant role for the carrier transport in HBTs. The inclusion of static velocity overshoot effects into HBT simulation is desirable, since it improves the accuracy of the bias-dependent, small-signal simulation and the estimation of device speed, as shown in Chapter 5. The static velocity overshoot in the collector contributes significantly to the overall speed of the device.
Reliable MC results for the high-field hole transport are needed for the calibration of the material models, especially for III-V materials. Such data are very rare, due to the complicated structure of the valence band. Fortunately, the influence of the hot-hole transport is relatively insignificant for npn-HBTs. This typically leads to the application of constant relaxation times for holes.

HEMT Simulation. In a first approach, constant energy relaxation times have been proposed for the simulation of HEMTs in [64] in combination with the extended Hänsch mobility model (3.141) [64]. Table 3.37 compiles the constant energy relaxation times for different materials. The combination of (3.141) and constant energy relaxation times allows the simulation of overshoot effects.
In combination with the regular Hänsch mobility model (3.139), constant energy relaxation times do not allow the modeling of the static velocity overshoot effect. However, the introduction of the hot-carrier dynamics is decisive for the speed evaluation of HEMTs. Further, in the extended Hänsch model (3.141), a lower bound for $T_\nu \geq T_\mathrm{L}$ must be introduced to avoid numerical shortcomings: The model yields non-physical results if arbitrary values for β are applied and carrier

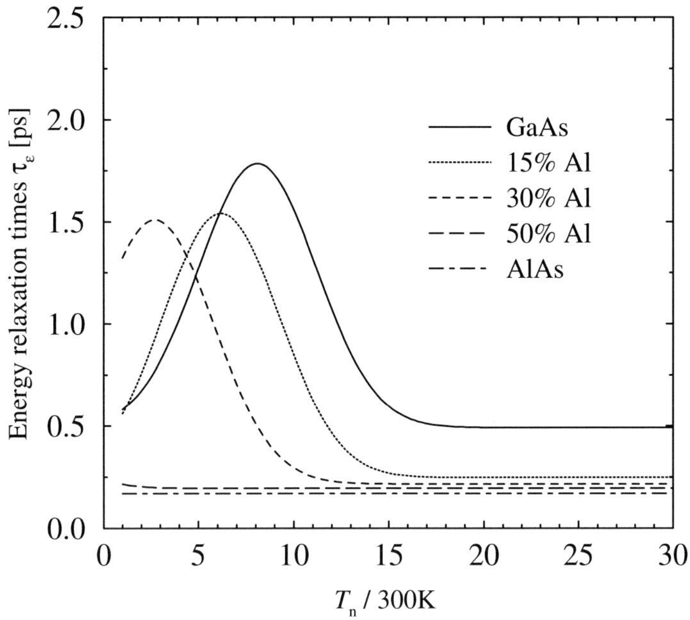

Fig. 3.65. Energy relaxation time as a function of electron temperature for different material compositions in $Al_xGa_{1-x}As$ at 300 K.

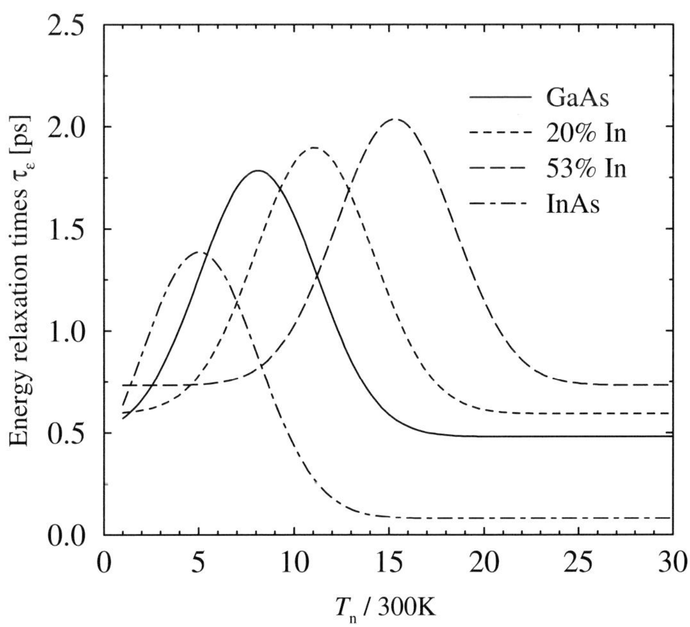

Fig. 3.66. Energy relaxation time as a function of electron temperature for different material compositions in $In_xGa_{1-x}As$ at 300 K.

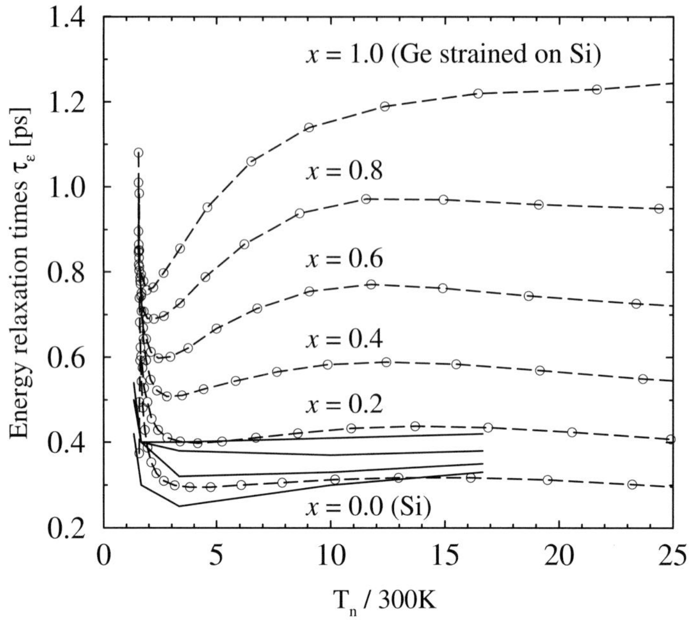

Fig. 3.67. Energy relaxation time as a function of electron temperature for different material compositions in $Si_{1-x}Ge_x$ at 300 K: MC simulation data.

cooling occurs, i.e. $T_\nu < T_L$. Thus it is very desirable to use energy-dependent relaxation times in combination with the mobility model of (3.139).

Table 3.37. Constant energy relaxation times fitted for III-V HEMTs

Material	$\tau_{\epsilon,n}$ [ps]	$\tau_{\epsilon,p}$ [ps]
$Al_{0.2}Ga_{0.8}As$	0.1	0.1
$In_{0.25}Ga_{0.75}As$	0.17	0.1
$In_{0.53}Ga_{0.47}As$	0.17	0.1
$In_{0.52}Al_{0.48}As$	0.1	0.1

Given the specific hydrodynamic transport model of Subsection 3.1.3 and the mobility model (3.139), the results from HEMT device simulation are compared to one- and two-dimensional MC simulations of the same device. It is found that the parameters from Table 3.35 and Table 3.36 allow neither the simulation of correct electron temperatures nor the correct velocities. This result of the hydrodynamic model is discussed by Grasser et al. in [184, 187] and attributed to the finite number of moments considered. Comparing the values of Table 3.37 and Table 3.35, the constant values obtained in previous work are significantly smaller than the minimum values of the energy relaxation times extracted from the bulk MC simulations [147]. In this case, the transport parallel to hetero-interfaces, which is naturally most relevant for HEMTs, and the quantization in the channel are considered to be responsible. MC simulation results for silicon including the quantization at the SiO_2 interface with $\tau_{\epsilon,n} = 0.1$ ps have been reported in [427]. This value for $\tau_{\epsilon,n}$ is significantly smaller than the bulk value in silicon $\tau_{\epsilon,n} \approx 0.3$ ps [149]. Additionally, the occurrence of Real Space Transfer (RST) modifies the high-field transport in the channel and has to be considered by the hydrodynamic model [390]. Ultimate verification of carrier temperatures and velocities is only achieved by one- and two-dimensional device MC simulations which include quantization and the effect of RST. However, modeling of quantization effects in thin channels and their impact on the energy relaxation times is a complicated issue. Qualitatively, it is generally assumed that the energy relaxation times reduce with the layer thickness, but the exact quantitative estimation is still a subject of research. Thus a scaling factor is used in combination with (3.147) to model the energy-dependent relaxation times in the simulations of HEMTs. The scaling parameter gives a relation between the bulk material values in Table 3.35 and the channel material values in Table 3.37. A general down-scaling factor of 8 is used for all HEMTs simulated in this work, which gives good agreement with our own MC simulations.

3.6 Generation and Recombination

The simulation of heterojunction bipolar and high electron mobility transistors depends strongly on the accurate modeling of generation and recombination effects. For HBTs, this is a rather trivial remark. HEMTs are generally considered unipolar devices, but phenomena such as frequency dispersion [388], gate currents [120, 25], and electroluminescence experiments give a strong hint of the influence of both carrier types and, in particular, their generation and recombination.

Carrier Lifetime. The total carrier lifetime observed experimentally is composed from its constituents in an inverse law:

$$\frac{1}{\tau_{\mathrm{tot}}} = \frac{1}{\tau_{\mathrm{SRH}}} + \frac{1}{\tau_{\mathrm{AUG}}} + \frac{1}{\tau_{\mathrm{DIR}}} + \frac{1}{\tau_{\mathrm{SURF}}}. \qquad (3.150)$$

A typical situation for the carrier lifetimes is given in Fig. 3.68, where three major contributions are assumed. These include doping-dependent Shockley-Read-Hall recombination, Auger recombination, and direct recombination. Additional effects can contribute, such as interface and surface recombination, resulting in additional τ_{SURF} in (3.150). The major generation and recombination mechanisms contributing to the total lifetime are discussed in the following.

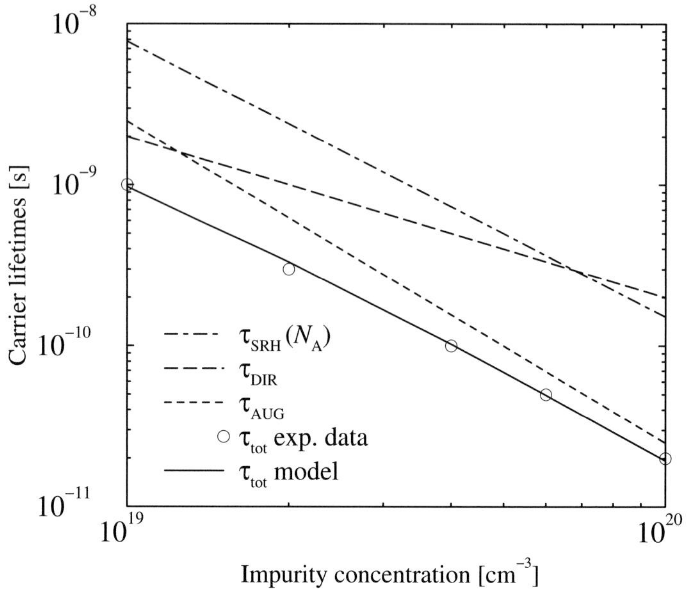

Fig. 3.68. Contributions (SRH, Auger, direct) to the total lifetime τ_{tot} in heavily C-doped GaAs as a function of impurity concentration.

3.6.1 Static Shockley-Read-Hall Recombination

Carrier generation in space charge regions and recombination in, for example, high injection regions are modeled using the well-known Shockley-Read-Hall (SRH) equations

$$R^{\mathrm{SRH}} = \frac{n \cdot p - n_i^2}{\tau_{\mathrm{p}} \cdot (n + n_1) + \tau_{\mathrm{n}} \cdot (p + p_1)}, \qquad n_i^2 = n_1 \cdot p_1 . \tag{3.151}$$

The quantity n_i is the intrinsic concentration and the auxiliary variables n_1 and p_1 are defined by

$$n_1 = N_{\mathrm{C}}(T_{\mathrm{L}}) \cdot \exp\left(\frac{-E_{\mathrm{C}} + E_{\mathrm{T}}}{k_{\mathrm{B}} \cdot T_{\mathrm{L}}}\right) \tag{3.152}$$

$$p_1 = N_{\mathrm{V}}(T_{\mathrm{L}}) \cdot \exp\left(\frac{-E_{\mathrm{T}} + E_{\mathrm{V}}}{k_{\mathrm{B}} \cdot T_{\mathrm{L}}}\right). \tag{3.153}$$

The recombination rate has its maximum for the trap energy level E_{T} located in the middle of the bandgap. The variables N_{C} and N_{V} are the carrier effective densities of states (see Section 3.3). The dependence on the lattice temperature T_{L} is given by the variables $n_1(T_{\mathrm{L}})$ and $p_1(T_{\mathrm{L}})$, and the recombination lifetimes for electrons and holes by $\tau_{\mathrm{n}}(T_{\mathrm{L}})$ and $\tau_{\mathrm{p}}(T_{\mathrm{L}})$. The thermal carrier velocities at 300 K are calculated using the relation

$$v_{\mathrm{th},\nu} = \sqrt{\frac{3 \cdot k_{\mathrm{B}} \cdot T_{\mathrm{L}}}{m_{\nu}}}, \quad \nu = n, p \tag{3.154}$$

to obtain the recombination times at 300 K:

$$\tau_{\mathrm{n},300} = \frac{1}{\sigma_{\mathrm{T,n}} \cdot N_{\mathrm{T}} \cdot v_{\mathrm{th,n},300} + s_{\mathrm{n}}/y} \tag{3.155}$$

$$\tau_{\mathrm{p},300} = \frac{1}{\sigma_{\mathrm{T,p}} \cdot N_{\mathrm{T}} \cdot v_{\mathrm{th,p},300} + s_{\mathrm{p}}/y}. \tag{3.156}$$

The recombination times are modeled using traps of donor, acceptor, or neutral type, as a function of trap density N_{T}, and the trap-capture cross-sections for electrons and holes, $\sigma_{\mathrm{T,n}}$ and $\sigma_{\mathrm{T,p}}$. The SRH effect is modeled independent of the electrical field. The effect of surface recombination is included by using non-zero surface recombination velocities for electrons s_{n} and holes s_{p}. The effect is stronger with decreasing distance to the surface y. Finally, the lattice temperature dependence is included in (3.160) and (3.161).

Doping Effects. The lifetimes in (3.155) and (3.156) are considered independent of the doping, as the number of traps N_{T} is considered independent of the doping. In typical modern semiconductor devices, most of all in bipolar devices, doping concentrations exceed several $10^{19}\mathrm{cm}^{-3}$ in GaAs [516], SiGe [49], and InGaAs [224], so that additional traps are introduced. Approximations of the Scharfetter relation [440, 448] are often used to model the experimentally-observed doping-dependence. This work uses:

$$\tau_{\mathrm{n,dop}} = \tau_{\mathrm{max,n}} \cdot \left(\frac{N_{\mathrm{A}} + N_{\mathrm{D}}}{N_{\mathrm{ref,n}}} \right)^{-\gamma} \tag{3.157}$$

$$\tau_{\mathrm{p,dop}} = \tau_{\mathrm{max,p}} \cdot \left(\frac{N_{\mathrm{A}} + N_{\mathrm{D}}}{N_{\mathrm{ref,p}}} \right)^{-\gamma} \tag{3.158}$$

Several other relations have been proposed to describe the doping dependence [448]. Model parameters are given in Table 3.38. Parameter values for GaAs are obtained from least square fits of the data given in [516] for the minority lifetime in p-doped layers with $N_{\mathrm{A}} \geq 10^{19}$ cm^{-3}. The parameter values for $\mathrm{In_{0.53}Ga_{0.47}As}$ and $\mathrm{In_{0.52}Al_{0.48}As}$ in [436] compile the data for both n and p-materials without regard to the minority/majority situation. The data for $\mathrm{In_{0.53}Ga_{0.47}As}$ in [92], also used in [113], compile the minority situation of n- and p-type material.
A serious shortcoming of (3.157) and (3.158) is that they cannot be used in conjunction with the dynamic SRH model. An elegant solution is to continue using the physics-based equations which account for trap dynamics (3.155)(3.156), but introduce the doping-dependence by modification of the trap concentration N_{T}.

$$N_{\mathrm{T,dop}} = N_{\mathrm{T}} \cdot \left(\frac{N_{\mathrm{A}} + N_{\mathrm{D}}}{N_{\mathrm{ref}}} \right)^{\gamma} \tag{3.159}$$

This combined relation employs the model parameters given in Table 3.38.

Table 3.38. Parameter values for the doping dependence of the SRH model

Material	$N_{ref,n}$ [cm^{-3}]	$N_{ref,p}$ [cm^{-3}]	$\tau_{max,n}$ [ns]	$\tau_{max,p}$ [ns]	γ -	Refs. -
Si	1e16	1e16	10000	3000	1	[153, 227]
SiGe	1e16	1e16	30000	10000	1	[297, 452]
GaAs	-	9e17	500	-	1.72	[516]
$In_{0.53}Ga_{0.47}As$	1e17	1e17	100	100	1	[436]
$In_{0.53}Ga_{0.47}As$	9e17	8e17	0.3	10	1	[92, 113]
$In_{0.52}Al_{0.48}As$	1e17	1e17	100	100	1	[436]

Trap-Assisted Band-To-Band Tunneling. The effect of trap-assisted band-to-band tunneling (TBB) is accounted for by field enhancement factors Γ_n and Γ_p, which are modeled after [438].

$$\tau_n(T_L) = \left(\frac{300\ \mathrm{K}}{T_L}\right)^{3/2} \cdot \frac{\tau_{n,300}}{1+\Gamma_n} \tag{3.160}$$

$$\tau_p(T_L) = \left(\frac{300\ \mathrm{K}}{T_L}\right)^{3/2} \cdot \frac{\tau_{p,300}}{1+\Gamma_p} \tag{3.161}$$

Since TBB is relevant only for materials with indirect bandgap, this effect is often neglected for technologically important direct bandgap III-V materials. The parameters for the SRH recombination model are summarized in Table 3.39. The exact trap energies have to be verified with analysis of the growth process given in the references and the processes applied to the semiconductors. In this work, midgap traps of $E_T = 0.8$ eV are used for AlGaAs. For InGaAs, $E_T = 0.3$ eV is typically assumed. The lifetimes obtained by (3.155)-(3.161), i.e. the function (3.155) of N_T, $v_{th,\nu}$, and $\sigma_{T,\nu}$, can be controlled by measurements of the minority lifetime in material of the same growth type, as given in [8]. In $Al_xGa_{1-x}As$, though a direct semiconductor for $x \leq 0.42$, the lifetime is dominated by SRH recombination. The data in [391] is given for undoped AlGaAs samples and the lifetimes remain in the 10 ns range for $x \leq 0.42$. The lifetimes for $Al_xGa_{1-x}As$ for $x \geq 0.3$ depend strongly on the growth and can be varied accordingly between 1 ns and 10 ns [391, 509]. Typical values are given for the surface recombination velocity. The surface recombination velocity for $Al_{0.4}Ga_{0.6}As$/GaAs interface [375] is supplied in Table 3.39. In general, lifetimes are strong functions of the processing conditions [210], so that the values given in Table 3.39 are to be considered default values. If the parameters $\sigma_{T,\nu}$, N_T, and $v_{\nu,300}$ cannot be separated, at least the experimental lifetime is made available. A great amount of data is available for the InGaP/GaAs heterojunctions in HBTs; however, for InGaP itself, SRH is mostly neglected with respect to the recombination in highly doped p-GaAs.

Table 3.39. Parameter values for the Shockley-Read-Hall recombination model

Material	N_T [cm^{-3}]	E_T-E_V [eV]	$\sigma_{T,n}$ [cm^2]	$\sigma_{T,p}$ [cm^2]	s_n [cm/s]	s_p [cm/s]	$\tau_{n,300}$ [ns]	$\tau_{p,300}$ [ns]	Refs.
Si	1e13	0.56	1e-15	1e-15	1e2	1e2-1e5	1-1e6	0.1-1e4	[254, 448] [121, 520]
Ge	1e13	0.33	1e-15	1e-15	1e2-1e4	1e2-1e4	>2e5	>2e5	[210, 297]
GaAs	1e14	0.2-0.8	1e-16	-	1e4	-	1-500	-	[394, 516]
	-	-	-	-	-	-	-	200	[391]
InP	1e13	0.1-0.5	1.7e-13	-	-	-	10	3e3	[253, 297] [393]
GaP	-	-	-	-	2e6	-	4-158	-	[126, 170]
GaSb	-	-	-	-	-	-	10	600	[483, 512]
GaN	-	-	-	-	-	-	6e6	-	[535]
AlN	-	-	-	-	-	-	3.5e7	-	[535]
$Si_{0.85}Ge_{0.15}$	≤1e14	0.44	1e-16	1e-16	-	-	0.11	-	[452]
$Si_{0.80}Ge_{0.20}$	≤1e14	0.4	1e-16	1e-16	-	-	≈ 1000	≈1000	[478, 482]
$Si_{0.77}Ge_{0.23}$	≤1e14	0.38	1e-16	1e-16	-	-	22	-	[452]
$Al_{0.17}Ga_{0.83}As$	1e15	0.2-0.8	5e-14	5e-15	-	-	8	8	[391, 395]
$Al_{0.40}Ga_{0.60}As$	-	-	-	-	210	-	-	-	[375]
InGaAs	<1e15	0.17-0.5	1e-15	-	-	-	100	100	[40, 436]
InAlAs	<1e15	-	-	-	-	-	100	100	[436]
InGaP	2e16	0.4	1e-14	1e-13	900	-	-	-	[375]
$Al_{0.75}Ga_{0.25}N$	-	-	-	-	-	-	2.5e7	-	[535]

3.6.2 Dynamic Shockley-Read-Hall Recombination

Frequency dispersion is one of the major concerns in both silicon-based and III-V-based RF devices, especially for broadband applications. The precise modeling of dynamic trap-related phenomena is therefore an important issue.

The Physical Model in the Dynamic Simulation. The present model is known as Shockley-Read-Hall theory. It is based on the numerical solution of the time-dependent basic semiconductor equations

$$\mathrm{div}(\varepsilon \cdot \mathrm{grad}\ \psi) = \mathrm{q} \cdot (n(t) - p(t) - n_\mathrm{T}(t) + N_\mathrm{A} - N_\mathrm{D}), \tag{3.162}$$

$$\mathrm{div}\ \mathbf{J}_\mathrm{n} = \mathrm{q} \cdot \left(R_{\mathrm{eff,n}}(t) + \frac{\partial n(t)}{\partial t} \right), \tag{3.163}$$

$$\mathrm{div}\ \mathbf{J}_\mathrm{p} = -\mathrm{q} \cdot \left(R_{\mathrm{eff,p}}(t) + \frac{\partial p(t)}{\partial t} \right), \tag{3.164}$$

where $R_{\mathrm{eff,n}}$ and $R_{\mathrm{eff,p}}$ are the effective (net) recombination rates for electrons and holes respectively, and $\mathrm{q} \cdot n_\mathrm{T}(t)$ is the total trapped charge. Trap-related quantities do not appear in the total-current-continuity equation, which is solved at the beginning of each time step

$$\mathrm{div}\left(\varepsilon \cdot \mathrm{grad}\,\frac{\partial \psi}{\partial t}\right) = \mathrm{div}\,\mathbf{J}_\mathrm{n} + \mathrm{div}\,\mathbf{J}_\mathrm{p}. \tag{3.165}$$

The model allows arbitrary interface and bulk trap distributions in both the energy and position space. For an individual single-level interface (or bulk) trap with density N_T located at the position x_i at energy level E_T, the trap dynamics are described by the equations

$$R_\mathrm{n} = \frac{N_\mathrm{T} - n_\mathrm{T}}{\tau_\mathrm{cap,n}}, \quad R_\mathrm{p} = \frac{n_\mathrm{T}}{\tau_\mathrm{cap,p}}, \quad G_\mathrm{n} = \frac{n_\mathrm{T}}{\tau_\mathrm{em,n}}, \quad G_\mathrm{p} = \frac{N_\mathrm{T} - n_\mathrm{T}}{\tau_\mathrm{em,p}}. \tag{3.166}$$

The time constants $\tau_{\mathrm{cap},\nu}$ denote the capture time constant of the carrier ν, and $\tau_{\mathrm{em},\nu}$ the emission time constant of the carrier ν. R_n is the electron capture rate, R_p the hole capture rate, G_n and G_p the electron and hole emission rate. The occupied trap density is denoted by n_T. The electron and hole capture and emission time constants are given by

$$\tau_\mathrm{cap,n}^{-1} = v_\mathrm{th,n} \cdot \sigma_\mathrm{T,n} \cdot n_\mathrm{s} \tag{3.167}$$

$$\tau_\mathrm{cap,p}^{-1} = v_\mathrm{th,p} \cdot \sigma_\mathrm{T,p} \cdot p_\mathrm{s} \tag{3.168}$$

$$\tau_\mathrm{em,n}^{-1} = v_\mathrm{th,n} \cdot \sigma_\mathrm{T,n} \cdot N_\mathrm{C} \cdot \exp\left(\frac{-E_\mathrm{C} + E_\mathrm{T}}{k_\mathrm{B} \cdot T_\mathrm{L}}\right) \tag{3.169}$$

$$\tau_\mathrm{em,p}^{-1} = v_\mathrm{th,p} \cdot \sigma_\mathrm{T,p} \cdot N_\mathrm{V} \cdot \exp\left(\frac{-E_\mathrm{T} + E_\mathrm{V}}{k_\mathrm{B} \cdot T_\mathrm{L}}\right). \tag{3.170}$$

$v_\mathrm{th,n}$ and $v_\mathrm{th,p}$ are the average thermal velocities towards the interface and $\sigma_\mathrm{T,n}$ and $\sigma_\mathrm{T,p}$ the average capture cross-sections for electrons and holes. N_C, E_C, N_V, and E_V are the effective densities of states and energy band edges in the conduction and valence band, respectively. n_s and p_s are the electron and hole surface concentrations at the position x_i. In this model, the conduction and valence band are reduced to single energy levels. It is assumed that the capture cross-sections associated with the emission and capture processes remain constant and equal to each other as if in equilibrium when (3.166), (3.167), and (3.169) are applied to non-equilibrium conditions. The static values are given in Table 3.39. The trap-dynamics equation is given by

$$\frac{dn_\mathrm{T}}{dt} = R_\mathrm{n} - G_\mathrm{n} - R_\mathrm{p} + G_\mathrm{p} = -\frac{n_\mathrm{T}}{\tau} + \frac{F_\mathrm{o} \cdot N_\mathrm{T}}{\tau} \tag{3.171}$$

where F_o is the steady-state occupancy function and τ an effective time constant which determines the actual trap dynamics.

$$\tau^{-1} = \tau_\mathrm{cap,n}^{-1} + \tau_\mathrm{em,n}^{-1} + \tau_\mathrm{cap,p}^{-1} + \tau_\mathrm{em,p}^{-1} \tag{3.172}$$

$$F_\mathrm{o} = \tau \cdot (\tau_\mathrm{cap,n}^{-1} + \tau_\mathrm{em,p}^{-1}) \tag{3.173}$$

If several trap levels exist in the forbidden gap, the trap-trap transitions are not accounted for in (3.171). Such an approach is proper for interface traps. A more detailed discussion about trap-dynamics equations and the numerical methods can be found in [196]. In the steady-state the implemented equations reduce exactly to the well-known Shockley-Read-Hall trap-assisted recombination model.

3.6.3 Trap-Assisted Band-to-Band Tunneling

Trap-assisted band-to-band tunneling (TBB) is a process which can be modeled closely to the SRH process, including the dependence on the electric field. It occurs in highly doped semiconductors, e.g. in HBT base layers. The trap-assisted band-to-band tunneling model describes the modification of carrier generation/recombination in high-field regions due to the Poole Frenkel effect. The model modifies the SRH lifetimes $\tau_{\mathrm{n,p}}$ (see Section 3.6.1) by the field enhancement factors Γ_{n} and Γ_{p}. Trap-assisted band-to-band tunneling is most significant for indirect semiconductors. Two models are typically used: the Hurkx and the Schenk models.

The Hurkx Trap-Assisted Band-to-Band Tunneling Model. In the Hurkx model [503], the enhancement factors $\Gamma_{\mathrm{n,p}}$ are modeled as a function of the local electric field. The implementation uses different equations for the high-field and low-field ranges. The equation for electrons at low-field reads:

$$\Gamma_{\mathrm{n}} = \sqrt{\pi} \cdot C_1 \cdot \exp\left(\frac{C_1^2}{3}\right) \cdot \left(2 - \operatorname{erfc}\left(\frac{C_1^2 - C_2}{2 \cdot C_1}\right)\right) \quad \text{for } C_1 \leq C_2 \tag{3.174}$$

and at high-fields:

$$\begin{aligned} \Gamma_{\mathrm{n}} = & \sqrt{\pi \cdot C_1} \cdot C_2^{1/4} \cdot \exp\left(-C_2 + C_1 \cdot \sqrt{C_2} + \frac{1}{3} \cdot C_1 \cdot C_2^{3/2}\right) \\ & \times \operatorname{erfc}\left(C_1^{1/2} \cdot C_2^{1/4} - C_1^{-1/2} \cdot C_2^{3/4}\right) \qquad \text{for } C_1 > C_2. \end{aligned} \tag{3.175}$$

The expressions for C_1 and C_2 are given by

$$C_1 = \frac{|\mathbf{E}| \cdot \mathrm{q} \cdot \hbar}{\sqrt{8 \cdot m_{\mathrm{n}}^* \cdot k_{\mathrm{B}}^3 \cdot T_{\mathrm{L}}^3}} \tag{3.176}$$

$$C_2 = \begin{cases} 0 & \text{for } n \geq n_i \cdot \exp\left(\dfrac{E_{\mathrm{g}}}{2 \cdot k_{\mathrm{B}} \cdot T_{\mathrm{L}}}\right) \\ \dfrac{E_{\mathrm{g}}}{2 \cdot k_{\mathrm{B}} \cdot T_{\mathrm{L}}} - \ln \dfrac{n}{n_i} & \text{for } n_{\mathrm{t}} \leq n \leq n_i \cdot \exp\left(\dfrac{E_{\mathrm{g}}}{2 \cdot k_{\mathrm{B}} \cdot T_{\mathrm{L}}}\right) \\ \dfrac{E_{\mathrm{g}}}{2 \cdot k_{\mathrm{B}} \cdot T_{\mathrm{L}}} - \ln \dfrac{n_{\mathrm{t}}}{n_i} & \text{for } n \leq n_{\mathrm{t}} \end{cases} \tag{3.177}$$

with

$$n_\mathrm{t} = n_i \cdot \exp\left(\frac{E_\mathrm{T}^*}{k_\mathrm{B} \cdot T_\mathrm{L}}\right). \tag{3.178}$$

Similar expressions exist for Γ_p, and for p_t:

$$p_\mathrm{t} = n_i \cdot \exp\left(-\frac{E_\mathrm{T}^*}{k_\mathrm{B} \cdot T_\mathrm{L}}\right). \tag{3.179}$$

The model requires one additional parameter, which denotes the effective electron tunneling mass m_n^*. It amounts to $\approx 0.25 \cdot m_0$ for silicon [223]. The trap energy E_T^* in this formulation refers to the Fermi energy of the intrinsic semiconductor. For other materials, where no direct experimental confirmation is available to extract the tunneling mass, the effective electron mass is used as an approximation. This is plausible, as stated in [223]. The light hole masses are used in the expressions for Γ_p. Since the model describes a modified SRH process, similar assumptions hold for the specific materials as discussed in the previous paragraph.

The Schenk Trap-Assisted Band-to-Band Tunneling Model. The derivation of the second trap-assisted band-to-band tunneling model similar to [227, 438] is based on multi-phonon interaction. The field enhancement factors are calculated using the equation

$$\begin{aligned} \Gamma_\mathrm{n,p} = &\left(1 + \frac{C_1^{3/2} \cdot \sqrt{E_\mathrm{T} - E_0}}{E_0 \cdot C_2}\right)^{-1/2} \cdot \frac{C_1^{3/4} \cdot (E_\mathrm{T} - E_0)^{1/4}}{2 \cdot \sqrt{E_\mathrm{T} \cdot E_0}} \cdot \\ &\left(\frac{C_1}{k_\mathrm{B} \cdot T_\mathrm{L}}\right)^{3/2} \times \exp\left(-\frac{E_\mathrm{T} - E_0}{C_2} + \frac{C_2 - k_\mathrm{B} \cdot T_\mathrm{L}}{2 \cdot C_2}\right. \\ &+ \frac{E_\mathrm{T} + 0.5 \cdot k_\mathrm{B} \cdot T_\mathrm{L}}{C_2} \cdot \ln\left(\frac{E_\mathrm{T}}{\epsilon_\mathrm{R}}\right) \\ &\left. - \frac{E_0}{C_2} \cdot \ln\left(\frac{E_0}{\epsilon_\mathrm{R}}\right) + \frac{E_\mathrm{T} - E_0}{k_\mathrm{B} \cdot T_\mathrm{L}} - \frac{4}{3} \cdot \left(\frac{E_\mathrm{T} - E_0}{C_1}\right)^{3/2}\right) \end{aligned} \tag{3.180}$$

The expressions for C_1, C_2, and E_0 are given by

$$C_1 = \hbar\theta = \left(\frac{\mathrm{q}^2 \cdot \hbar^2 \cdot |\mathbf{E}|^2}{2 \cdot m_\mathrm{n,p}^*}\right)^{1/3} \tag{3.181}$$

$$C_2 = \hbar\omega_0 \tag{3.182}$$

$$E_0 = 2 \cdot \sqrt{\epsilon_\mathrm{F}} \cdot \left(\sqrt{\epsilon_\mathrm{F} + E_\mathrm{T} + \epsilon_\mathrm{R}} - \sqrt{\epsilon_\mathrm{F}}\right) + \epsilon_\mathrm{R}. \tag{3.183}$$

The quantities ϵ_F and ϵ_R are expressed by

$$\epsilon_\mathrm{R} = R_\mathrm{H} \cdot \hbar\omega_0 \tag{3.184}$$

$$\epsilon_\mathrm{F} = \frac{(2 \cdot \epsilon_\mathrm{R} \cdot k_\mathrm{B} \cdot T_\mathrm{L})^2}{C_1^3}. \tag{3.185}$$

In addition to the effective tunneling masses m_n^*, the model requires three parameters: the effective phonon energy $\hbar\omega_0$, the so-called Huang-Rhys factor R_H representing the electron-phonon coupling, and E_T which denotes the trap energy relative to the reference energy E_0. Table 3.40 gives values for silicon, an indirect semiconductor, and for direct semiconductors.

Table 3.40. Parameter values for the Schenk trap-assisted band-to-band tunneling model

Material	$\hbar\omega_0$ [eV]	R_H [-]	E_T [eV]	m_n^* [eV]	References
Si	0.068	3.5	E_V+0.55	0.26	[259, 438, 439]
GaAs	0.033	6.0	E_V+0.71	0.067	[2]
GaN	0.066	6.5	E_V+0.60	0.2	[421]

In the case of hydrodynamic simulation, the effects of carrier cooling are handled as described in Section 3.6.1. In both TBB models, the trap level E_T is assumed to be in the middle of the bandgap. Both models evaluate the field enhancement factors in the same way, independent of generation or recombination.

3.6.4 Direct Band-to-Band Tunneling

The direct band-to-band tunneling model(BB) describes the carrier generation in high-field regions based on the Zener effect without any influence of local traps. The BB tunneling process describes the field emission of valence electrons leaving back holes. This occurs especially at pn-junctions with strongly bent band edges due to high fields (above 8×10^5 V/cm) and high doping levels (above 10^{19} cm^{-3}) at both sides of the junction. If the magnitude of the electric field increases, the trap-assisted band-to-band tunneling process is replaced by the BB process. A typical application for the model is at the base-collector junction in HBTs for high collector voltages. Two BB models are available which again are named after Hurkx and Schenk.

The Hurkx Direct Band-to-Band Tunneling Model. The first model is based on [223, 503]. If a pn-junction is biased in reverse direction ($n \cdot p < n_i^2$) generation of carrier pairs occurs. In forward direction ($n \cdot p > n_i^2$) recombination occurs. The generation/recombination rate is calculated by

$$R^{\mathrm{BB}} = A \cdot \frac{n \cdot p - n_i^2}{(n + n_i) \cdot (p + n_i)} \cdot \left(\frac{|\mathbf{E}|}{E_0}\right)^{\sigma} \cdot \exp\left(-\frac{B}{|\mathbf{E}|} \cdot \left(\frac{E_\mathrm{g}}{E_{\mathrm{g},300}}\right)^{3/2}\right) \tag{3.186}$$

using $E_0 = 1$ V/cm, $\sigma = 2.5$ for indirect and $\sigma = 2.0$ for direct semiconductors, $A = 4 \times 10^{14}$ cm^{-3}s^{-1}, and $B = 1.9 \times 10^6$ V/cm.

The Schenk Direct Band-to-Band Tunneling Model. The second direct band-to-band tunneling model is similar to those in [227, 438]. The model evaluates the generation rates depending on the local electric field and assumes isotropic material. The BB generation rate is calculated by

$$R^{\mathrm{BB}} = A \cdot |\mathbf{E}|^{3.5} \cdot \left(\frac{n \cdot p - n_i^2}{(n + n_i) \cdot (p + n_i)} \right) \times \left(\frac{(F_c^{\mp})^{-\frac{3}{2}} \cdot \exp\left(-\frac{F_c^{\mp}}{|\mathbf{E}|}\right)}{\exp\left(\frac{h \cdot \omega}{k_{\mathrm{B}} \cdot T_{\mathrm{L}}}\right) - 1} + \frac{(F_c^{\pm})^{-\frac{3}{2}} \cdot \exp\left(-\frac{F_c^{\pm}}{|\mathbf{E}|}\right)}{1 - \exp\left(\frac{-h \cdot \omega}{k_{\mathrm{B}} \cdot T_{\mathrm{L}}}\right)} \right). \tag{3.187}$$

The critical field strengths are

$$F_c^{\pm} = B \cdot (E_{\mathrm{g}}(T_{\mathrm{L}}) \pm h \cdot \omega)^{3/2}. \tag{3.188}$$

The upper sign in (3.187) refers to tunneling generation ($n \cdot p < n_i^2$) and the lower sign to tunneling recombination. The quantity $h\omega$ denotes the transverse accoustic phonon energy. The default model parameter values used in DESSIS for silicon are: $A = 9 \times 10^{20}$ V^{-2}cm^{-1}s^{-1}, $B = 2.15 \times 10^{7}$ Vcm^{-1}eV$^{-3/2}$, and $h\omega = 18.6$ meV [439].

3.6.5 Direct Carrier Generation/Recombination Model

In semiconductors with a direct bandgap such as GaAs, InAs, InP, GaN, and the corresponding direct ternary compounds, direct generation/recombination is very important. This has also been confirmed by numerous optical experiments for devices. The effect is modeled using the following equation

$$R^{\mathrm{DIR}} = C^{DIR} \cdot (n \cdot p - n_i^2). \tag{3.189}$$

Model parameter values for C^{DIR} at 300 K are compiled in Table 3.41 and Table 3.42.
As several publications have shown [8], the lifetime τ in AlGaAs is dominated by SRH recombination. For InGaAs, the data to determine C^{DIR} is taken from [212]. For the indirect semiconductors silicon and GaP, the radiative recombination is relatively inefficient, as given in Table 3.41.

3.6.6 Auger Generation/Recombination

The Auger generation/recombination processes make up one subclass of the three-particle processes. The related three-particle process, impact ionization, is discussed in the next section. For HEMTs, Auger effects are discussed in the context of the off-state breakdown [25], for HBTs they are important due to the high doping in the base layer. The band-to-band Auger recombination/generation rate R^{AUG} is modeled according to (3.190).

Table 3.41. Model parameters for direct recombination in *basic materials*

Material	C^{DIR} [cm^3/s]	References
Si	1.8e-15, 1.1e-14	[169, 289]
Ge	3e-14, 6.4e-14, 5.25e-15	[367, 527, 297]
GaAs	0.5e-10–7.2e-10	[8, 39, 486, 496, 516, 529]
InAs	2.1e-11, 8.5e-11	[513, 529]
InP	2.0e-10, 1.26e-9	[8, 529]
GaP	3.0e-15, 5.4e-14	[289, 513]
GaN	2.58e-10, 1.1e-8, 6.6e-10, 1.5e-11	[128, 318, 352, 409, 573]
AlN	4e-11	[128]
InN	2.21e-10, 2.7e-10	[128, 573]
GaSb	1.0e-10, 1.1e-10, 2.4e-10	[297, 367, 573]
InSb	5e-11,1.91e-11	[297, 573]

Table 3.42. Model parameters for direct recombination in *alloy materials*

Material	x	C^{DIR} [cm^3/s]	Refs.	Material	x	C^{DIR} [cm^3/s]	Refs.
Si/$Si_{1-x}Ge_x$	0.3	9e-14	[118]	InGaP	0.5	5e-11	[479]
AlGaAs	0.38	1.4e-10	[9]	AlGaN	0.2	3e-11	[128]
InGaAs	0.53	4e-10, 9.6e-11	[212, 574]	InGaN	0.2	≈4e-11	[128]
InAlAs	0.52	1.4e-10	-	InAlN	0.2	4e-11	[128]

$$R^{\mathrm{AUG}} = \left(C_{\mathrm{n}}^{\mathrm{AUG}} \cdot n + C_{\mathrm{p}}^{\mathrm{AUG}} \cdot p\right) \cdot \left(n \cdot p - n_i^2\right). \tag{3.190}$$

Parameter values can be found in Table 3.43. A weak temperature dependence of R^{AUG} with a exponential fit is suggested in [449]. The carrier concentrations prevailing for the extraction of the values reported are also given. Comparing different direct III-V semiconductor materials, Auger coefficients C_ν^{AUG} increase for decreasing bandgap, as was shown by Riech et al. in [422]. Since bandgap and effective mass are related, a systematic methodology to obtain unknown parameter values is to plot the Auger coefficients as a (universal) function of effective mass, as suggested in [318].

The temperature dependence of the coefficients in silicon [222] can, for example, be fitted by a simple power law with α= 0.5-0.6 using:

$$C_\nu^{\mathrm{AUG}}(T_{\mathrm{L}}) = C_{\nu,300K}^{\mathrm{AUG}} \cdot \left(\frac{T_{\mathrm{L}}}{300\ K}\right)^\alpha \tag{3.191}$$

or in an polynonmial expansion, as given in (3.192).

$$C_\nu^{\mathrm{AUG}}(T_{\mathrm{L}}) = C_{\nu,300K}^{\mathrm{AUG}} + K_1 \cdot \left(\frac{T_{\mathrm{L}}}{300\ K}\right) + K_2 \cdot \left(\frac{T_{\mathrm{L}}}{300\ K}\right)^2 \tag{3.192}$$

In Table 3.43, well-known values are given for silicon [469]. The value for germanium is higher than the value for silicon due to the smaller bandgap of germanium. The value for $Si_{1-x}Ge_x$ varies between the silicon and germanium values

Table 3.43. Model parameters for Auger generation/recombination at 300 K

Material	Comp. x	C_n^{AUG} [cm^6/s]	C_p^{AUG} [cm^6/s]	n,p [cm^{-3}]	References
Si	-	1.0e-31	2.28e-31	-	[297, 469]
	-	3.7e-31	-	nid	[222]
Ge	-	1e-30	-	-	[297]
GaAs	-	2-6e-30	1.6e-29	p, 1e19	[422, 514, 516]
	-	-	4e-30	-	[32]
InAs	-	1e-26	2.2e-27	-	[422]
	-	3e-27	-	-	[167]
InP	-	1.7e-33	9e-31	-	[167]
InP	-	9e-31	9e-31	-	[297]
GaP	-	1e-30	1e-30	-	[297]
GaN	-	1e-30, 1.4e-31	1e-31	-	[83, 318]
GaSb	-	5e-30	5e-30	-	[297, 483]
InSb	-	5e-26	-	-	[297]
$Si_{1-x}Ge_x$	0.3	-	6.25e-30	4e18	[118]
$Al_xGa_{1-x}As$	0.0	1.9e-31	12e-31	5e18	[498]
	0.1	1.2e-31	8.5e-31	5e18	[498]
	0.2	0.7e-31	6.1e-31	5e18	[498]
$In_xGa_{1-x}P$ (bulk)	0.5	3e-30	3e-30	-	[297]
$In_xGa_{1-x}As$ (QW)	0.53	7-9e-29	-	1.9e18	[453]
$In_xGa_{1-x}As$ (bulk)	0.53	2e-27	-	1.9e18	[453]
$In_xGa_{1-x}As$	0	-	6.5e-30	degenerate	[168]
$In_xGa_{1-x}As$	0.1	-	1.5e-29	degenerate	[168]
$In_xGa_{1-x}As$	0.2	-	3.6e-29	degenerate	[168]
$In_xGa_{1-x}As$	0.5	-	3.8e-29	degenerate	[168]
$In_xGa_{1-x}As$	1	-	2.2e-27	degenerate	[168]
$In_xAl_{1-x}As$	0.52	1.4-3e-28	-	-	[341]

as a function of x [118]. For ternary III-V semiconductors, parameter values are available for T_L= 300 K for most of the materials and are compiled in Table 3.43. A material-dependent model is presented in [498] for $Al_xGa_{1-x}As$ as a function of temperature T_L. The behavior of C_ν^{AUG} in AlGaAs is found to be similar to that of C_ν^{AUG} in silicon. However, due to the different underlying band-to-band processes, the temperature dependence cannot be generalized as a material property alone. For narrow quantum wells, a distinction is made for $In_{0.53}Ga_{0.47}As$ in Table 3.43, and the rates are significantly modified depending on the thickness of the quantum well (QW). The value given is determined for a 11 nm wide well [453]. Further, a consistent calculation is given in [168] that supplies the material dependence as a function of indium content x. No Auger coefficients are found for AlN and InN, to the best of the authors' knowledge.

3.6.7 Impact Ionization

Impact ionization (II) as a high-field generation process plays a dominant role in the understanding of HBT and HEMT device degeneration.

Impact Ionization for the Drift-Diffusion Transport Model. In the DD approximation the generation rates are usually calculated as:

$$G_\nu^{II} = \alpha_\nu \cdot \frac{|J_\nu|}{q}, \qquad \nu = \mathrm{n}, \mathrm{p}. \tag{3.193}$$

G_ν is introduced into the continuity equation as:

$$R^{II} = -G_\mathrm{n} - G_\mathrm{p}. \tag{3.194}$$

In a very straightforward model, e.g. for use in large area diodes, impact ionization rates can be modeled according to

$$\alpha_\mathrm{n} = \alpha_\mathrm{n}^\infty \cdot \exp\left(-\left(\frac{E_\mathrm{n}^\mathrm{crit}\cdot|\mathbf{J}_\mathrm{n}|}{\mathbf{E}\cdot\mathbf{J}_\mathrm{n}}\right)^{\beta_\mathrm{n}}\right) \tag{3.195}$$

and

$$\alpha_\mathrm{p} = \alpha_\mathrm{p}^\infty \cdot \exp\left(-\left(\frac{E_\mathrm{p}^\mathrm{crit}\cdot|\mathbf{J}_\mathrm{p}|}{\mathbf{E}\cdot\mathbf{J}_\mathrm{p}}\right)^{\beta_\mathrm{p}}\right). \tag{3.196}$$

This approach is commonly used to compile the different sets of experimental data. A number of sets, obtained from various measurement and MC approaches, are summarized in Table 3.44.

Silicon, Germanium, and Silicon-Germanium. Experimental data for silicon is found in great variety [107, 181, 293, 434, 526]. Electron and hole data for pure germanium can be found in [117].
Very advanced descriptions have been suggested for II in silicon FETs [183] and for different crystal orientations [539]. For SiGe, a great number of publications is available including measurements [294] and MC simulations for silicon [183, 246] and SiGe [563, 564] of the different material compositions and for different strain situations [140]. Some of these data are compiled in Table 3.44.

III-V Semiconductors. Fig. 3.69, Fig. 3.70, Fig. 3.71, and Fig. 3.72 compile several measurements and sets of MC data for III-V materials to highlight the properties of impact ionization for several bulk materials.
Fig. 3.69 shows the generation for GaAs as a function of the inverse electric field using the fit function of data in [78]. In Fig. 3.70, fit functions for the measured coefficients for $In_{0.53}Ga_{0.47}As$ are shown for data taken from [377, 521]. Earlier data in [396, 397] deviate significantly. In order to sort the sets of data, the extraction procedure is additionally named in Table 3.44. The direct comparison of

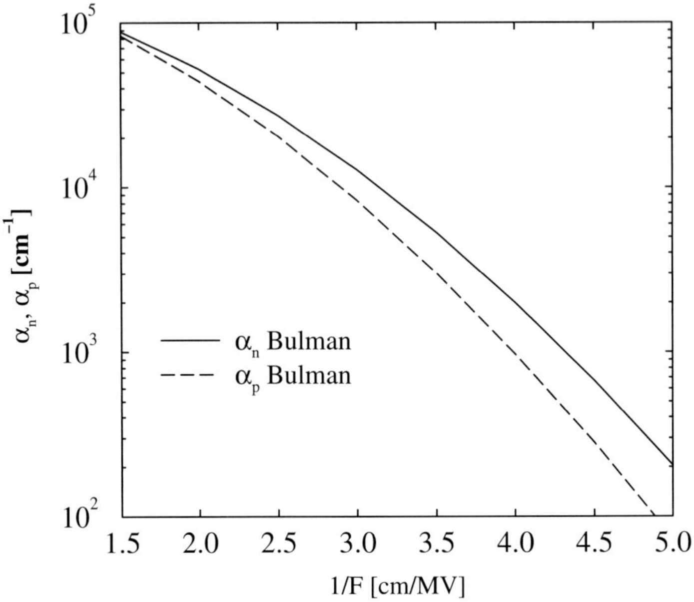

Fig. 3.69. Fit function of the measured impact ionization coefficients as a function of the inverse field for GaAs at 300 K.

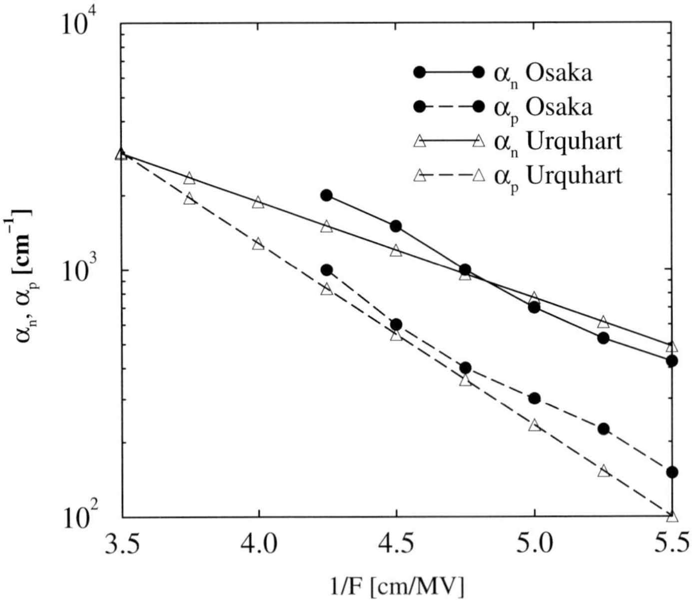

Fig. 3.70. Measured impact ionization coefficients as a function of the inverse field for $In_{0.53}Ga_{0.47}As$ at 300 K.

Table 3.44. Impact ionization parameters for *basic materials*[1]

Material T_L, N_D	E [V/cm]	α_n^{∞} [cm^{-1}]	E_n^{crit} [V/cm]	β_n [-]	α_p^{∞} [cm^{-1}]	E_p^{crit} [V/cm]	β_p [-]	Extraction	Refs.
Si	1.75-6e5	7.03e5	1.23e6	1	1.582e6	2.03e6	1	D	[526]
Ge	-	4.9e5	7.9e5	1	2.15e5	7.1e5	1	D	[116]
GaAs	-	2.99e5	6.848e5	1.6	2.215e5	6.57e5	1.75	D	[77]
	-	2.2e5	6.66e5	1.52	2.16e5	7.03e5	1.56	D	[404]
AlAs	-	3.69e5	1.43e6	1.33	1.99e5	1.21e6	1.56	D	[404]
InAs									
77 K	5e4-3e5	1.09e5	9.8e4	1	-	-	-	MC	[147]
300 K	5e4-2.5e5	2.46e5	9.16e4	1	-	-	-	MC	[147]
300 K	5e3-1e5	1.25e5	3.6e4	1	-	-	-	MC	[67]
InP									
1.2e15	2.4-3.6e5	1.12e7	3.11e6	1	4.79e6	2.55e6	1	D	[105]
3.0e16	3.6-5.6e5	2.93e6	2.64e6	1	1.62e6	2.11e6	1	D	[105]
1.2e17	5.3-7.7e5	2.32e5	8.46e5	2	2.48e5	7.89e5	2	D	[105]
GaN	2-4.05e6	1.4e7	1.48e7	1.40	2.95e5	5.17e6	2	MC	[370]
	3-3.75e6	8.9e6	1.05e7	2	-	-	-	MC	[262]
	2-3.3e6	1.14e7	2.389e7	1	-	-	-	MC	[79]
	$\geq$1e6	2.9e8	3.4e7	1	-	-	-	FET	[280]
AlN	3.3-5e6	8.87e6	3.759e7	1	-	-	-	MC	[79]
InN	7e5-2e6	6.20e5	3.54e7	1	-	-	-	MC	[30]
GaSb	2.85e4-5e4	1.07e7	4.03e5	1	7.16e5	4.02e5	1	D	[215]
InSb	50-150	-[2]	-	-	-	-	-	D	[531]

[1] D: pn-diode measurements; MC: Monte Carlo simulation; [2] model does not fit well the measurements

the model parameters is complicated by the fact that the parameters β are not always used for fitting and instead are set to one. A relatively stable database is available for GaAs. The value for AlAs is found from the fit of the material composition of $Al_xGa_{1-x}As$ for $0 \leq x \leq 0.6$. For InAs, the soft onset of impact ionization cannot be modeled very well with equations (3.195) and (3.196). Thus the data obtained from the fit is only valid for the range of the electric field indicated in Table 3.44. However, the low breakdown field and the positive temperature coefficient of the InAs parameters can be observed. The doping dependence can be evaluated for InP in Table 3.47. Impact ionization is reduced for higher impurity concentrations. Data are available for comparably higher fields for GaN than for the GaAs-based materials. The high critical fields extracted show the breakdown hardness of this material's system due to the large bandgap [68]. MC simulations are available in [79, 262, 370]. Impact ionization coefficients for AlN are given in [79]. Direction-dependent electron II data for InN is given in [30]. The data given in Table 3.44 holds for the Γ-A direction. Further, impact

Table 3.45. Impact ionization parameters for *alloy materials*[1]

Material	Comp. x	α_n^∞ [cm^{-1}]	E_n^{crit} [V/cm]	β_n [-]	α_p^∞ [cm^{-1}]	E_p^{crit} [V/cm]	β_p [-]	Extraction	Refs.
$Si_{1-x}Ge_x$(u)[2]	0.2	7.03e5	1.5e6	1.0	-	-	-	MC	[140]
$Si_{1-x}Ge_x$(s)	0.1	6.8e5	1.14e6	1.0	-	-	-	MC	[563]
	0.2	6.6e5	1.0e6	1.0	-	-	-	MC	[563]
	0.3	6.4e5	8.66e5	1.0	-	-	-	MC	[563]
AlGaAs	0.1	1.81e5	6.31e5	2	3.05e5	7.22e5	1.9	D	[425]
	0.15	2.17e5	7.74e5	1.533	2.51e5	8.60e5	1.516	D	[404]
	0.2	1.09e6	1.37e6	1.3	6.45e5	1.11e6	1.5	D	[425]
	0.3	2.21e5	7.64e5	2.0	2.79e5	8.47e5	1.9	D	[425]
	0.3	2.69e5	9.54e5	1.476	3.20e5	1.06e6	1.43	D	[404]
	0.4	1.74e7	3.39e6	1.0	3.06e6	2.07e6	1.2	D	[425]
	0.6	2.95e7	1.16e8	1.44	3.11e5	1.215e6	1.433	D	[404]
InGaAs	0.53	5.9e5	1.33e6	1	1e6	1.63e6	1	D	[377]
	0.53	6.9e4	9e5	1	1.15e6	1.7e6	1	D	[521]
InAlAs	0.52	7.36e4	5.62e5	2	1.57e4	4.88e5	2	D	[85]
	0.52	8.6e6	3.5e6	1	2.3e7	4.5e6	1	D	[544]
InGaP	0.48	4.57e5	1.413e6	1.73	4.73e5	1.425e6	1.65	D	[172]
	0.5	3.85e6	3.71e6	1	1.71e6	3.195e6	1	D	[157]
AlGaN	0.2	1.51e7	3.167e7	1	-	-	-	MC	[79]
In_xGa_{1-x}	x=0.67								[376]
$As_{1-y}P_y$	y=0.3	4.8e6	2.389e6	1	3.97e6	2.46e6	1	D	[377]

[1] MC: Monte Carlo; D: diode; [2]: u (unstrained); s (strained)

ionization is an important effect in GaSb, given its small bandgap (see Table 3.9). The data for GaSb are taken from [215].

Ternary III-V Compounds. Impact ionization parameters for different material compositions of ternary compounds from the above basic materials are listed in Table 3.45. Measurements are available for most of the compositions of $Al_xGa_{1-x}As$ and for $In_{0.52}Al_{0.48}As$. The data for InGaAs in [521] and [544] have been used for InP HEMT simulation in [490]. The following formulae are applied for those data sets which were presented consistently for various material compositions to model the material dependence of the impact ionization parameters:

$$\alpha_{\nu,AB}^\infty = x \cdot \alpha_{\nu,A}^\infty + (1-x) \cdot \alpha_{\nu,B}^\infty + x \cdot (1-x) \cdot \alpha_{\nu,AB}^0 \quad (3.197)$$

$$\beta_{\nu,AB} = x \cdot \beta_{\nu,A} + (1-x) \cdot \beta_{\nu,B} + x \cdot (1-x) \cdot \beta_{\nu,AB}^0 \quad (3.198)$$

$$E_{\nu,AB} = x \cdot E_{\nu,A} + (1-x) \cdot E_{\nu,B} + x \cdot (1-x) \cdot E_{\nu,AB}^0 . \quad (3.199)$$

The parameters extracted for $Al_xGa_{1-x}As$ are given in Table 3.46 and shown in Fig. 3.71. A systematic material dependence can be observed. A similar complete set of data is supplied in [79] for $Al_xGa_{1-x}N$.

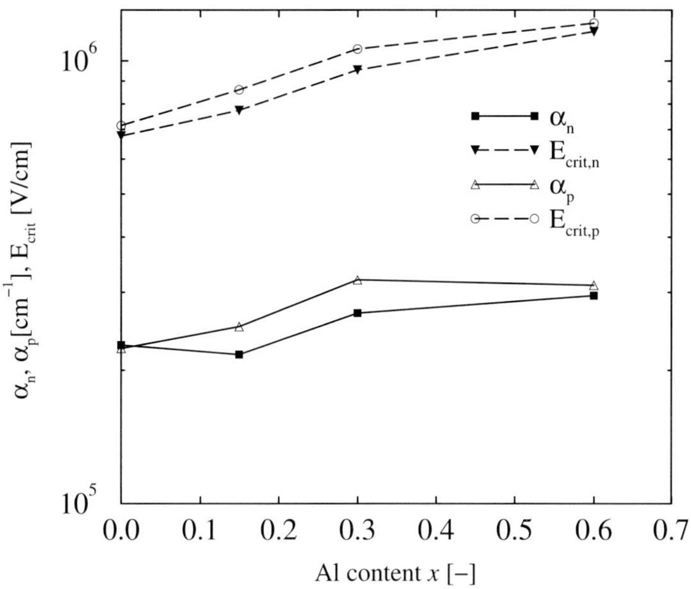

Fig. 3.71. Impact ionization parameters as a function of material composition for $Al_xGa_{1-x}As$ at 300 K.

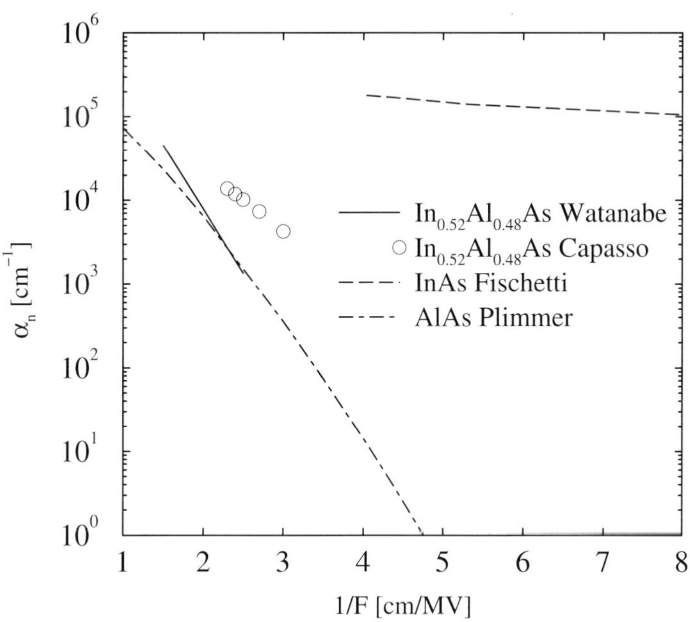

Fig. 3.72. Impact ionization coefficients as a function of material composition for $In_xAl_{1-x}As$ at 300 K.

Fig. 3.72 shows the data available for the typical HEMT barrier material $In_xAl_{1-x}As$. The experimental situation for $In_{0.52}Al_{0.48}As$ is also indicated. The plot is useful for the modeling of different material compositions in InAlAs/InGaAs HEMTs and InP HBTs. For InAs, MC data from [147] are given in Fig. 3.72. For AlAs, data from the extraction in Table 3.46 are assumed.
For the simulation of power HEMTs and HBTs, it is extremely desirable to model the temperature dependence of the impact ionization rates for $T_L > 300$ K in order to understand the interaction between electrically and thermally induced device and performance degeneration. Unfortunately, very little data are available to extract the dependence for the layer materials $In_{0.25}Ga_{0.75}As$ and $In_{0.48}Al_{0.52}As$. For $In_{0.53}Ga_{0.47}As$, a positive temperature coefficient is reported in [362] resulting in increasing impact ionization with rising lattice temperature T_L. Fig. 3.73 shows the temperature dependence for electron and holes in GaAs, where impact ionization decreases as a function of T_L. Fig. 3.74 gives the extracted model parameters from a consistent set of measurement data from [497] for InP as a function of temperature. The data for $\alpha_\nu{}^\infty$ and E_ν^{crit}, also presented in Table 3.47, show a systematic increase of the critical field as a function of T_L. α first rises and then drops above a certain T_L. However, the overall temperature coefficient of the impact ionization coefficients is negative for the temperatures observed. Measurement data for $Al_xGa_{1-x}As$ are published in [572]; very unfortunately, the paper contains several inconsistencies in data and comparisons applied. For GaAsSb, no data on impact ionization is available to the authors' knowledge. For $In_xGa_{1-x}As_yP_{1-y}$ impact ionization data are available for different material compositions x, y in [376, 377], as given in Table 3.45.

Table 3.46. Impact ionization parameters for bulk compound semiconductors

Material	Valid for x	$\alpha^0_{n,AB}$ [cm^{-1}]	$\beta^0_{n,AB}$ [-]	$E^0_{n,AB}$ [V/cm]	$\alpha^0_{p,AB}$ [cm^{-1}]	$\beta^0_{p,AB}$ [-]	$E^0_{p,AB}$ [V/cm]	Refs.
$Si_{1-x}Ge_x$(u)	0-1.0	0.0	0.0	2.3e6	-	-	-	[140]
$Si_{1-x}Ge_x$(s)	0-0.3	0.0	0.0	-9.0e5	-	-	-	[563]
$Al_xGa_{1-x}As$	0-0.6	5.1e4	1.44	1.75e5	4.45e4	-0.54	8.66e5	[404]
$Al_xGa_{1-x}N$	0-1.0	3.4126e7	0	2.491e7	-	-	-	[79]

u: unstrained, s: strained

Table 3.47. Measured impact ionization parameters for InP as a function of lattice temperature

T_L [K]	α_n [cm^{-1}]	E_n^{crit} [V/cm]	β_n -	α_p [cm^{-1}]	E_p^{crit} [V/cm]	β_p -
298	9.2e6	3.44e6	1	4.3e6	2.72e6	1
373	1.22e7	3.73e6	1	1.06e7	3.36e6	1
398	1.026e7	3.78e6	1	1.39e7	3.63e6	1
448	1.04e7	3.89e6	1	1.06e7	3.56e6	1

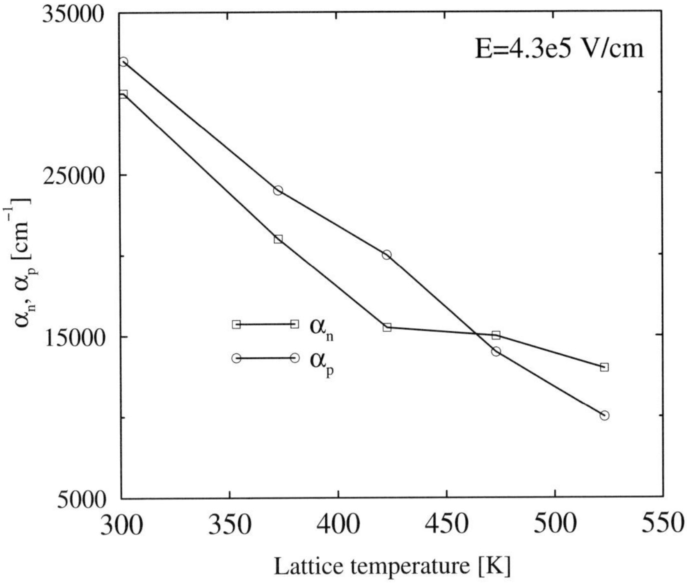

Fig. 3.73. Impact ionization coefficients as a function of lattice temperature for a given electric field in GaAs.

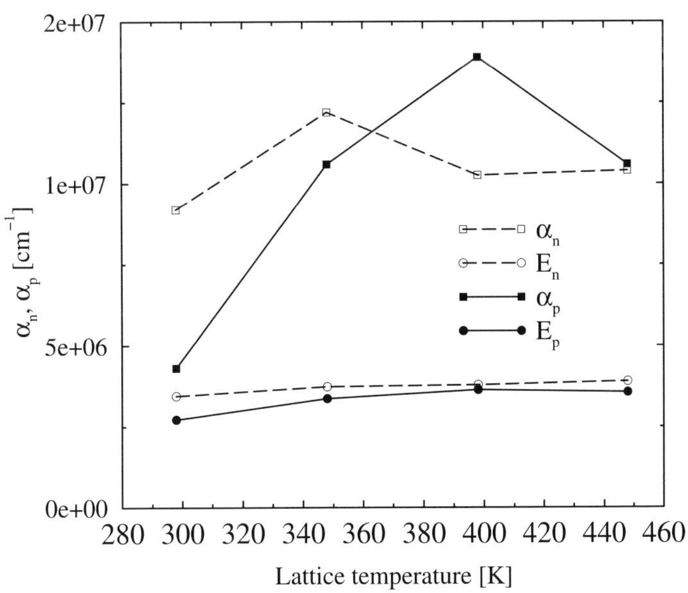

Fig. 3.74. Impact ionization parameters as a function of lattice temperature in InP.

Table 3.47 gives data sets for various lattice temperatures compiled in the form of (3.195) and (3.196). The following approach is also used to obtain a more systematic view of the temperature dependence for other materials. Chau and Pavlidis compiled available sets of data [95], as given in the previous section, into the Okuto-Crowell model [372]. Baraff proposed an approach [28] for estimation of impact ionization data at unknown temperatures and material compositions. A similar approach is used here for materials which have a temperature dependence similar to that of GaAs, $Al_xGa_{1-x}As$, and InP. Contrary to [95] this work does not include $In_xGa_{1-x}As$ for $x_{crit} \geq 0.53$ where the temperature coefficient is positive, as is shown in [120, 362]. The Okuto-Crowell impact ionization model is basically a modification of (3.195) and (3.196), where three parameters enter the formula:

$$\alpha_{n,p} = \left(\frac{q \cdot F}{E_i}\right) \exp\left\{0.217\left(\frac{E_i}{E_r}\right)^{1.14} - \sqrt{\left[0.217\left(\frac{E_i}{E_r}\right)^{1.14}\right]^2 + \left(\frac{E_i}{q \cdot F \cdot \lambda}\right)^2}\right\}. \tag{3.200}$$

E_i denotes a threshold energy; E_r is the average energy loss per scattering event; λ denotes the mean free path for the optical phonon scattering. The temperature dependence of (3.200) is introduced in (3.201) and (3.202).

$$E_r = E_{0,r} \tanh\left(\frac{E_{r,0}}{2k_B \cdot T_L}\right) \tag{3.201}$$

$$\lambda = \lambda_0 \tanh\left(\frac{E_{r,0}}{2k_B \cdot T_L}\right) \tag{3.202}$$

Model parameter values are compiled in Table 3.48. The data given include the modifications as stated by Chau and Pavlidis in [95]. Thus, for the materials used in this work, a systematic scheme is found to fill the blanks of the database as a function of temperature. This suggests a consistent approach to the lattice temperature dependence to support the modeling in the hydrodynamic II model given in the next section.

Simplified Hydrodynamic Model. In a straight forward approach, a simplified model can be applied for hydrodynamic transport. The generation rates are modeled as

$$G_n(T_n, T_L, n) = n \cdot A_n \cdot \exp\left(\frac{-B_n \cdot E_g}{k_B \cdot T_n}\right) \tag{3.203}$$

$$G_p(T_p, T_L, p) = p \cdot A_p \cdot \exp\left(\frac{-B_p \cdot E_g}{k_B \cdot T_p}\right). \tag{3.204}$$

The parameters B_ν are used to obtain a threshold energy for the materials similar to that given in the previous section as a function of the bandgap. This provides a material-dependent consistency. This model is numerically less complex than

Table 3.48. Impact ionization parameters for the Okuto-Crowell model

Material	Comp. x	$E_{\mathrm{i,n}}$ [eV]	$E_{\mathrm{i,p}}$ [eV]	$E_{\mathrm{r,0}}$ [meV]	λ_0 [Å]	E_r [meV]	λ [Å]	Refs.
Si	-	$E_g(T_L)$	$E_g(T_L)$	-	76/55	63	62/45	[496]
Ge	-	$E_g(T_L)$	$E_g(T_L)$	-	105	37	65	[496]
GaAs	-	$E_g(T_L)$+0.52	$E_g(T_L)$+0.7	43	60.65	29.3	50	[95]
AlAs	-	$E_g(T_L)$	$E_g(T_L)$	-	-	50.1	54.4	[513]
InAs	-	$E_g(T_L)$+0.03	-	-	-	29.3	39.2	[364]
InP	-	$E_g(T_L)$+0.16	$E_g(T_L)$-0.21	52.15	56.6	39.9	45.2	[95]
GaP	-	$E_g(T_L)$	$E_g(T_L)$	-	42	50	32	[496]
GaSb	-	$E_g(T_L)$+0.12	E_g	-	-	30	15	[189]
AlSb	-	E_g	E_g	-	-	39.5	15	[189]
GaN	-	$E_g(T_L)$+ 0.58	-	-	-	87	55	[79, 511]
$Si_{1-x}Ge_x$	0-0.3	$E_g(T_L) + 0.18$	$E_g(T_L)$	-	-	37-65[1]	65	[563]
$Al_xGa_{1-x}As$	0.1	$E_g(T_L)$+0.18	$E_g(T_L)$+0.31	45.12	53	31.7	40.9	[95]
	0.2	$E_g(T_L)$+0.51	$E_g(T_L)$+0.64	47.1	54.17	34	42.3	[95]
	0.3	$E_g(T_L)$+0.68	$E_g(T_L)$+0.86	48.06	55.6	35.1	44	[95]
	0.4	$E_g(T_L)$+0.69	$E_g(T_L)$+0.94	50.45	55.95	37.9	44.4	[95]
$In_xAl_{1-x}As$	0.52	$E_g(T_L)$+0.71	$E_g(T_L)$+1.31	45.37	45.45	32	32	[95]

[1] linear interpolation Si and Ge

Table 3.49. Parameter values in bulk III-V semiconductors for the simplified HD impact ionization model

Material	$A_n = A_p$ [1/s]	B_n [-]	B_p [-]
GaAs	2.7e12	1.36	1.50
$Al_{0.2}Ga_{0.8}As$	5e12	1.29	1.37
$In_{0.25}Ga_{0.75}As$	6.5e12	1.38	1.56
$In_{0.53}Ga_{0.47}As$	9e12	1.38	1.66
$In_{0.52}Al_{0.48}As$	4.2e12	1.47	1.87

the previous one. Using the parameters A_ν as an effective fit constant, impact ionization can be included into the simulation: the parameters are derived either from an approach similar to that in the previous section, or from the fit of gate currents. Typical values for the extracted parameters are shown in Table 3.49.

Hydrodynamic Impact Ionization Model. After the foregoing discussion of some principal properties of bulk material within the DD and a simplified HD approach, a more sophisticated impact ionization modeling is addressed in this section. The model by Quade et al.[410] is based on the assumption of a heated Maxwellian energetic distribution of the carriers. Since impact ionization is a generically a hydrodynamic effect based on the elevated carrier temperature, the generation rates for direct semiconductors are written as

$$G_{\mathrm{n}}(T_{\mathrm{n}},n) = n\cdot A\cdot\left(\frac{1}{2}\cdot\sqrt{u}\cdot\exp\left(\frac{-1}{u}\right)-\mathrm{erfc}\left(\frac{1}{\sqrt{u}}\right)\right) \tag{3.205}$$

and

$$G_{\mathrm{p}}(T_{\mathrm{p}},p) = p\cdot A\cdot\left(\frac{1}{2}\cdot\sqrt{u}\cdot\exp\left(\frac{-1}{u}\right)-\mathrm{erfc}\left(\frac{1}{\sqrt{u}}\right)\right). \tag{3.206}$$

For indirect semiconductors, however, the formula reads

$$G_{\mathrm{n}}(T_{\mathrm{n}},n) = n\cdot A\cdot\left(\left(1+\frac{u}{2}\right)\cdot\mathrm{erfc}\left(\frac{1}{\sqrt{u}}\right)-\frac{1}{2}\cdot\sqrt{u}\cdot\exp\left(-\frac{1}{u}\right)\right) \tag{3.207}$$

and

$$G_{\mathrm{p}}(T_{\mathrm{p}},p) = p\cdot A\cdot\left(\left(1+\frac{u}{2}\right)\cdot\mathrm{erfc}\left(\frac{1}{\sqrt{u}}\right)-\frac{1}{2}\cdot\sqrt{u}\cdot\exp\left(-\frac{1}{u}\right)\right) \tag{3.208}$$

using the carrier energy scaled in units of the threshold energy E_i:

$$u = \frac{k_{\mathrm{B}}\cdot T_{\nu}}{E_{\mathrm{i},\nu}}. \tag{3.209}$$

The two sets of equations, (3.205)and (3.206), and (3.207) and(3.208), are derived as two limiting cases from the same basic model as explained in [410]. The difference arises from the energies allowed after the impact ionization process in the case of a direct and indirect semiconductor. The symbol erfc() denotes the complementary error function 1 − erf(). Special care has to be taken for the implementation of the direct model to avoid numerical uncertainties that may arise due to often-encountered poor implementations of the complementary error function [407].
The lattice temperature dependence is modeled in the prefactor A [259] using four parameters. This allows the inclusion of the two different temperatures represented by GaAs and $\mathrm{In_{0.53}Ga_{0.47}As}$.

$$A(T_{\mathrm{L}},T_{\mathrm{n}},E_{\mathrm{i}}) = C_1\cdot\exp\left(C_2\cdot\frac{T_{\mathrm{L}}}{T_0}+\frac{E_{\mathrm{i}}\cdot u}{k_{\mathrm{B}}\cdot T_0}\cdot\left(C_3-C_4\cdot\frac{T_{\mathrm{L}}-T_0}{T_0}\right)\right) \tag{3.210}$$

E_{i} denotes the threshold energy at $T_{\mathrm{L}} = T_0 = 300$ K modeled as given above. For indirect semiconductors, the energy E_{i} approximately equals the bandgap, while for direct semiconductors there is a significant difference, as given in Table 3.50. The parameter extraction is achieved as follows: Basically, the impact ionization generation rates are found as a function of the corresponding carrier temperatures by one-dimensional MC simulation using the potential profile in a HEMT channel or in an HBT collector at high bias. The model parameters

Table 3.50. Parameter values for the HD impact ionization model for electrons and holes

Material	C_1 [1/s]	C_2 [-]	C_3 [-]	C_4 [-]	$E_{i,n}$ [eV]	$E_{i,p}$ [eV]
Si	1.05e8	3.823	0.34633	0.0922	$E_g(T_L)$	$E_g(T_L)$[259]
GaAs	1.3e8	3.8	0.35	0.1	$E_g(T_L)$+0.52	$E_g(T_L)$+0.72
$Al_{0.2}Ga_{0.8}As$	1.0e8	3.8	0.35	0.1	$E_g(T_L)$+0.51	$E_g(T_L)$+0.64
$In_{0.25}Ga_{0.75}As$	2.3e8	3.8	0.35	0.1	$E_g(T_L)$+0.41	$E_g(T_L)$+0.60
$In_{0.53}Ga_{0.47}As$	2.0e8	3.8	0.35	-0.1	$E_g(T_L)$+0.30	$E_g(T_L)$+0.51
$In_{0.52}Al_{0.48}As$	2.0e8	3.8	0.35	0.1	$E_g(T_L)$+0.71	$E_g(T_L)$+1.31

in (3.210) are then fitted to the generation rates obtained by the MC model as a function of carrier temperature. Thus, the model parameters contain all the assumptions made by the HD impact ionization model, which suggests G_ν to be a single valued function of the average carrier temperature T_ν. The values found useful are given in Table 3.50.
The data for the lattice-temperature dependence are obtained by comparison with the approach in the previous section, which suggests modeling the threshold energy. For $In_{0.53}Ga_{0.47}As$, the lattice temperature dependence is changed by the

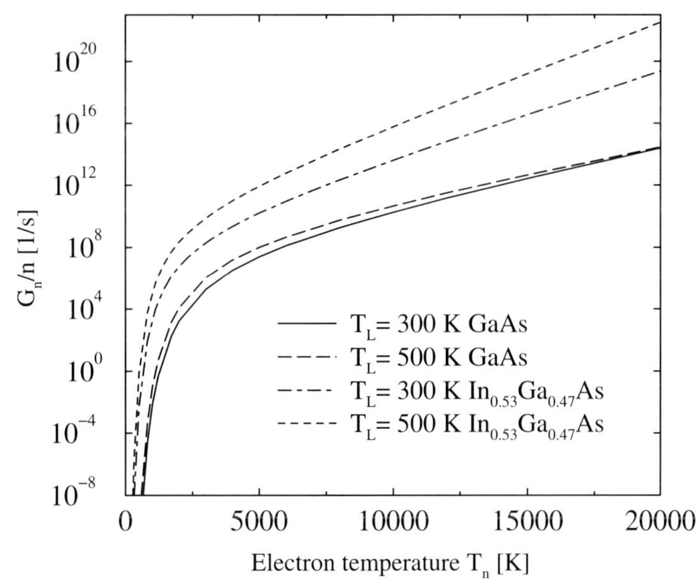

Fig. 3.75. Impact ionization rate as a function of carrier temperature for GaAs and $In_{0.53}Ga_{0.47}As$ with the lattice temperature T_L as a parameter.

sign of the parameter C_4, which leads to the increase of the impact ionization with the lattice temperature.
Fig. 3.75 shows the principal difference of impact ionization for GaAs, representative for AlGaAs and InAlAs for all material compositions, and for $In_{0.25}Ga_{0.75}As$. It can be seen that, for GaAs, the slope of G_ν drops with increasing lattice temperatures T_L. This leads to a crossing of the curves for carrier temperatures of about $T_\nu =$ 20000 K, i.e. for an average carrier energy of $\approx$ 2 eV. MC simulations of HEMTs show that carriers already reach energies $\geq$ 1 eV for the relatively low bias of $V_{DS} = 2$ V. For $In_{0.53}Ga_{0.47}As$, the opposite behavior is found in Fig. 3.75: the generation increases as a function of T_L, which causes hot-electron reliability problems. If material compositions other than those given in Table 3.50 are needed, they can be interpolated between the materials given.

4. RF Parameter Extraction for HEMTs and HBTs

This chapter addresses the extraction of RF quantities such as Y- and related parameters, derived quantities such as stability factors and cut-off frequencies, and, last but not least, small-signal equivalent circuit elements for various topologies. It further provides a view of the physical understanding of the bias dependence of small-signal equivalent circuit elements in HEMTs and HBTs in analytical models and TCAD approaches.

4.1 RF Parameter Extraction Methods

Several approaches can be used for the extraction of RF quantities. An overview is given in [291]. The simulation itself can be based on several approaches, e.g. Fourier decomposition or applying quasi-static or equivalent-circuit parameter models. These approaches require a transient simulation. Basically, many time steps are necessary to obtain sufficient time-domain accuracy to generate sufficient frequency-domain accuracy. These approaches are thus very CPU-time consuming, as only a limited number of time steps can be performed in a reasonable amount of CPU-time.

4.1.1 Transient Analysis

In this approach, small-signal sinusoidal voltages are applied to the device terminals in two independent simulations. A frequency of typically 5 GHz or 10 GHz is chosen for the simulation to avoid dispersion effects in a lower frequency range. Typical small-signal voltages used for the extraction are 1 mV for HBT and 10 mV for HEMT simulations. The intrinsic Y-parameters are extracted from the simulation results (see [291]) as

$$Y_{11} = \frac{\partial I_1}{\partial V_1}\bigg| V_2 = \text{const.} \qquad Y_{21} = \frac{\partial I_2}{\partial V_1}\bigg| V_2 = \text{const.} \tag{4.1}$$

$$Y_{12} = \frac{\partial I_1}{\partial V_2}\bigg| V_1 = \text{const.} \qquad Y_{22} = \frac{\partial I_2}{\partial V_2}\bigg| V_1 = \text{const.} \tag{4.2}$$

The indices denote the ports 1 and 2 of the two-port device, 1 being the input and 2 the output. In the case of a HEMT in common source configuration, I_1 is the gate current I_{G}, and I_2 denotes the drain current I_{D}. This approach is computationally expensive, as two full transient simulations have to be completed. Depending on the topology, the intrinsic small-signal equivalent circuit elements are calculated from the intrinsic Y-parameters: directly for FETs, or after conversion to intrinsic Z-parameters for HBTs. The elements accounting for parasitic effects, i.e. inductances and capacitances, result from standard extraction procedures such as open/short thru-test-structures and are included in the calculation. In this way the extrinsic S-parameters can be obtained for all frequencies and are suitable for comparison to the measurements in the same frequency range and for the same microwave reference plane.

4.1.2 Sinusoidal Steady-State Analysis

A second method uses a specialized solver, as proposed in [291, 382]. The direct small-signal simulation mode is based on the S^3A (sinusoidal steady-state analysis) approach presented in [291], which is rigorously correct, regarded to be computationally inexpensive, and very accurate. The high accuracy is due to the formal linearization of the device. The numerical effort is minimized since the calculations are performed directly in the frequency domain. The advantage of such an approach is that for one bias situation only one additional simulation step is necessary to obtain the full RF information at a specific frequency, i.e. the simulation is much more efficient than transient approaches.

4.1.3 Further Extraction Approaches

Other approaches apply quasi-static approximations. A very general approach for FETs (including HEMTs) is the extraction of capacitances from DC simulations

$$C_{\mathrm{g,tot}} = \frac{\partial Q_G(V_{\mathrm{DS}}, V_{\mathrm{GS}})}{\partial V_{\mathrm{DS}}} + \frac{\partial Q_G(V_{\mathrm{DS}}, V_{\mathrm{GS}})}{\partial V_{\mathrm{GS}}}. \tag{4.3}$$

By stepping the voltage between two contacts only, e.g. varying only V_{GS}, the change of total gate charge Q_G and thus the total gate capacitance $C_{\mathrm{g,tot}}$ can be extracted. By varying only V_{DS} at a constant V_{GS}, the gate drain capacitance C_{gd} can be extracted. From

$$C_{\mathrm{gs}} = C_{\mathrm{g,tot}} - C_{\mathrm{gd}} \tag{4.4}$$

the gate source capacitance C_{gs} is finally available [64]. Another simplified approach has been demonstrated in [398] for HBTs. By applying small steps of V_{BE}, f_{T} can be calculated by

$$f_{\mathrm{T}} = \frac{g_m}{2\pi \cdot C_{\mathrm{in}}} \tag{4.5}$$

$$g_{\mathrm{m}} = \frac{\Delta J_{\mathrm{C}}}{\Delta V_{\mathrm{BE}}} \tag{4.6}$$

$$C_{\mathrm{in}} = \frac{\Delta Q_{\mathrm{s}}}{\Delta V_{\mathrm{BE}}}. \tag{4.7}$$

C_{in} represents the input capacitance and g_{m} the transconductance, while Q_{s} represents the storage charge. Typical analytical models for HEMTs and HBTs are discussed in the next section.

4.2 Contributions to the Small-Signal Equivalent Circuit Elements and their Bias-Dependence

4.2.1 Analytical HEMT Models

The intrinsic eight-element small-signal equivalent circuit [34] (see Fig. 4.1) is commonly used for the extraction of small-signal equivalent circuit elements for MESFETs and HEMTs. The extraction of parasitic elements from measurements is typically performed according to standard procedures [34, 419]. Contrary to the classical separation of the intrinsic device and the parasitic environment, Fig. 4.1 additionally shows a simulator shell, which includes the intrinsic shell and the parasitic ohmic resistances. This is useful since the semiconductor contributions to the resistances are a genuine part of the simulation domain. However, parasitic capacitances and inductances can be obtained from gate width dependent measurements, so they can be considered extrinsic for the physical simulation. Extended equivalent circuits have been suggested, e.g. in [419], to include leakage and impact ionization phenomena in InAlAs/InGaAs HEMTs.
Fig. 4.2 shows such an extended equivalent circuit for a millimeterwave HEMT. In the input circuit, two resistances, R_{pgs} and R_{pgd}, are added to account for the gate leakage. In the output circuit, three additional elements are added: g_{mim} introduces a V_{DS} dependent contribution to g_{m} due to impact ionization, and the combination of C_{im} and R_{im} present an additional time constant explained below. Genetic algorithms can be used in order to determine small-signal equivalent circuit elements [419]. This circuit, usually used for noise modeling, can be applied to extract breakdown information from measured S-parameters at high V_{DS} bias in combination with standard procedures.

Treatment of the Parasitic Elements.

Line Resistors. In Fig. 4.1 a de-embedding concept is indicated by the dashed line. As outlined above, the line resistances R_{S}, R_{G}, and R_{D} are completely included in the simulation shell, since the source and drain resistors are partly defined by the semiconductor parts of the device and the semiconductor/metal alloy at the ohmic contacts. Both inductances and capacitances are considered extrinsic. The resistances R_{S} and R_{D} are split into two contributions which have to be evaluated for each process and layout.

$$R_{\text{S}} = R_{\text{Ssemi/alloy}} + R_{\text{SM}}, \quad R_{\text{D}} = R_{\text{Dsemi/alloy}} + R_{\text{DM}} \tag{4.8}$$

To account for the metal part above the semiconductor/alloy/metal transition, an additional line resistor is added in the simulation. It is defined by the conductivity of the metal used and the geometry of the contact. The contributions

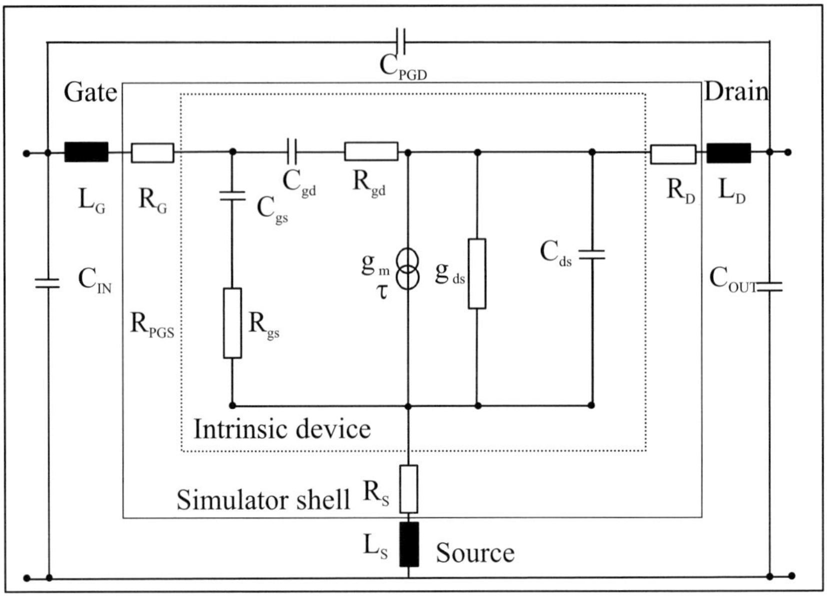

Fig. 4.1. Standard Π-type small-signal equivalent circuit of a HEMT.

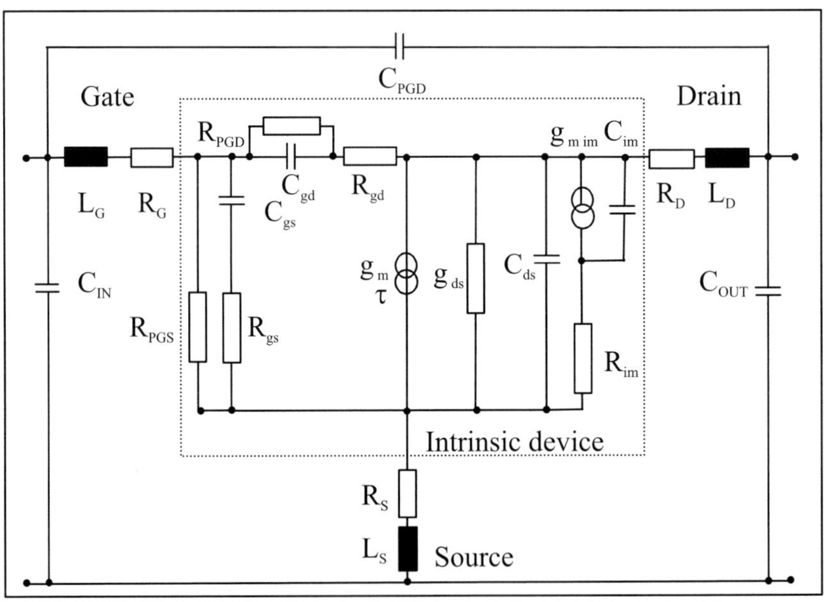

Fig. 4.2. Extended small-signal equivalent circuit of a HEMT including gate leakage and impact ionization.

R_{SM} and R_{DM} to R_S and R_D account for the metal parts and are extracted by measurements of a special short structure, or alternatively by gate forward transistor measurements [34, 419]. DC measurements of the ohmic resistances can be performed as a reference. The extracted drain resistance can be bigger than the source resistance in agreement with measurements: at the drain side the electrons have to pass the heterojunction also for symmetrical devices, which leads to an additional contribution. The different ohmic contact situations are addressed in Section 6.4. Because of their influence on noise performance, the gate resistances R_G have to be known very well. They are dominated by the metal part of the gate and thus are determined by the width of the single gate-finger. A line resistor R_G, which basically accounts for the gate processing technology, is specified in the Schottky contact to obtain realistic intrinsic voltages in the simulation.
Moreover, R_G is an important parameter for the analysis of the output conductance, since for simulations including self-heating R_G shows a strong dependence on the gate temperature. The gate metal is one of the hottest parts of the device. The increase in temperature leads to changes of the overall gate resistance. This results in effects such as the negative output conductance observed in measurements. To allow for the comparison with experimental data, R_G is modeled

$$R_G(T_L) = R_G(300\ \mathrm{K})\cdot(1+\alpha\cdot(T_L-300\ \mathrm{K})) \tag{4.9}$$

where α is the temperature coefficient of the resistivity of the metal. R_{SM} and R_{DM} are modeled equivalently. Typical values for the resistivity and α are compiled in Table 4.1. For the gate composite metal a value of $\alpha = 4\times10^{-3}\mathrm{K}^{-1}$ is typically assumed .

Table 4.1. Values for the resistivities of various IC metalizations

Metal	$\rho_{300\ \mathrm{K}}$ [Ω·cm]	α [K^{-1}]	References
Cu	1.71e-6	3.9e-3	[287, 546]
Au	2.20e-6	3.4e-3	[287, 328]
Ag	1.55e-6	3.8e-3	[328, 546]
Al	2.81e-6	3.9e-3	[328, 546]
Ti	5.54e-6	4.0e-3	[328]
W	5.60e-6	4.5e-3	[546]

Parasitic Capacitances, Inductances, and Parasitic Line Delay. The capacitance C_{IN} is positioned in parallel to C_{gs}, C_{OUT} is parallel to C_{ds}, and C_{pgd} parallel to C_{gd}. Open structures of a given layout are measured for various gate widths where the inner transistor is physically removed. To separate parasitic capacitances, alternatively, measurements of a transistor in pinch-off are evaluated [34]. Using the values obtained for the parasitic capacitances C_i and measuring a short structure, the metal parts of R_G and R_D as well as the parasitic inductances L_i can be determined from the S-parameters. Alternatively, the determination of the L_i can be replaced by forward measurements of a HEMT at $V_{DS} = 0$ V at different current levels [34]. A gate width dependent port extension τ_{port} is added to account for the delay of a coplanar line which is additionally part of the transistor layout in the measurement. For overall justification, examples given in the next two chapters show the comparison of simulation and extraction.

Contributions to the Transconductance g_m. In the one-dimensional linear charge control model from [24] the transconductance g_m is described as

$$g_m = \frac{\varepsilon_r \cdot v_{eff} \cdot W_g}{d_{eff}} \cdot \frac{1}{\sqrt{1 + \left(\frac{n_c}{n_{sheet}}\right)^2}} \tag{4.10}$$

where ε_r is the relative dielectric constant, v_{eff} the effective carrier velocity, W_g the gate width, and d_{eff} an effective gate-to-channel separation. n_{sheet} is the sheet charge density and n_c a unity reference sheet concentration defined as

$$n_c = \frac{E_{crit} \cdot l_g \cdot c_{eff}}{q} \tag{4.11}$$

where E_{crit} is a critical field for the onset of a "saturated" behavior and c_{eff} is here defined to be an effective channel capacity assuming a fixed distance between channel charge and gate in a simple capacitor model.

As can be seen from (4.10), both a higher sheet density n_{sheet} and a higher effective carrier velocity v_{eff} lead to higher g_m. This explains the different properties of the material systems AlGaAs/InGaAs and InAlAs/InGaAs. For example, InAlAs/InGaAs transistors yield both a higher sheet density and higher effective carrier velocities than AlGaAs/InGaAs HEMTs. Furthermore, the difference of DC and RF g_m can be explained due to carrier generation/recombination resulting in changes of the sheet densities n_{sheet}. To this point, neither velocity saturation nor parasitic charge modulation are included in the model. A term called modulation efficiency ME describing velocity saturation can be derived defining an intrinsic delay time τ_i as

$$\tau_i = \frac{C_{g,tot}}{g_m} = \frac{l_g + L_{od}}{v_{eff}} \cdot \frac{1}{ME} \tag{4.12}$$

with $C_{g,tot}$ the total gate capacitance. The considerations of (4.12) are purely one-dimensional, i.e. the effects of an additional V_{DS} bias variation are not taken into account. They can be included using an extension of the symmetric space charge region considered in (4.12) by a length L_{od}. This length describes the drain side extension of the space charge region. Parasitic charge modulation is further included by setting an upper limit towards the modulation efficiency ME. The model of (4.12) further assumes a saturated effective carrier velocity v_{eff}, which is independent of V_{DS}. This V_{DS} dependence of the effective carrier velocities can be understood with the help of two-dimensional device simulation.

Contributions to the Input Capacitance C_{gs}. The basic contribution to C_{gs} arises from the interaction of the channel charge and the gate metal. Additionally, fringe capacitances [15] are added due to coupling between the semiconductor caps, the ohmic metal, and the gate metal. The capacitance C_{gs} is therefore written as

$$C_{gs}(V_{GS}, V_{DS}) = C_{spaceDC}(V_{GS}, V_{DS}) + C_{fringe} + C_{traps}(V_{GS}, V_{DS}, f) \tag{4.13}$$

The additional term C_{traps} is added to emphasize the effect of the generation/recombination of additional carriers when applying a third terminal voltage. The difference between RF and DC g_m due to the activation of traps and carrier generation/recombination in a dynamic process with time constants of the order of μs is generally acknowledged [388]. Above frequencies of excitation of a few MHz this results in the stable, time-invariant equilibrium which is extracted in RF measurements [365]. Above a few MHz it is further independent of the frequency, yet not of the applied bias. It is very desirable to reduce generation/recombination. The inclusion of these effects into device modeling [441] is helpful in explaining the dependence of device properties on V_{DS} bias. The bias dependence of C_{gs} is further strongly related to the shape and the carrier charge concentration distribution of the space charge region. The compact model assumes a charge at an effective distance d_{eff}, yet the center of this charge varies as a function of V_{DS}.

Contributions to C_{gd}. The gate-drain capacitance C_{gd} is expressed as suggested in [102].

$$C_{gd}(V_{DS}) = C_{met} + C_{pass} + C_{semi}(V_{DS}) \tag{4.14}$$

The contributions of the gate and drain metal periphery C_{met} and the coupling of the gate metal with the semiconductor caps through the SiN passivation C_{pass} do not depend on V_{DS}. The dependence of C_{gd} on V_{DS} arises mostly from the coupling of gate and drain through the barrier and channel layer, represented by C_{semi}.

Contributions to the Output Capacitance C_{ds}. The drain-source capacitance C_{ds} originates from the capacitive coupling of source and drain and represents the intrinsic output capacitance in the standard model of Fig. 4.1. Bias-dependent contributions to C_{ds} arise from buffer and channel layers depending on the overall carrier distribution. Since a one-dimensional model does not consider a V_{DS} voltage, the analysis remains for a two- or three-dimensional model. C_{ds} can be written as

$$C_{ds}(V_{DS}) = \sum C_{ds}^{j} \tag{4.15}$$

where the index j represents all layers between the gate metal and the semi-insulating substrate. Thus the elements g_m, C_{gs}, C_{gd}, and even C_{ds} find a relatively neat physical interpretation in compact models. A physical interpretation can be found for the other four elements, but unfortunately the non-ambiguous transformation of four complex S-parameters into eight small-signal equivalent circuit elements leads to rather abstract quantities with respect to the physical interpretation.

Contributions to the Output Conductance g_{ds}. The output conductance g_{ds}, and its inverse, the channel resistance R_{ds}, are subject to a number of influences that must be carefully separated. Special care has to be taken when comparing DC g_{ds} and RF g_{ds}. DC g_{ds} values obtained from the output characteristics can appear to be negative. A first effect to be separated is the effect of self-heating and the temperature dependence of the parasitic resistances, especially the gate resistance R_G. Self-heating leads to a rise of R_G and thus to a reduction of the internally applied V_{GS} bias, so self-heating causes an internal bias shift [320]. A second effect is the influence of carrier generation/recombination. As reported especially for GaN HEMTs [132], but also for pseudomorphic HEMTs [388], the occurrence of traps leads to a hysteresis in the DC output characteristics, an effect which is not necessarily seen in RF parameters. The RF value of g_{ds} extracted from measured data is therefore always positive.
When extending one-dimensional charge control models using a compact current model, a saturated velocity model leads to a constant DC output conductance as a function of V_{DS} in the so-called saturation region [152]. Introducing first order compact approaches [23] for impact ionization leads to an exponential increase of the drain current I_D as a function of V_{DS} bias.

Contributions to R_{gs} and R_{gd}. The ohmic resistances R_{gs} and R_{gd} represent a rather macroscopic and concentrated form of the carrier transport near the gate contact. Due to the strongly locally varying carrier concentrations in the depleted barrier layer ($n \leq 10^{10}$ cm^{-3}) and in the channel ($n \geq 10^{18}$ cm^{-3}), and due to carrier temperatures near the gate of $T_\nu \geq 10 \cdot T_L$, a direct physical interpretation is not suitable. The transport situation is extremely non-linear and thus non-ohmic. However, the information contained in the extraction is to be understood in a heuristic sense. Since both a current I_G and the voltages V_{GS} and V_{GD} are available, the effective impedances R_{gs} and R_{gd} can be calculated. Especially R_{gs} does supply a high amount of information, for example, for compact noise modeling [419].

Contributions to τ. If the effective carrier velocity v_{eff} used in the compact models for the transconductance g_m is interpreted as bias-dependent group velocity, a somewhat physical interpretation of τ can be found. In this case τ includes the phase information of this group velocity. This explains, on the one hand, why the absolute values of τ are in the order of magnitude of extracted delay times, and, on the other hand, why a direct interpretation as the delay time of the device is not appropriate. The typical delay times are extracted from the inverse of f_T values for analog or from ring oscillators for digital applications.

4.2.2 Extracted Device Quantities: Invariants

The current gain cut-off frequency f_T is defined as

$$|h_{21}|_{f=f_T} = 1. \tag{4.16}$$

This definition of f_T has various approximations in terms of intrinsic small-signal equivalent circuit elements. The most important approximations for HEMTs are (assuming the parasitic inductances and capacitances to be stripped off) according to Baeyens [24]

$$f_T = \frac{g_{mi}}{2\pi \cdot [C_{gs} + C_{gd}] \cdot [1 + g_{ds} \cdot (R_S + R_D)] + C_{gd} \cdot g_{mi} \cdot (R_S + R_D)} = \\ = \frac{1}{2\pi \cdot \tau_T} \tag{4.17}$$

and even simpler in (4.18), neglecting the parasitic resistances and the output conductance:

$$f_T = \frac{g_{mi}}{2\pi \cdot (C_{gs} + C_{gd})}. \tag{4.18}$$

On the contrary, the inclusion of the pad capacitance C_{pad} and the fringe capacitance C_{fringe} from (4.13) in (4.17) yield an extrinsic charging time and extrinsic f_T respectively.

$$\tau_{ext} = \frac{1}{2\pi \cdot f_{T,ext}} = \tau_T + \frac{C_{pad}}{g_{m,ext}} + \frac{C_{fringe}}{g_{mi}}. \tag{4.19}$$

The intrinsic transconductance g_{mi} and the extrinsic transconductance $g_{m,ext}$ are then related by:

$$g_{mi} = \frac{g_{m,ext}}{1 - g_{ds,ext} \cdot (R_S + R_D) - g_{m,ext} \cdot R_S} \tag{4.20}$$

where $g_{ds,ext}$ is the extrinsic output conductance. Using (4.17)-(4.19), f_T is precisely defined for different levels of de-embedding.
The quantity f_{max} can be defined in several ways, depending on the invariants used for its definition [195]. First, defining the unilateral power gain U allows for the highest values of f_{max} in any device representing the maximum gain in a lossless reciprocal de-embedding [195].

$$U(f) = \frac{\left|\frac{S_{21}}{S_{12}} - 1\right|^2}{2 \cdot k \cdot \left|\frac{S_{21}}{S_{12}}\right| - 2 \cdot \mathrm{Re}\left(\frac{S_{21}}{S_{12}}\right)} \tag{4.21}$$

Second, the maximum available gain (MAG) and the maximum stable gain (MSG) can be used to define f_{max}

$$\mathrm{MAG}(f) = \left|\frac{S_{21}}{S_{12}}\right| \cdot (k \pm \sqrt{k^2 - 1}) \tag{4.22}$$

with the constraint for the stability factor $k \geq 1$. $f_{\max}$ is then determined as

$$|\mathrm{MAG}|_{f=f_{\max}} = 1. \tag{4.23}$$

The MAG drops with a slope of −20 dB/dec as a function of frequency near $|\mathrm{MAG}| = 1$. Kurokawa's stability factor k is defined by the S-parameters as [195]

$$k(f) = \frac{1 - |S_{11}|^2 - |S_{22}|^2 + |S_{11}S_{22} - S_{12}S_{21}|^2}{2 \cdot |S_{12}||S_{21}|}. \tag{4.24}$$

The Maximum Stable Gain (MSG) is used when $k \leq 1$ and defined as

$$\mathrm{MSG} = \left|\frac{S_{21}}{S_{12}}\right|. \tag{4.25}$$

For HEMTs, $f_{\max}$ can be expressed in several ways, as outlined for example in [24].

4.2.3 Analytical HBT Models

For the extraction of S-parameters for HBTs, the same methodologies can be used as stated for HEMTs above. However, generally the extraction of small-signal equivalent circuit elements is more complicated. A direct extraction of

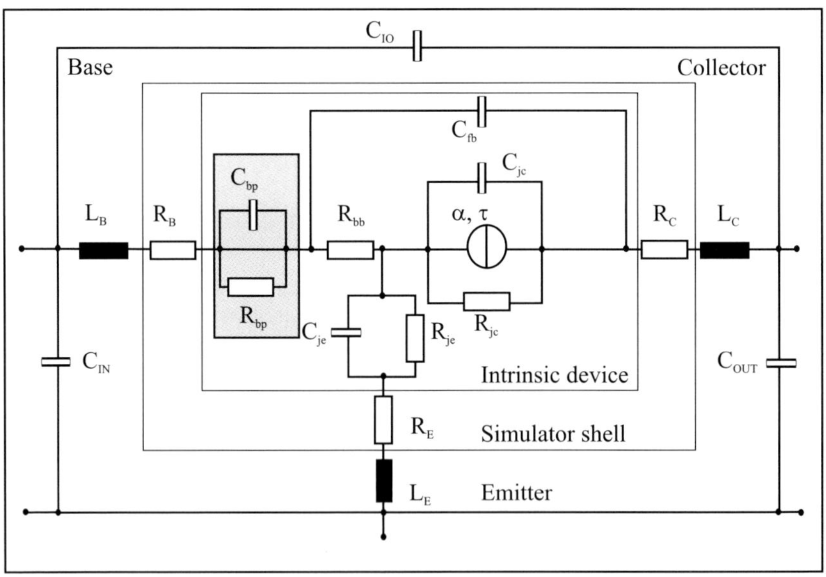

Fig. 4.3. T-type eight-element small-signal equivalent circuit of a HBT.

the small-signal equivalent circuit elements from the intrinsic Y-parameters at one frequency is practically impossible, as the compact extraction method needs the information obtained at various measured frequencies [432]. For transient simulation, when only one frequency is available per simulation, this approach becomes even more demanding on computational time for the HBT.
A T-like eight-element small-signal equivalent circuit model [400, 432] is commonly used for the analysis of the bias dependence of the small-signal equivalent circuit elements (see Fig. 4.3) in HBTs. Ohmic resistances are completely accounted for in the TCAD simulations, as denoted in Fig. 4.3 by distinguishing the intrinsic transistor and the simulator shell. The extrinsic S-parameters are calculated using the same typical values for the extrinsic parasitic elements (inductances and capacitances) as extracted from the underlying technology [446] for compact modeling.
The model topology is a standard approach commonly used in large-signal models [431] except from the additional non-linear input circuit shaded in grey. This approach is only used for modeling in the early stages of process development, when ohmic contacts are still immature and show an additional capacitive behavior.

Contributions to R_{bb}. R_{bb} denotes the internal base resistance from a high-frequency-oriented point of view [522] and can be written as

$$R_{bb} = R_{Con} + R_{spread} + R_{gap}. \tag{4.26}$$

R_{Con} denotes the contact resistance for the npn-transistors between the metal and the highly p-doped base layer. R_{spread} denotes the resistance which is due to a spreading of the carriers in the base layer. R_{gap} denotes the ohmic equivalent of the resistance through the semiconductor of the non-metalized region between emitter and base.

Contributions to $C_{bc,tot}$. The overall base-collector capacitance $C_{bc,tot}$ is one of the most important equivalent circuit elements. It is usually considered to be composed of two contributions

$$C_{bc,tot} = C_{jc} + C_{fb} \tag{4.27}$$

The internal base-collector capacitance C_{jc} adds up to the external base-collector capacitance, called C_{fb} here, and is given in Fig. 4.3. For high-speed devices, C_{jc} is a speed-limiting element and a main target quantity for device optimization. The quantitative separation of an internal contribution to C_{jc} and the external contribution C_{fb} is a very delicate matter. For SHBTs, the qualitative separation is possible as follows [447]. For low currents the internal base-collector capacitance rises with the collector current due to the spreading of the collector current and the base push-out. The second contribution C_{fb} drops with the current due to the effect of velocity modulation in III-V semiconductors with non-homogenous velocity-field characteristics. The overall behavior thus is determined by relative and absolute contributions during the device optimization process which can mask each other. Thus device simulation provides a valuable means of determining the internal contribution to $C_{\mathrm{bc,tot}}$.

Contributions to α and τ. Within the topology of Fig. 4.3, α denotes the active amplification as a current gain and a phase factor τ. The amplification α can be written

$$\alpha = \frac{\mathrm{q} \cdot I_{\mathrm{E}}}{k_{\mathrm{B}} \cdot T_{\mathrm{L}}}. \tag{4.28}$$

The equation shows that a high emitter current I_{E} is a prerequisite to obtaining high amplification, or, for device optimization, a low emitter resistance R_{E}. Further, (4.28) explains the attractivity of rising currents and current densities for both Si and III-V devices for extremely high-speed operation, which interferes with device reliability [224].

The Cut-off Frequencies f_{T} and f_{max}. The current gain cut-off frequency f_{T} in HBTs can be expressed as

$$\frac{1}{2\pi \cdot f_{\mathrm{T}}} = \tau_{\mathrm{ec}} = \tau_{\mathrm{eb}} + \tau_{\mathrm{b}} + \tau_{\mathrm{c}} + \tau_{\mathrm{sc}}. \tag{4.29}$$

In (4.29), τ_{eb} denotes the emitter-base transit time, τ_{b} the base transit time, and τ_{sc} the space charge transit time due to the base push-out and the collector space charge regions at elevated collector-emitter voltages. Each of these, with further approximations from equivalent circuit elements [447], can be optimized individually by TCAD device simulation. If the contributions to f_{T} in (4.29) are rewritten, we can analyze not only the overall f_{T} from the h_{21} extrapolation obtained from AC simulation, but also the individual contributions in (4.29).

$$\tau_{\mathrm{c}} = (R_{\mathrm{C}} + R_{\mathrm{E}}) \cdot C_{\mathrm{bc,tot}} \tag{4.30}$$

From the detailed geometrical, velocity, and charge carrier distribution information we obtain in a TCAD simulation, further contributions in (4.29) can be analyzed. The approximations are similar to those in [447].

$$\tau_{\mathrm{eb}} = (C_{\mathrm{eb,dep}} + C_{\mathrm{bc,tot}}) \cdot \frac{k_{\mathrm{B}} \cdot T_{\mathrm{L}}}{\mathrm{q} \cdot I_{\mathrm{C}}} \tag{4.31}$$

$C_{\mathrm{eb,dep}}$ is the emitter-base depletion capacitance.

$$\tau_{\mathrm{b}} = \frac{X_{\mathrm{b}}^2}{2 \cdot D_{\mathrm{n}}} \tag{4.32}$$

X_{b} is the (effective) base width and D_{n} is the effective diffusivity in the base,

$$\tau_{\mathrm{sc}} \approx \frac{W_{\mathrm{C}}}{2 \cdot v_{\mathrm{eff}}} \tag{4.33}$$

while W_C is the collector depletion width and v_{eff} an effective carrier velocity. (4.31)-(4.33) contain a number of these effective quantities, which can be better judged by TCAD for the given bias situation than by first order (bulk) approximations. Further, the set of equations (4.29)-(4.33) stresses the importance of the optimization of the base-collector capacitance C_{jc}.
Similar to HEMTs, several approximations can be found for the cut-off frequency f_{max} in HBTs as a function of the small-signal equivalent circuit elements. The most common approximation

$$f_{max} = \sqrt{\frac{f_T}{8\pi \cdot R_{bb} \cdot C_{jc}}} \qquad (4.34)$$

demonstrates the critical influence of the intrinsic base-collector capacitance C_{jc} on f_{max}. Further, the base resistance R_{bb} needs to be reduced, typically by extremely high base doping.
With these extraction procedures and the simulation of S-parameters, highly accurate small-signal equivalent circuit elements can be extracted with a minimum of computational effort, especially with the AC solver method. Simulation results and a comparison to measurements are presented in the next two chapters.

5. Heterojunction Bipolar Transistors

5.1 General Considerations

Heterojunction Bipolar Transistors (HBTs) are an advanced development of the Bipolar Junction Transistors (BJTs). The basic principles of operation of bipolar transistors are explained in detail elsewhere, e.g. in [305, 496, 570].
Typically HBTs employ a heterojunction (a junction between two different semiconductor materials) for the emitter-base junction. Using a wider bandgap emitter material or lower bandgap base material introduces an energy barrier which prevents the injection of minority carriers in the emitter while preserving high injection of majority carriers from the emitter into the base. Thus a high current gain can also be maintained at a higher base doping concentration, which means a reduced base resistance and improved high-frequency performance [69].
Some devices, which are often referred to as Double Heterojunction Bipolar Transistors (DHBTs), employ a second heterojunction for the base-collector junction. For Si HBTs this is inevitable because of the lack of an emitter material with a wider bandgap than that of Si. Therefore a lower bandgap material (SiGe) is used in the base, while Si is used in the emitter and the collector. In III-V HBTs the second heterojunction is usually introduced on purpose to improve the breakdown performance, as demonstrated later in this chapter.
The heterojunctions can be either graded or abrupt, depending on whether a gradual change of the material composition or abrupt change of the material is introduced. The former is not a 'true' heterojunction and is governed by diffusion processes. In the latter, both the current flow and the carrier heat flow are dominated by tunneling-supported thermionic field-emission processes.
This chapter provides industrially relevant examples of AlGaAs/GaAs, InGaP/GaAs, and InAlAs/InGaAs/InP SHBTs, as well as Si/SiGe/Si and InP/InGaAs/InP DHBTs. All HBTs are fabricated as npn-structures because of the better high-frequency performance and thus wider popularity. The simulations have been performed mostly with the device simulator MINIMOS-NT indexMINIMOS-NT and have been compared to available measurements or, in some cases, to results from a commercial device simulator [227]. Benchmark structures with exact layer specifications are given in Appendix A and can be used for comparison and calibration of different device simulators.

5.2 SiGe HBTs

SiGe HBTs are currently undergoing a rapid development challenging the high-speed domain, which has been restricted to III-V-based semiconductor devices so far. The devices employ strained SiGe layers in the base to achieve better transport properties in the base and thus better performance.

5.2.1 Device Fabrication

There are various techniques for epitaxial growth of strained SiGe layers. The very first SiGe HBT [229] was produced by Molecular Beam Epitaxy (MBE). This technique allows for very accurate control of layer thicknesses down to a few angstroms. However, the most popular technique today is the cost-efficient Chemical Vapor Deposition (CVD) [50, 206, 233]. In all approaches a low-temperature epitaxial process is used to grow high-quality boron-doped strained SiGe layers. The germanium content can have a constant 'box' profile or vary with position and have a triangular or trapezoidal shape. Undesired out-diffusion of boron from the base is typically avoided by introduction of undoped spacers below and above or by addition of small contents ($< 1\%$) of carbon, resulting in SiGe:C (SiGeC) layers [134]. Quite often polysilicon emitters are used to suppress hole injection, to improve the common-emitter current gain, or to preserve emitter efficiency, while scaling lateral geometries and junction depths [22]. A more detailed discussion on fabrication elements common for all industrial technologies can be found in [106].

5.2.2 Process Simulation

It is often favorable to use process simulation to achieve accurate device structures and realistic doping profiles. Commercial software products, such as TSUPREM [504] or DIOS [227], are capable of handling silicon-based materials, including strained SiGe.

As demonstrated in an example later in this section, such simulation can allow analysis of a variation of process parameters, such as energy and dose for ion implantation. The devices under investigation are polysilicon-emitter double-base SiGe HBTs epitaxially grown by a CVD process. An implanted n-well is used, similar to the one used in the standard CMOS technology. The buried layer is connected via sinker to the collector contact. The germanium content has a triangular shape. The base-emitter junction is formed by Rapid Thermal Processing, which causes out-diffusion of arsenic from the polysilicon emitter layer into the crystalline silicon.

The process simulation starts from the blank wafer and ends with the final device. It reflects real device fabrication as accurately as possible. The implant profiles as well as annealing steps are calibrated to one-dimensional SIMS profiles. To save computational resources the simulation domain covers only one half of the real device, which is symmetric, and the collector-sinker is not included in the structure.

5.2.3 Device Simulation and Calibration to Measurements

Applying a global calibration scheme, good agreement can be achieved both with experimental DC results (forward and output characteristics) and with high-frequency data. It is possible to extract various sets of small-signal equivalent circuit parameters as well as related figures of merit by means of simulation with MINIMOS-NT.

Large Emitter Area Devices for Surface-Independent Calibration. Submicron HBTs often suffer from surface-related effects. To obtain a model calibration independently of such effects, five large area 400 × 400 μm^2 $Si_{1-x}Ge_x$ HBTs have been fabricated with the same geometry and layer specification as shown in Fig. 5.1. The devices differ only in the Ge content x in the base, which is constant for a given device and varies between 16% and 28%. The following layer sequence applies for all devices: The emitter consists of an antimony-doped $n^+ = 2 \times 10^{20}$ cm^{-3} contact region and an $n = 10^{18}$ cm^{-3} region. The $Si_{1-x}Ge_x$ base is boron-doped ($p^+ = 5 \times 10^{19}$ cm^{-3}) and has thin undoped spacers on its top and bottom. The collector is antimony-doped ($n^- = 2 \times 10^{16}$ cm^{-3}) and is followed by an intrinsic buffer and the substrate.
The device with the lowest Ge content in the base, $x = 16\%$, is analyzed first. Fig. 5.2 presents results of several consecutive simulations (lines) compared to measured data at $V_{BC} = 0$ V (symbols). They are presented as an example of calibration of forward Gummel plot characteristics. The first simulation is performed with the default models and the device is defined according to the specification (dashed lines). Thus important effects such as BGN and SRH recombination are still not taken into account. Note the significant disagreement between simulation and measurement. Comparatively good agreement for I_C is observed when the BGN is accounted for in the simulation (dot-dashed lines).
As stated in [408], the spacers are used to prevent out-diffusion of boron from the base to the emitter in the following thermal processing. In the final device, the SiGe spacers contain boron and are parts of the real base layer. In the next simulation, gradual boron doping profiles varying from 5×10^{13} cm^{-3} at the spacer/emitter and spacer/collector interfaces to 5×10^{19} cm^{-3} at the base/spacer interfaces are included. The only remaining step is to include SRH recombination in order to match the base current. Finally good agreement for the complete bias range is achieved (solid lines).
The next simulations are performed with the remaining four devices, which have higher germanium content in the base. Fig. 5.3 compares simulated and measured Gummel plots for the devices with $x = 16\%$, 22%, and 28% Ge. All simulations

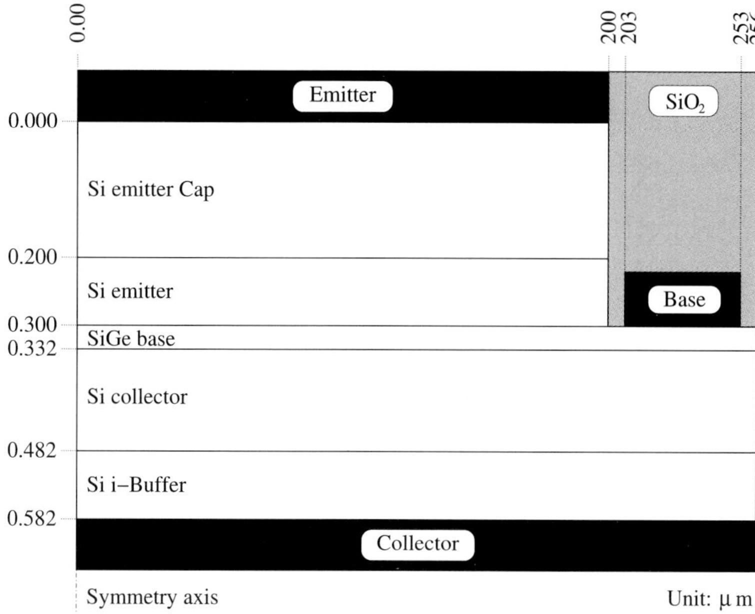

Fig. 5.1. Layer structure of the five simulated SiGe HBTs.

are performed with the same set of models without adjusting the model parameters. The only exception is made for the concentration of traps in the base, because with the higher Ge content more traps are also introduced. Thus good agreement is finally achieved.

A Test-HBT with a 0.4 μm Emitter Length. The Monte Carlo simulation results and the SiGe HBT from [359] have been employed for the calibration of models used in hydrodynamic simulations. Fig. 5.4 shows the device structure. The emitter consists of an $n^+ = 10^{19}$ cm^{-3} contact region and an $n = 5 \times 10^{18}$ cm^{-3} region. The 20 nm $Si_{0.8}Ge_{0.2}$ base ($p^+ = 10^{19}$ cm^{-3}) is followed by the collector ($n = 2 \times 10^{17}$ cm^{-3}) and the subcollector ($n^+ = 10^{19}$ cm^{-3}) layers. The Monte Carlo simulation results have been compared to drift-diffusion and hydrodynamic simulation results obtained with MINIMOS-NT for the electron concentration, the electrostatic potential, the electron velocity, and the electron temperature distribution in the device (see Fig. 5.5). Such comparisons are useful not only to verify the validity of the models for SiGe and the hydrodynamic transport model, but also to analyze why and where the hydrodynamic model should be used, keeping in mind the higher computational effort. Mixed-mode device/circuit simulations of a five-stage Current-Mode Logic (CML) ring oscillator containing ten SiGe HBTs (Fig. 5.6) have been presented in [185]. Simulations using the drift-diffusion transport model deliver lower inverter delay times compared to simulations with the hydrodynamic model (see Fig. 5.7). This is mainly

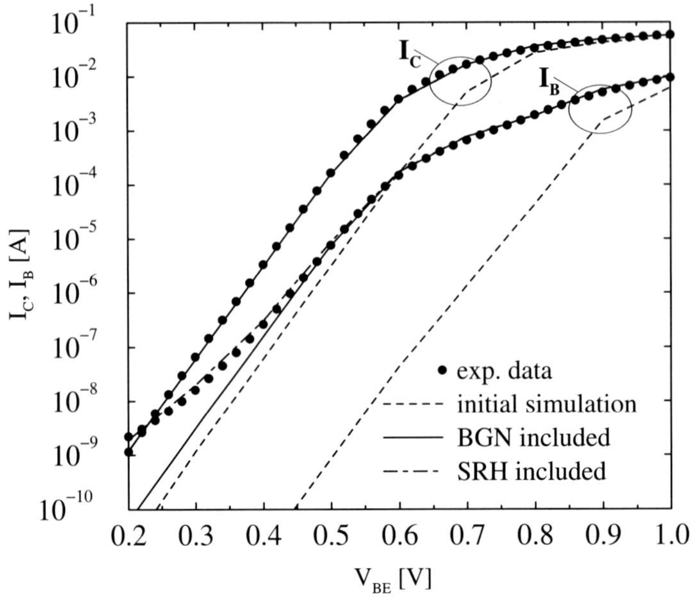

Fig. 5.2. Forward Gummel plots at $V_{\mathrm{BC}} = 0$ V: Study of different effects in a $\mathrm{Si}_{0.84}\mathrm{Ge}_{0.16}$ HBT.

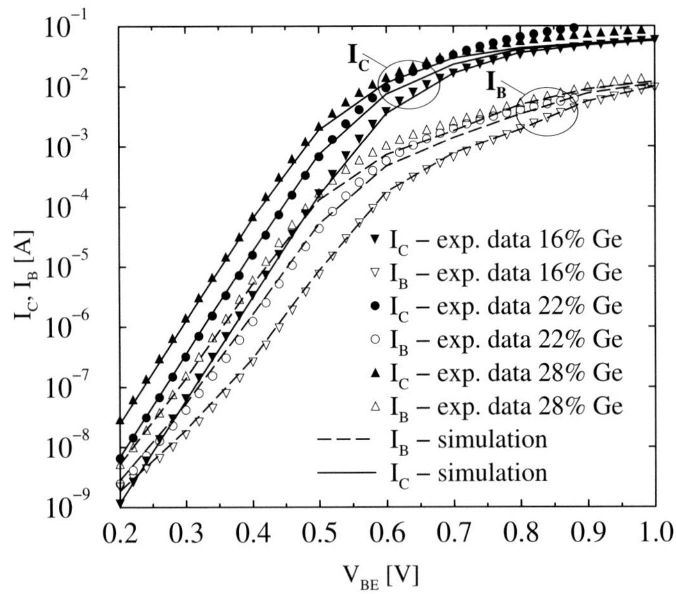

Fig. 5.3. Forward Gummel plots at $V_{\mathrm{BC}} = 0$ V: Comparison of measurement and simulation for different material compositions in $\mathrm{Si}_{1-x}\mathrm{Ge}_x$.

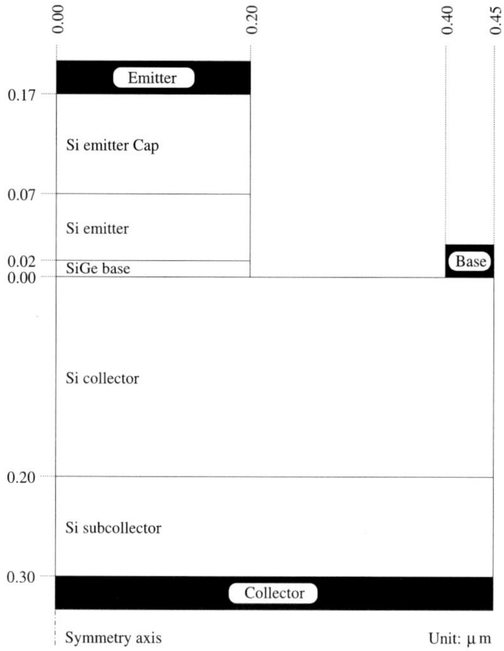

Fig. 5.4. Layer structure of a 0.4 μm SiGe HBT.

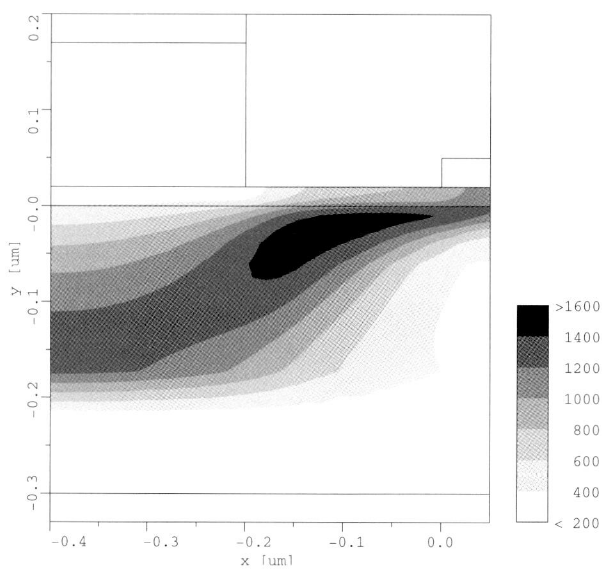

Fig. 5.5. Simulated electron temperature distribution (given in [K]) in the SiGe HBT at $V_{BE} = 0.87$ V and $V_{CE} = 1$ V.

due to the velocity overshoot in the base-collector space charge region which cannot be modeled using drift-diffusion simulations [359]. Furthermore, since hydrodynamic simulations deliver higher currents, the overall speed of the circuit is higher. The error of the drift-diffusion simulation in the inverter delay is in the range of 20% compared to the hydrodynamic simulation, thus proving the necessity of a hydrodynamic model.

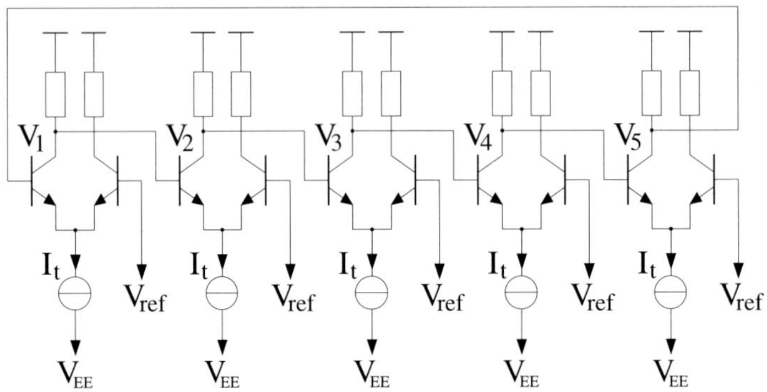

Fig. 5.6. CML ring oscillator circuit.

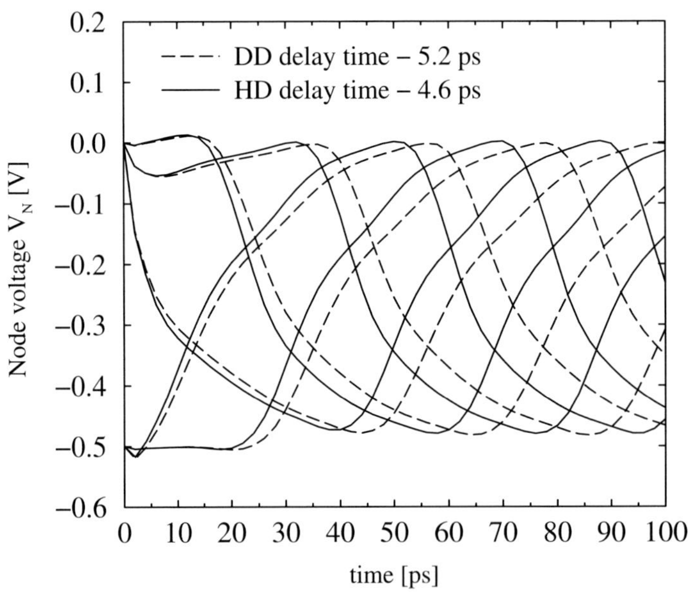

Fig. 5.7. Comparison of the simulated drift-diffusion and hydrodynamic transient response of the CML ring oscillator.

Analyzed Industrial SiGe HBT Structures. SiGe HBTs from an industrial vendor are analyzed in this section. The devices are part of proven 0.8 μm and 0.35 μm BiCMOS technologies which include a CMOS process and a high-performance analog-oriented HBT module [242]. The applications reach from circuits for mobile communication to high-speed networks.
The methodology for characterization and optimization of SiGe HBTs involves process calibration, device calibration employing two-dimensional device simulation, and automated Technology Computer Aided Design (TCAD) optimization.

SiGe HBT from the 0.8 μm Technology Node. The influence of the selectively-implanted-collector (SIC) doping on device performance has been studied in order to obtain an optimized doping profile for specific requirements (high speed or high breakdown voltage). For that purpose, four SiGe HBT structures with emitter areas of $6 \times 0.8\ \mu\text{m}^2$ have been investigated both experimentally and by means of process simulation, followed by two-dimensional device simulation. The simulated device structure with the phosphorus SIC implant is shown in Fig. 5.8. This allows the use of a very precise simulation grid in areas of interest (Fig. 5.9).
The only process step in which the four HBTs (hereafter referred to as Dev. 1, Dev. 2, Dev. 3, Dev. 4) differ is the combination of energy and dose used for the SIC implants, as summarized in Table 5.1. The resulting phosphorus doping profiles in vertical cuts under the emitter windows of the four devices are shown in Fig. 5.10.

Table 5.1. Summary of key SiGe HBT process and device parameters

Device	Energy [keV]	Dose [cm^{-2}]	f_T [GHz]	BV_CE0 [V]	$f_\text{T} \times BV_\text{CE0}$ [GHz·V]
Dev.1	480	7e12	32	4.0	128
Dev.2	480	3e13	40	3.7	148
Dev.3	300	7e12	33	3.1	102
Dev.4	300	3e13	42	2.3	97

A comparative Monte Carlo simulation of ion implantation [219] of phosphorus in silicon and SiGe has been performed to examine the accuracy of the process simulation with respect to SiGe (see Fig. 5.11).
A comparative device drift-diffusion simulation with MINIMOS-NT and DESSIS has been performed to cross-check the accuracy of the physical models. Both device simulators correctly reproduce the measured forward Gummel plot at 300 K (see Fig. 5.12) with default models. The slight increase of collector current I_C with dose and energy at high bias is due to the differences in the base push-out effect. f_T is extracted by small-signal AC analysis.
The only fitting parameters used in the simulation are the contribution of bandgap narrowing to the conduction band (here about 80% and 20% for donor

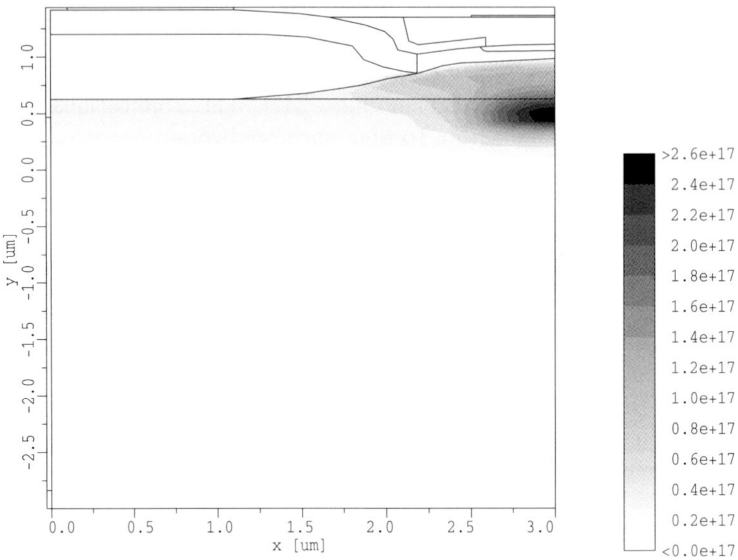

Fig. 5.8. Simulated device structure and phosphorus collector implant [cm^{-3}] in the 0.8 μm SiGe HBT.

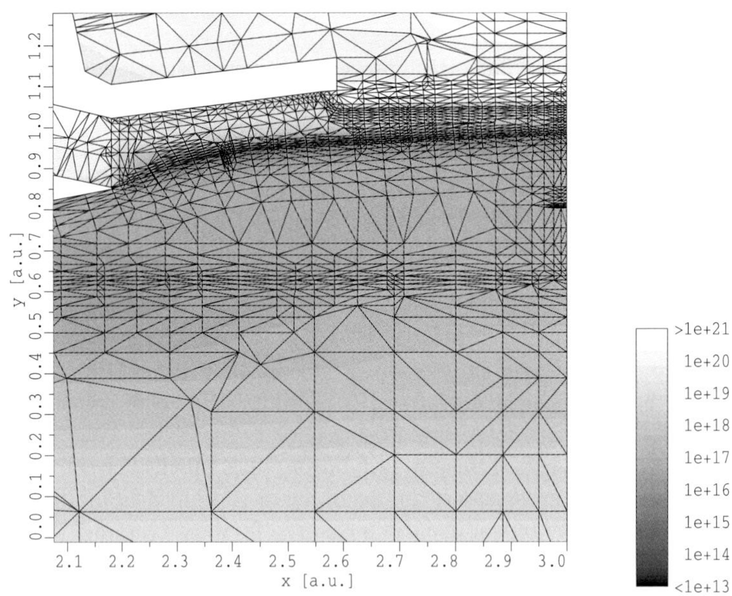

Fig. 5.9. Gridding example from the active device area.

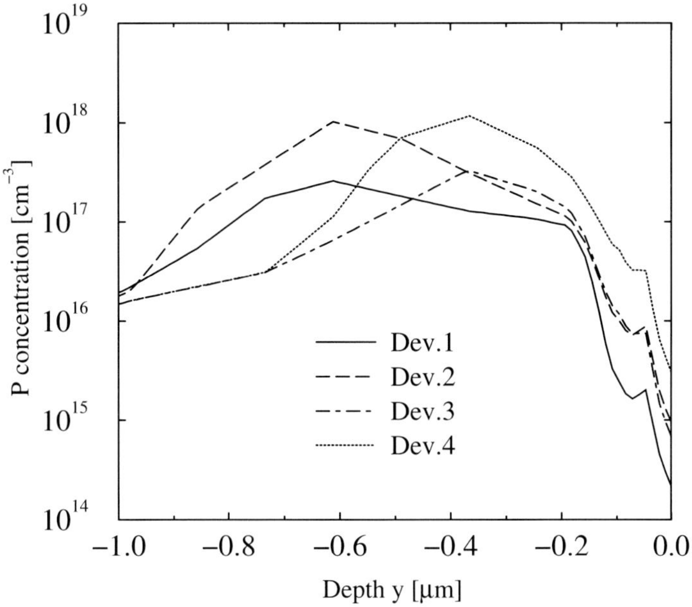

Fig. 5.10. Phosphorus doping profiles under the emitter contact in four devices.

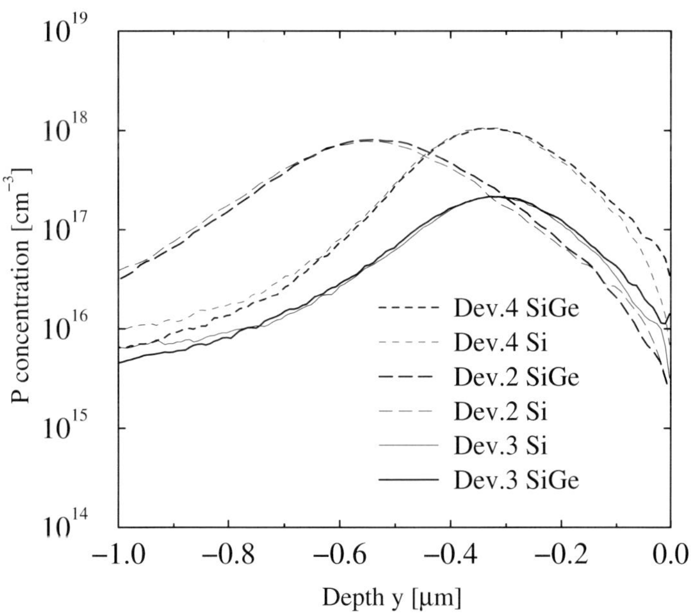

Fig. 5.11. Comparative Monte Carlo simulation of ion implantation of phosphorus in Si and SiGe.

and acceptor doping respectively), and the concentration of traps in the Shockley-Read-Hall recombination model (here 10^{13} cm^{-3}).
However, as can be seen in Fig. 5.13 and Fig. 5.14, both DESSIS and MINIMOS-NT failed to explain the experimentally observed similarity in peak f_T between Dev. 1 and Dev. 3 and between Dev. 2 and Dev. 4.
An automated device calibration of the SIC implant has been performed within a TCAD framework [484]. It turned out that a 50% increase in phosphorus in the collector of the two low-dose devices (Dev. 1 and Dev. 3) gives an acceptable qualitative agreement using the drift-diffusion simulation.
It is known that with shrinking device dimensions non-local effects, such as velocity overshoot, become more pronounced. Neglecting these effects can be a reason for underestimating f_T [247]. To resolve the disagreement of the drift-diffusion simulations and the f_T measurements, simulations with the hydrodynamic transport model are performed, which improve the results quantitatively (see Fig. 5.15).
Fig. 5.16 shows the electron velocity overshoot over the greater part of the base region. The simulated velocity in the hydrodynamic simulation is about twice the saturation velocity limit in the drift-diffusion case (10^7 cm/s). This correlates to the higher electron energy in the collector (see Fig. 5.17) and explains the increase of f_T in comparison to drift-diffusion simulations (see Fig. 5.13 and Fig. 5.14). The good agreement of f_T at low currents is very important, since HBTs typically operate at much lower frequencies than at the maximum f_T. Simulations prove that in this range optimizations of the SIC implant do not have an influence on f_T, i.e. the base-emitter capacitance but not the base-collector capacitance is dominating. The maximum f_T has been found to have a stronger dependence on the dose than on the energy of the implants.
Furthermore, the important figure-of-merit $f_T \times BV_{CE0}$ (see Table 3.6) reaches a maximum for high SIC implant energies (deep implant) and high SIC doses. We found that the higher f_T for high-dose/low-energy SIC implants is due to a smaller base width and a delayed onset of the base push-out effect due to the higher collector doping.

SiGe HBT from the 0.35 μm Technology Node. The investigated SiGe HBTs from the next generation have emitter areas of 12×0.4 μm^2. The device structure with the phosphorus SIC implant is shown in Fig. 5.18.
The device structure with the phosphorus SIC implant is shown in Fig. 5.18 as resulting from process simulation with DIOS.
MINIMOS-NT is employed for two-dimensional device simulation. The following important physical effects are properly modeled (see Chapter 3) and accounted for in the simulation in order to get good agreement with measured forward (Fig. 5.19) and output (Fig. 5.20) characteristics using the concise set of models and parameters: stress-induced bandgap reduction due to Ge content; doping-dependent bandgap narrowing; anisotropic electron minority mobility in strained

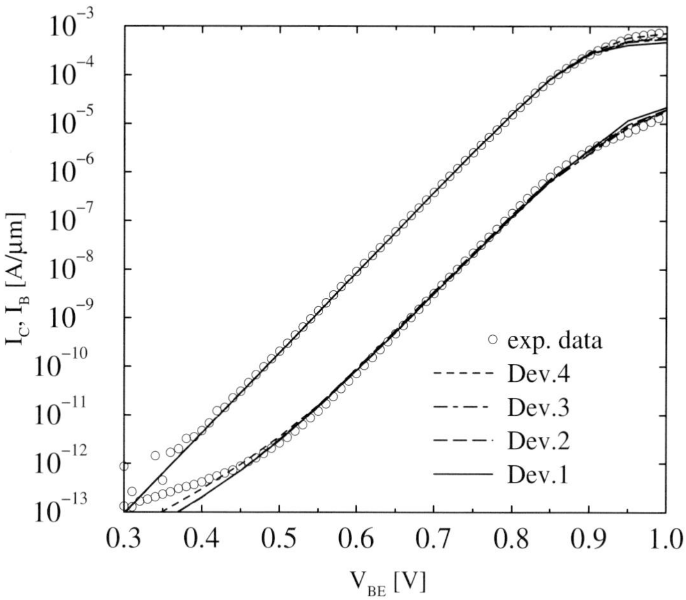

Fig. 5.12. Forward Gummel plots at $V_{\mathrm{BC}} = 0$ V: Comparison of measurement and simulation.

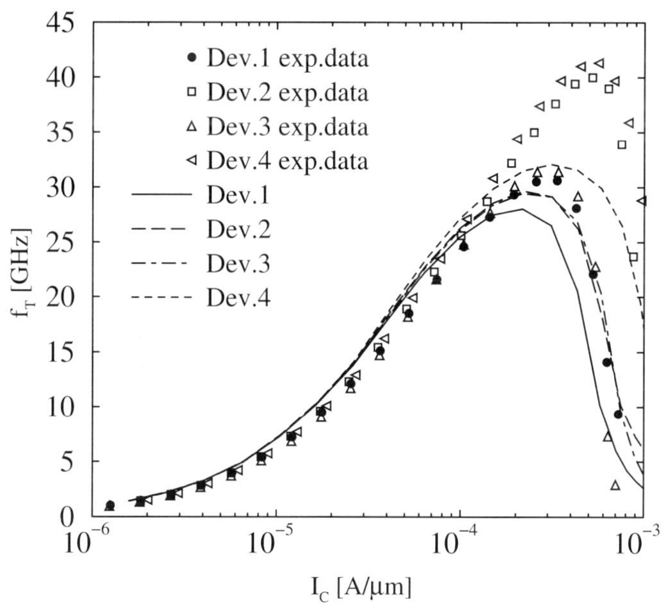

Fig. 5.13. Current gain cut-off frequency f_{T} as a function of I_{C} at $V_{\mathrm{CE}} = 1.5$ V: Comparison of measurement and drift-diffusion simulation with the device simulator DESSIS.

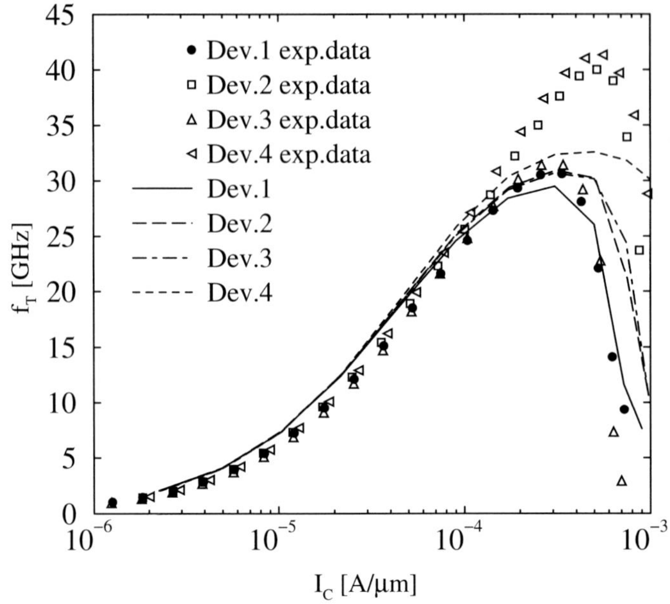

Fig. 5.14. Current gain cut-off frequency f_T as a function of I_C at $V_{CE} = 1.5$ V: Comparison of measurement and drift-diffusion simulation with the device simulator MINIMOS-NT.

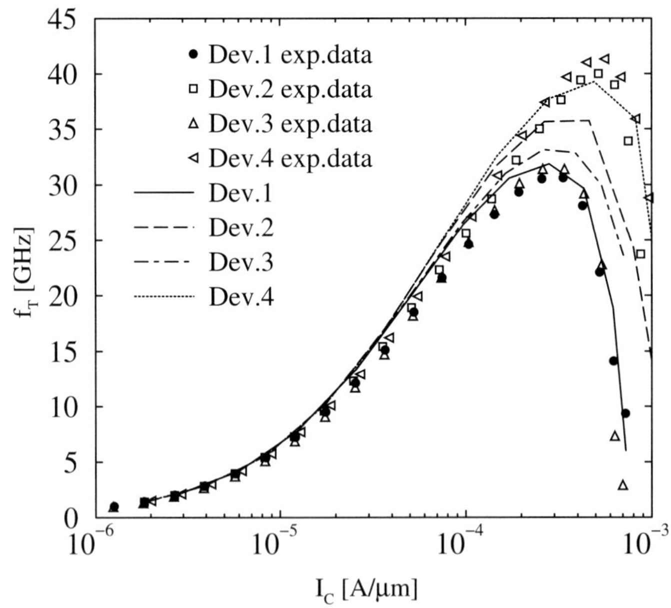

Fig. 5.15. Current gain cut-off frequency f_T as a function of I_C at $V_{CE} = 1.5$ V: Comparison of measurement and hydrodynamic simulation with MINIMOS-NT.

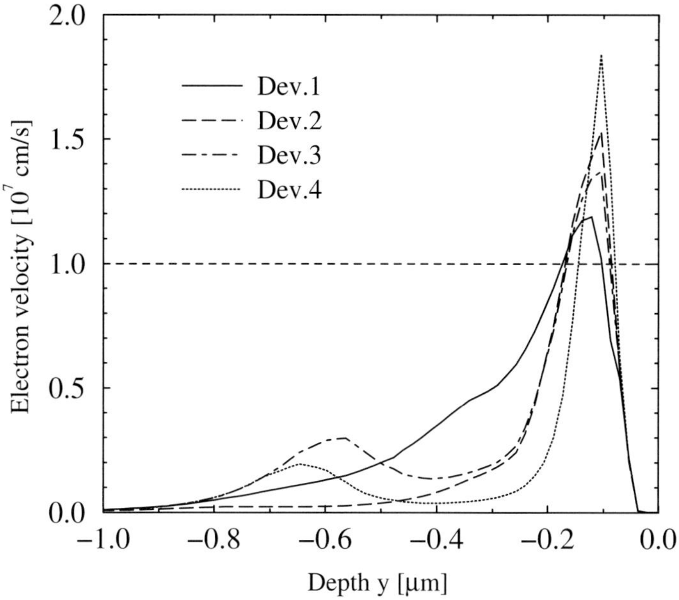

Fig. 5.16. Simulated electron velocity overshoot in the base-collector space charge region at $V_{\mathrm{CE}} = V_{\mathrm{BE}} = 0.88$ V.

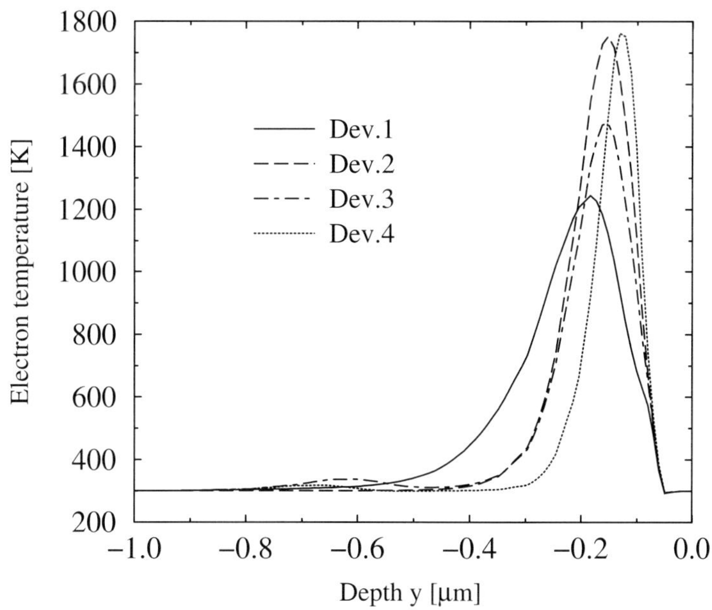

Fig. 5.17. Simulated electron temperature distribution in the four devices at $V_{\mathrm{CE}} = V_{\mathrm{BE}} = 0.88$ V.

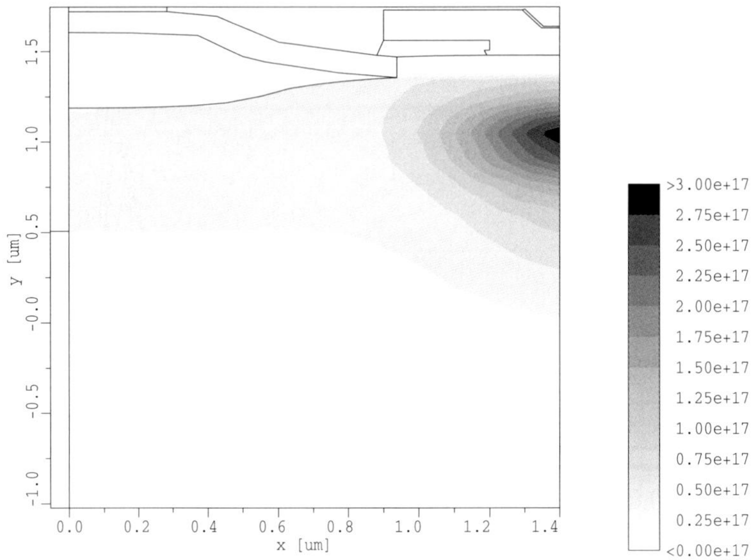

Fig. 5.18. Simulated device structure and phosphorus collector implant [cm^{-3}] in the 0.4 μm SiGe HBT.

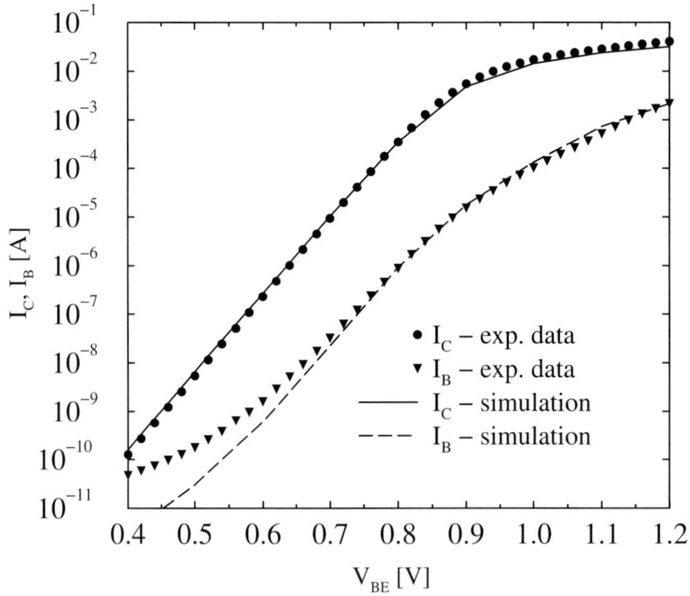

Fig. 5.19. Forward Gummel plots at $V_{\mathrm{BC}} = 0$ V: Comparison of measurement and simulation at 300 K.

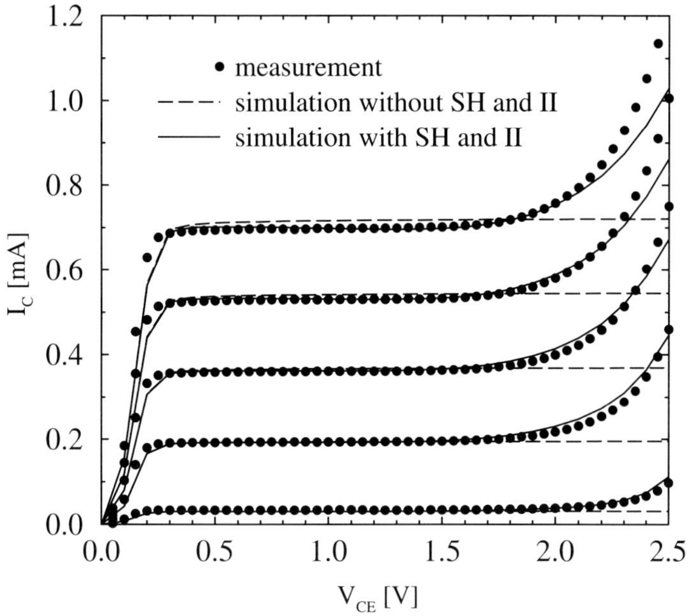

Fig. 5.20. Simulated output characteristics with and without self-heating (SH) and impact ionization (II) and comparison to measurements. I_B is stepped by 0.4 μA from 0.1 μA to 1.7 μA.

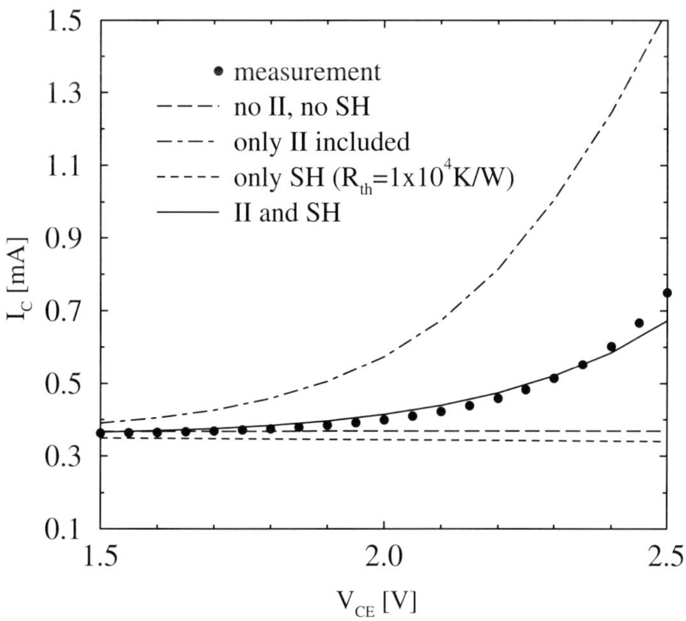

Fig. 5.21. Output characteristics at $I_B = 0.9$ μA: The increasing I_C at high V_{CE} reveals the interplay between self-heating (SH) effect and impact ionization (II) generation.

SiGe; Shockley-Read-Hall, Auger, and surface recombination; impact ionization (II) generation; and self-heating (SH).
The only fitting parameters used in this simulation are the contributions of bandgap narrowing to the conduction band, the trap charge density in the Shockley-Read-Hall recombination model (here 10^{14} cm^{-3}), the velocity recombination for holes in the polysilicon contact model [569] used at the emitter contact, and the substrate thermal resistance.
A closer look at the increasing collector current I_C at high collector-to-emitter voltages V_{CE} and constant base current I_B, stepped by 0.4 μA from 0.1 μA to 1.7 μA, reveals the interplay between self-heating and impact ionization (see Fig. 5.21). While impact ionization leads to a strong increase of I_C, self-heating decreases it. In fact, both I_C and I_B increase due to self-heating at a given bias condition. As the change is higher for I_B, in order to maintain I_B at the same level, V_{BE} and, therefore, I_C decrease.

5.2.4 RF Simulations

A proper DC calibration is an important prerequisite for AC simulation. Note that it is necessary for AC simulations to take the complete device structure into account, in order to consider the capacitances between collector and substrate C_{CS}, as well as between base and collector C_{BC}.
The accuracy of the simulated (intrinsic) Y-parameters is checked by calculating the row and column sums of the admittance matrix, which have to be zero according to Kirchhoff's laws. The simulation yields errors of about 10^{-16} A/V for typical matrix entries of 10^{-3} A/V. The transformation from Y-parameters to intrinsic S-parameters is completely analytical. In this example, the pad parasitics are accounted for in the measurement environment. Thus, measured and simulated intrinsic S-parameters can be directly compared without transformation to extrinsic parameters.
Fig. 5.22 and Fig. 5.23 compare simulated and measured S-parameters at V_{CE} = 1 V and current densities J_C = 28 kA/cm^2 and J_C = 76 kA/cm^2 in the frequency range between 50 MHz and 31 GHz. For the same device we calculated the matched gain and the short-circuit current gain h_{21} in order to extract the figures-of-merit f_T and f_{max} found at the respective unity-gain intersections. Matched gain in this case means the unilateral gain, assuming a perfect match at the output. Fig. 5.24 and Fig. 5.25 show the comparison of simulated results and the corresponding measurement data. While the measurement covers a range up to 31 GHz, the simulation is extended to frequencies beyond the unity-gain point. The peak value of the f_T-curve in Fig. 5.24 corresponds exactly to the frequency of the respective intersection in Fig. 5.25. The effect of the introduction of anisotropic electron mobility is demonstrated in Fig. 5.24. In addition, results obtained by a commercial device simulator (DESSIS [227]), using default models and parameters, are included for comparison.

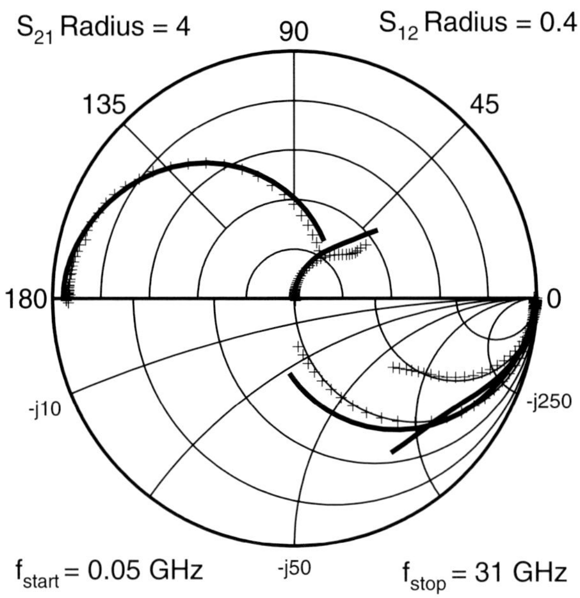

Fig. 5.22. Simulated (–) and measured (+) S-parameters in a Smith-polar chart from 50 MHz to 31 GHz at $V_{\mathrm{CE}} = 1$ V and $\mathrm{J_C} = 28$ kA/cm^2.

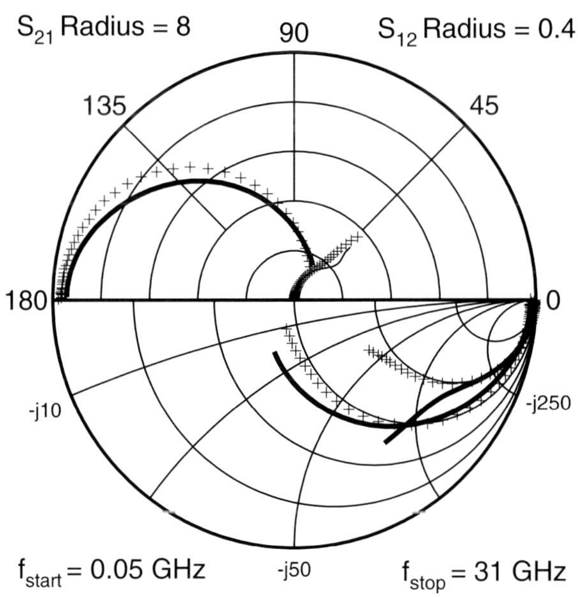

Fig. 5.23. Simulated (–) and measured (+) S-parameters in a Smith-polar chart from 50 MHz to 31 GHz at $V_{\mathrm{CE}} = 1$ V and $\mathrm{J_C} = 76$ kA/cm^2.

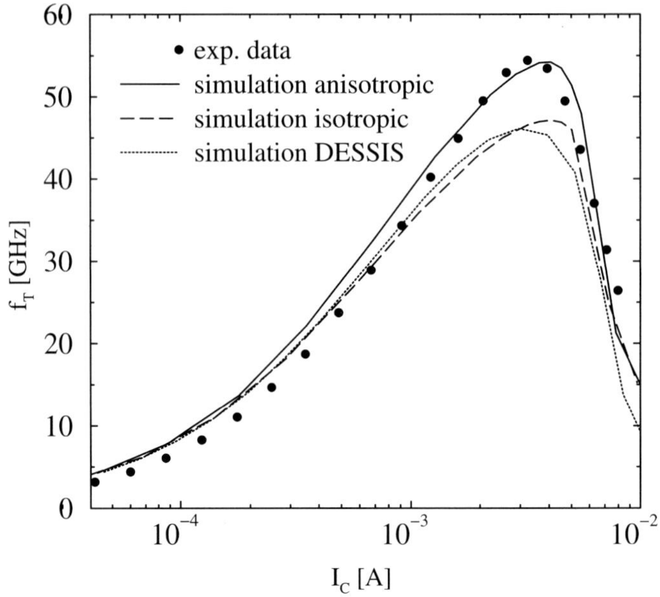

Fig. 5.24. Simulated (anisotropic: solid line; isotropic: dashed line) and measured (circles) current gain cut-off frequency f_{T} as a function of collector current I_{C} at $V_{\mathrm{CE}} = 1$ V.

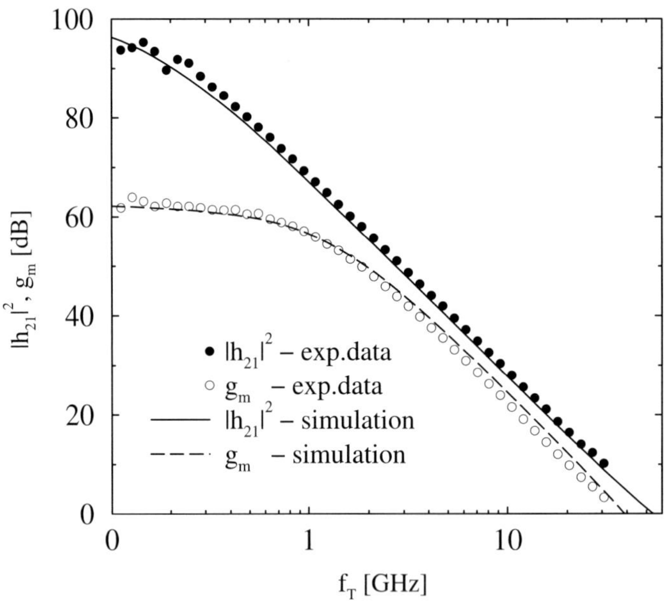

Fig. 5.25. Simulated and measured short-circuit current gain h_{21} and matched gain as a function of frequency at $V_{\mathrm{CE}} = 1$ V and $\mathrm{J_C} = 76$ kA/cm^2.

5.3 High-Power GaAs HBTs

Results of two- and three-dimensional simulations of Heterojunction Bipolar Transistors (HBTs) on GaAs substrate are presented in this section. An overview of the physical models used and comparisons with experimental data are given. In particular, the electrical behavior of AlGaAs/GaAs and InGaP/GaAs one-finger high-power HBTs with emitter areas of $3 \times 30\ \mu\text{m}^2$ is studied at several ambient temperatures. Typical operation temperatures for the devices under consideration range from 300 K up to 380 K [446].

5.3.1 Device Fabrication

All HBT structures mentioned here are epitaxially grown by MOCVD on GaAs substrates. In the first device, later on referred to as Dev. 1, the emitter cap is formed by an n^+-InGaAs/n-GaAs layer. The emitter consists of 80 nm silicon-doped n-$Al_{0.28}Ga_{0.72}As$ with 20 nm graded layers on its top and bottom. The 90 nm GaAs base is carbon-doped ($p^+ = 3 \times 10^{19}$ cm^{-3}). It is followed by a 700 nm GaAs collector ($n^- = 2 \times 10^{16}$ cm^{-3}) and a 700 nm GaAs subcollector ($n^+ = 5 \times 10^{18}$ cm^{-3}). The simulated device structure is shown in Fig. 5.26. In two devices (Dev. 2 and Dev. 3), the graded AlGaAs layer next to the base-to-emitter junction is replaced by a 20 nm InGaP ledge layer ($n = 4 \times 10^{17}$ cm^{-3} and 3×10^{17} cm^{-3}, respectively). Both devices differ in the base carbon doping concentration ($p^+ = 3 \times 10^{19}$ cm^{-3} vs. $p^+ = 4 \times 10^{19}$ cm^{-3}) and layer thickness (80 nm vs. 120 nm). A 40 nm InGaP emitter ($n = 4 \times 10^{19}$ cm^{-3}) HBT with 120 nm base thickness is the last device type (Dev. 4) considered. The simulated InGaP HBT structure is shown in Fig. 5.27. All HBTs are processed by etching a double mesa structure and passivated by Si_3N_4.

5.3.2 DC Simulation Results

One concise set of model parameters is used in all simulations. The extrinsic parasitics, such as interconnect resistances, are accounted for in the simulation by adding lumped resistances of 1 Ω to all electric contacts.

Forward and Reverse Gummel Plots. Fig. 5.28 compares the simulated and measured forward Gummel plots for Dev. 1 at 293 K. The agreement at moderate and high voltages is good, which is typical for operation of this kind of devices. The measured leakage currents at $V_{BE} < 1$ V cannot be reproduced by simulation, despite the fact that several generation/recombination mechanisms, such as SRH recombination, surface recombination, and band-to-band tunneling are taken into account.
Encouraging results are also achieved for Dev. 2. Fig. 5.29 includes the simulated Gummel plot at 373 K to demonstrate the ability of MINIMOS-NT to reproduce the thermal device behavior correctly. Fig. 5.30 and Fig. 5.31 show comparisons

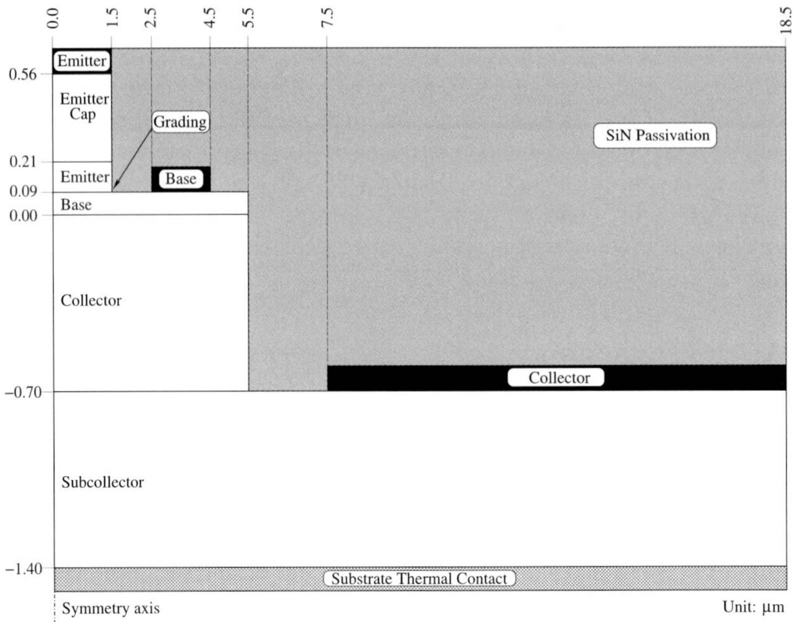

Fig. 5.26. Simulated device structure of an AlGaAs/GaAs HBT.

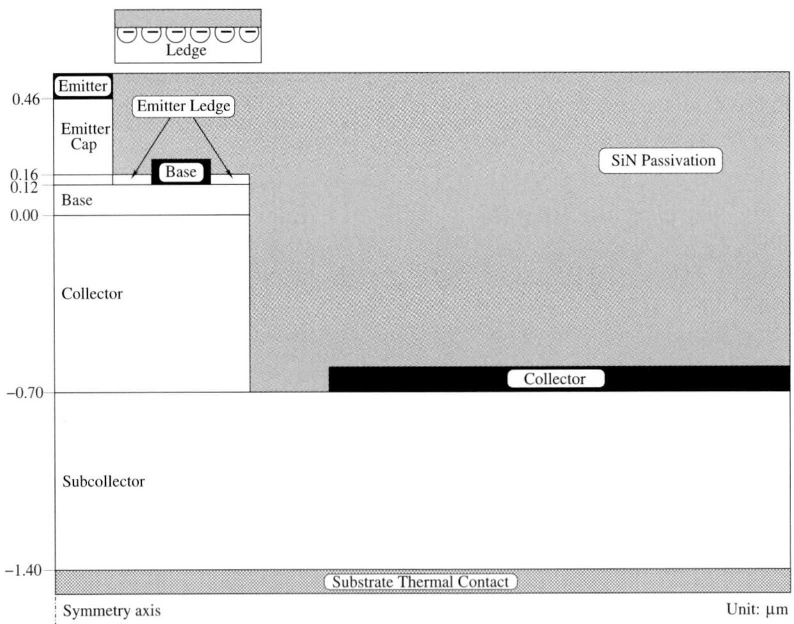

Fig. 5.27. Simulated device structure of an InGaP/GaAs HBT.

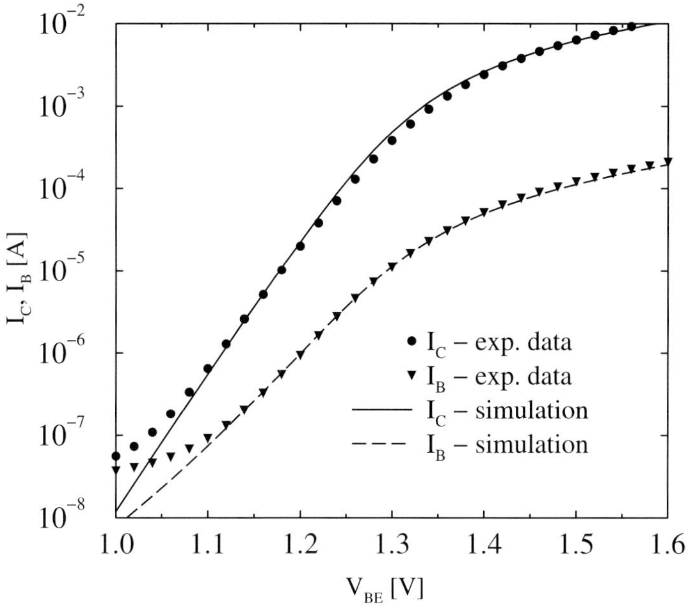

Fig. 5.28. Forward Gummel plots at $V_{BC} = 0$ V for Dev. 1: Comparison of simulation and measurement at 293 K.

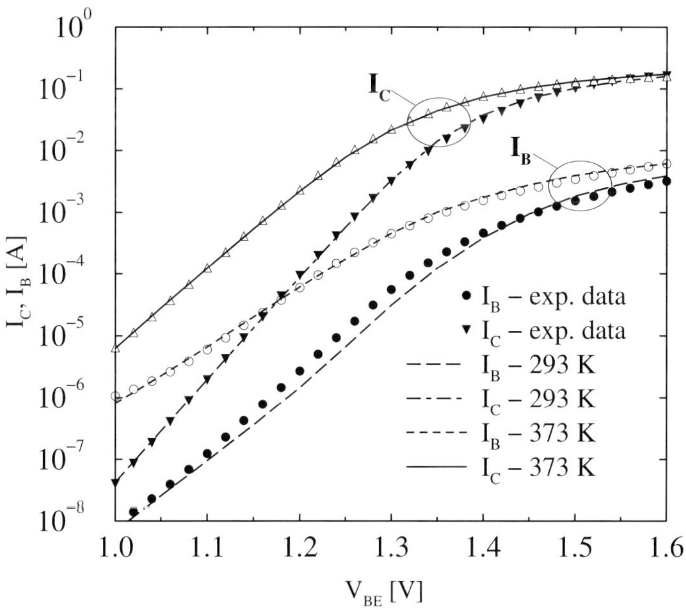

Fig. 5.29. Forward Gummel plots at $V_{BC} = 0$ V for Dev. 2: Comparison of simulation and measurement at 293 K and 373 K.

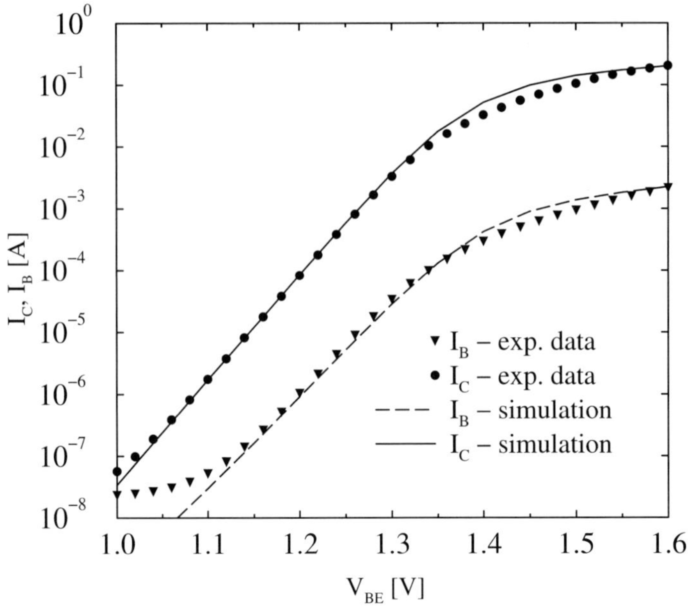

Fig. 5.30. Forward Gummel plots at $V_{\mathrm{BC}} = 0$ V for Dev. 4: Comparison of simulation and measurement at 293 K.

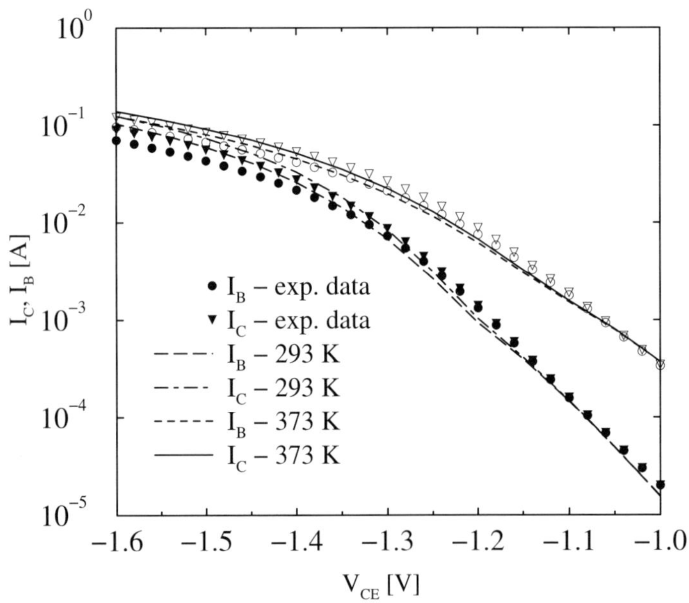

Fig. 5.31. Reverse Gummel plots at $V_{\mathrm{BE}} = 0$ V for Dev. 4: Comparison of simulation and measurement at 293 K and 373 K.

with measured data for Dev. 4. Note that very good agreement for the reverse Gummel plots is also achieved at higher temperatures. These results verify the temperature dependence of the physical models and therefore are a good prerequisite for simulation of self-heating effects.

Simulation with Self-Heating. In order to account for self-heating effects, the lattice heat flow equation (3.18) is solved self-consistently with the energy transport equations (3.6) and (3.7). An additional substrate thermal contact is introduced. The thermal heat flow density is calculated by (3.39) using a thermal contact resistance R_{TH} with a measured value of about 400 K/W [61, 350]. The thermal dissipation through the emitter and base contacts is neglected, therefore a Neumann boundary condition is assumed. This is a worst case assumption, as the emitter contact is often used for cooling the device through air-bridges.

Simulation of Output Characteristics. Achieving realistic results in simulation of HBT output characteristics is a difficult problem. This is especially true for the power devices considered in this work. As stated in [305], power dissipation increases with collector-to-emitter voltage V_{CE}, gradually elevating the junction temperature above the ambient temperature.
This leads to gradually decreasing collector currents I_{C} for constant base currents I_{B} applied or gradually increasing I_{C} for constant base-to-emitter voltages V_{BE} applied. For example, Fig. 5.32 compares simulated and measured output device characteristics for Dev. 4 with constant I_{B}, stepped by 0.1 mA from 0.1 mA to 0.5 mA. Fig. 5.33 compares simulated output device characteristics to measurements for constant $V_{\mathrm{BE}} = 1.4$ V to 1.45 V using a 0.01 V step. A good agreement is achieved by simulation including self-heating effects. Fig. 5.34 presents the peak lattice temperature $T_{\mathrm{L,max}}$ in the device as a function of V_{CE}. For Dev. 3, the temperature reaches as much as 400 K for the specified thermal resistance. As already stated in [384], such lattice temperatures significantly change the material properties of the device and, consequently, its electrical characteristics. This confirms the necessity of having exact material parameters for the full temperature range before including self-heating effects.
Hydrodynamic simulation is needed to account for non-local effects. The electron temperature (see Fig. 5.35) is used as an input to the hydrodynamic mobility model and therefore allows simulation of the electron velocity overshoot in the collector space charge region. The carrier temperature also influences the current flux across the heterointerfaces: the higher the carrier temperature, the more carriers are able to surmount the energy barrier. Considering the nature of the simulated devices (including graded and abrupt heterojunctions) and the high electron temperatures observed at maximum bias (above 2500 K - see Fig. 5.35), a thermionic-field emission interface model (3.71)–(3.74) is used in conjunction with the hydrodynamic transport model.
The resulting lattice temperature distribution in the device at $V_{\mathrm{CE}} = 6.0$ V and $V_{\mathrm{BE}} = 1.45$ V is shown in Fig. 5.36. The heat generated at the heterojunctions flows out of the device in the direction of the substrate heat sink. In the opposite

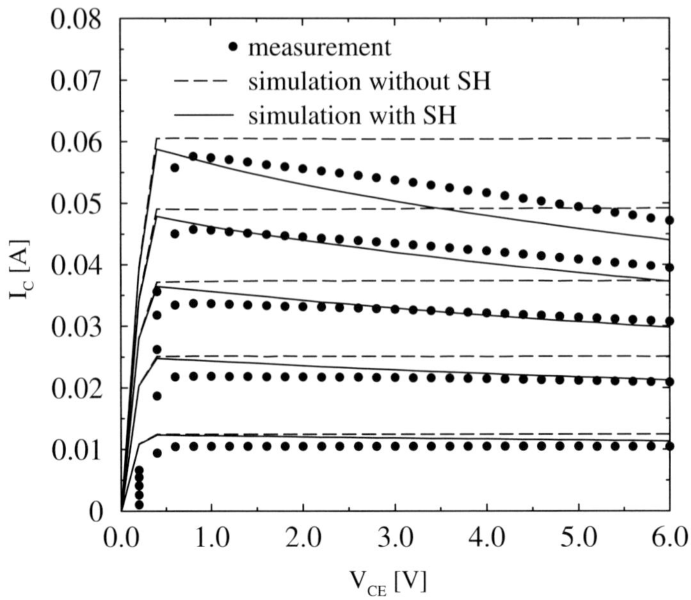

Fig. 5.32. Output characteristics for Dev. 4: Simulation with and without self-heating compared to measurements at constant $I_B = 0.1 - 0.5$ mA.

direction the heat cannot leave the device and therefore the emitter finger is heated significantly up to 400 K. The simulation shown is of practical interest. It demonstrates the necessity of using both a thermal shunt (e.g. an air-bridge) at the emitter contact and of reducing the substrate thickness as much as possible in order to remove the heat.

5.3.3 S-Parameter Simulation

After having a correct description of the DC device behavior including thermal aspects, small-signal analysis is needed to connect DC and RF device operation. The following subsections compare measured S-parameters to results from transient or AC simulations.

As discussed in Chapter 4, there are several approaches to computing bias-dependent S-parameters, e.g. [16, 416], applying quasi-static or equivalent-circuit parameter models. These approaches employ transformations in the time domain to extract S-parameters. These methods are both more CPU-time consuming (the steady-state has to be reached for each bias and frequency) and more inaccurate (only a limited number of time-steps in reasonable CPU-time, equivalent-circuit approximation, as discussed in Chapter 4) compared to sinusoidal steady-state analysis (S^3A) [291].

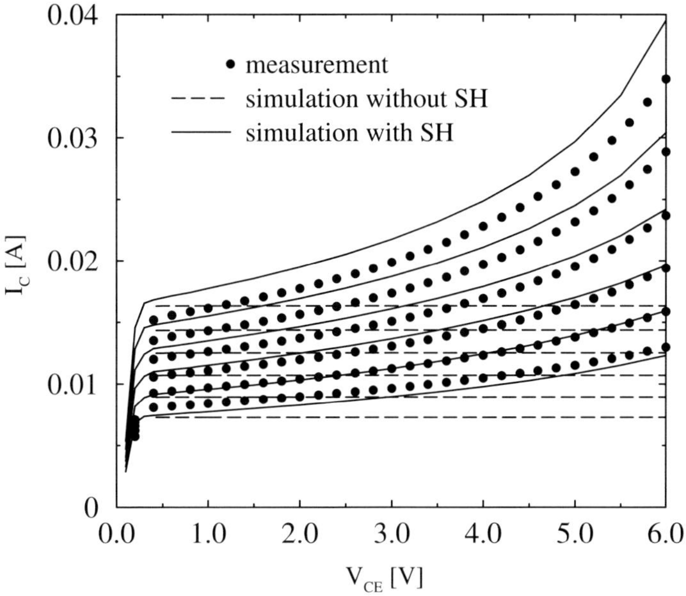

Fig. 5.33. Output characteristics for Dev. 3: Simulation with and without self-heating compared to measurements for constant $V_{\rm BE} = 1.4 - 1.45$ V.

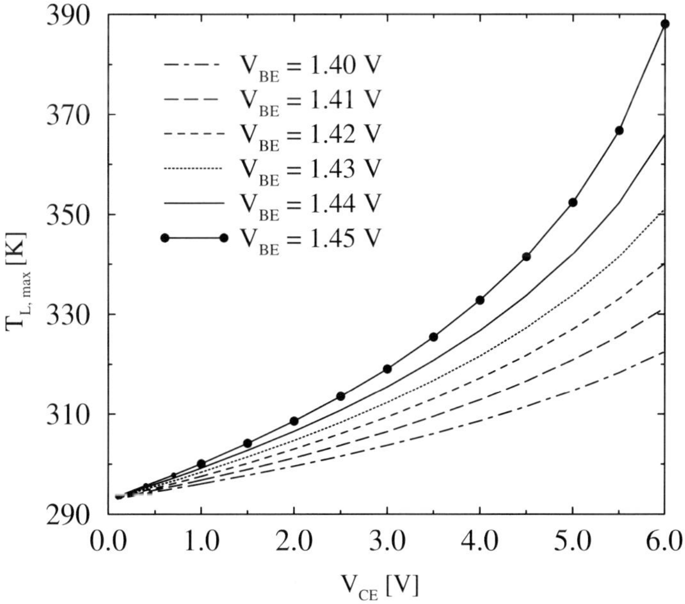

Fig. 5.34. Peak lattice temperature $T_{\rm L,max}$ (given in K) in the device as a function of $V_{\rm CE}$ in Dev. 3.

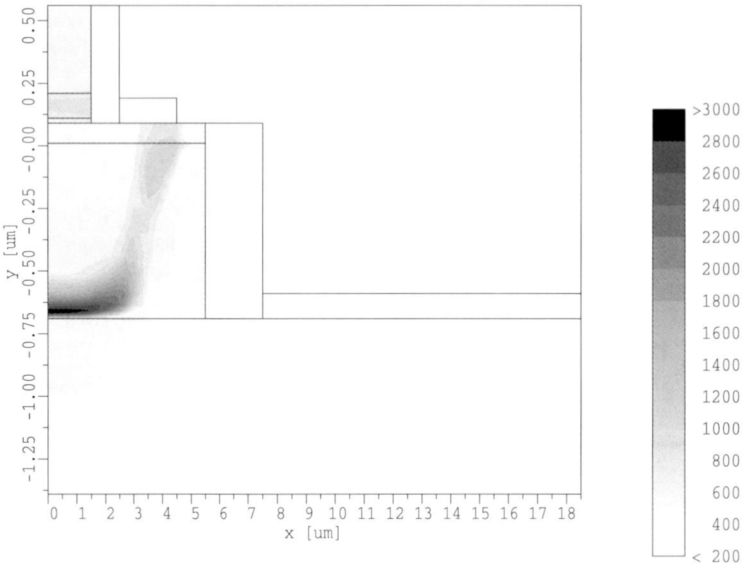

Fig. 5.35. Simulated electron temperature distribution (given in [K]) at $V_{\mathrm{CE}} = V_{\mathrm{BE}} =$ 1.6 V.

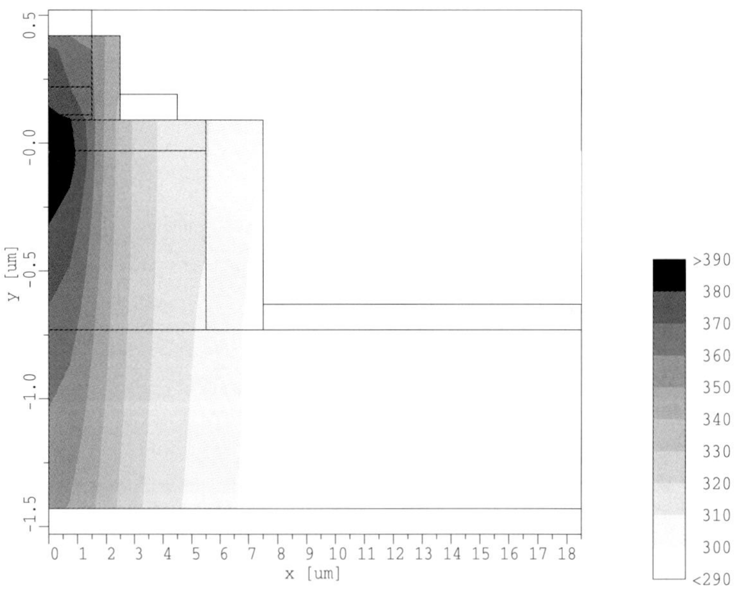

Fig. 5.36. Simulated lattice temperature distribution (given in [K]) at $V_{\mathrm{CE}} = 6.0$ V and $V_{\mathrm{BE}} = 1.45$ V: A substrate thermal contact with $R_{\mathrm{TH}} = 400$ K/W is added.

Transient Simulation of AlGaAs/GaAs HBT. In the following example, small-signal sinusoidal voltages are applied to the device terminals of $3 \times 30\ \mu m^2$ AlGaAs/GaAs HBT with an InGaP ledge (Dev. 3) in two independent simulations. The intrinsic Y-parameters are extracted from the transient simulation results as shown in [291]. A frequency of 5 GHz is chosen for the simulation to avoid dispersion effects in a lower frequency range. The intrinsic Y-parameters are converted to Z-parameters and from those the intrinsic small-signal equivalent circuit elements are calculated. The T-like eight-element small-signal equivalent circuit model [400] is used, as shown in Fig. 4.3. The ohmic resistances are completely accounted for in the simulation. The S-parameter calculation uses the same typical values for the extrinsic parasitic elements (inductances and capacitances) as extracted from the underlying technology [446] for compact modeling. From this set of parameters, extrinsic S-parameters are calculated for the frequency range of 0 to 20 GHz, suitable for comparison to the measurements in the same frequency range.
The combined Smith-polar chart in Fig. 5.37 shows a comparison of the simulated and measured S-parameters at $V_{CE} = 3$ V and $I_C = 22$ mA in the frequency range between 0 and 20 GHz. Due to the large emitter area, the limitations of the eight-element intrinsic circuit, and the fact that the physical extraction procedure includes no adjustment of parameters, perfect agreement cannot be expected. However, the small-signal extraction supplies valuable information about the modeling and the device itself.
Using the extraction above, it is found that the physical background of some important small-signal equivalent circuit elements, such as the base resistance R_{bb} and the emitter junction capacity C_{je}, can be improved. For example, neglecting the difference between majority and minority electron mobilities results in an underestimation of the electron mobility in the GaAs base of the HBT (see Section 3.4). This results in a significant overestimation of the extracted base resistance.

Small-Signal AC Simulation of InGaP/GaAs HBT. In a second approach, S-parameters are directly extracted from small-signal AC analysis of a one-finger $3 \times 30\ \mu m^2$ InGaP/GaAs HBT (Dev. 4). The two-port pad parasitic equivalent circuit (outer shell in Fig. 4.3) is used for analytical conversion of the simulated intrinsic Y-parameters to extrinsic S-parameters in order to compare simulation results directly to measurements. The pad capacitances considered read $C_{pBE} =$ 150 fF, $C_{pCE} = 75$ fF, and $C_{pBC} = 24$ fF; the parasitic inductances are $L_E =$ 1 pH, $L_B = 75$ pH, and $L_C = 50$ pH. These values result from measurements of open/short thru-test-structures [446].
The combined Smith-polar charts in Fig. 5.38, Fig. 5.39, and Fig. 5.40 show comparisons of simulated and measured S-parameters at $V_{CE} = 3$ V and $J_C =$ 2 kA/cm^2, $J_C = 8$ kA/cm^2, and $J_C = 15$ kA/cm^2, respectively, in the frequency range between 50 MHz and 10 GHz. For the same device, the matched gain and the short-circuit current gain h_{21} are calculated in order to extract the figures-of-

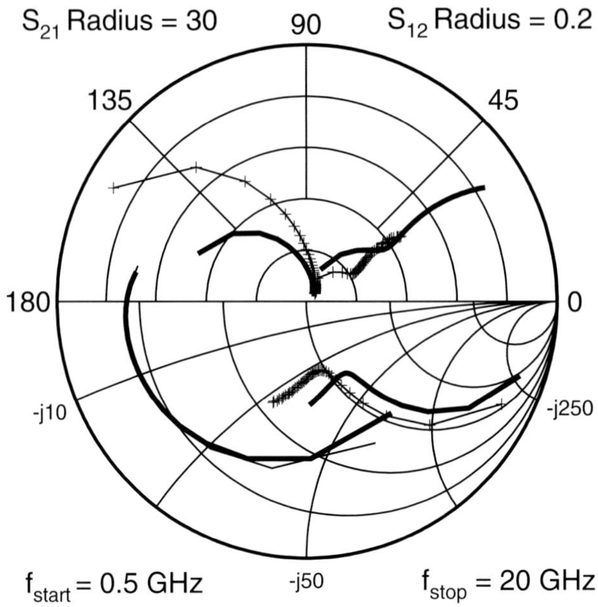

Fig. 5.37. Simulated (–) and measured (+) S-parameters in a Smith-polar chart from 0 to 20 GHz at $V_{\mathrm{CE}} = 3$ V and $J_{\mathrm{C}} = 25$ kA/cm^2.

merit f_{T} and $f_{\max}$ found at the respective unity-gain intersections. Fig. 5.41 shows the comparison of the simulated results and the corresponding measurement data.

Computational Effort. The S^{3}A AC simulation takes about 200 s CPU-time on a 2.4 GHz Linux Pentium machine for S-parameters computation with 20 frequency steps. For comparison, the conventional small-signal equivalent circuit approach takes about 590 s CPU-time on the same machine for 200 time steps at a given frequency to obtain sufficient accuracy for one frequency. To avoid a high number of time steps in the time domain for all frequencies required, only one frequency is used to extract equivalent circuit elements valid in a specific frequency range. The time required for such a post-processing of the transient simulation results to obtain the S-parameters at all frequencies is not included. Thus, the more accurate approach can speed up the frequency domain simulation by about 98% (taking one frequency into account). The good agreement with measured data and the reduction of the computational time demonstrate the quality and the efficiency of the small-signal AC approach. It enables further extensive optimization tasks, for example large signal extraction, with hundreds of runs in a reasonable amount of time. Thus two-dimensional physical simulation allows a direct relation between the material properties and the high-frequency device behavior.

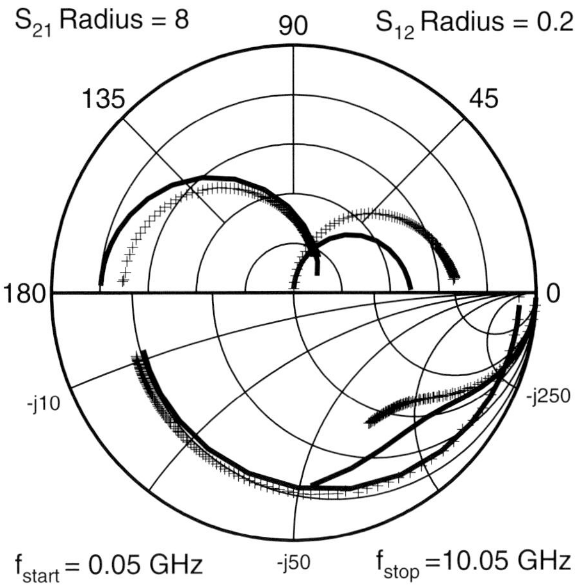

Fig. 5.38. Simulated (–) and measured (+) S-parameters in a Smith-polar chart from 50 MHz to 10 GHz at $V_{CE} = 3$ V and $J_C = 2$ kA/cm^2.

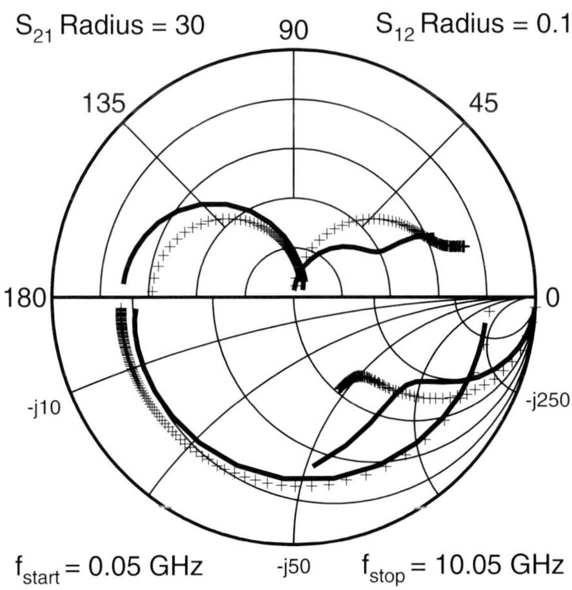

Fig. 5.39. Simulated (–) and measured (+) S-parameters in a Smith-polar chart from 50 MHz to 10 GHz at $V_{CE} = 3$ V and $J_C = 8$ kA/cm^2.

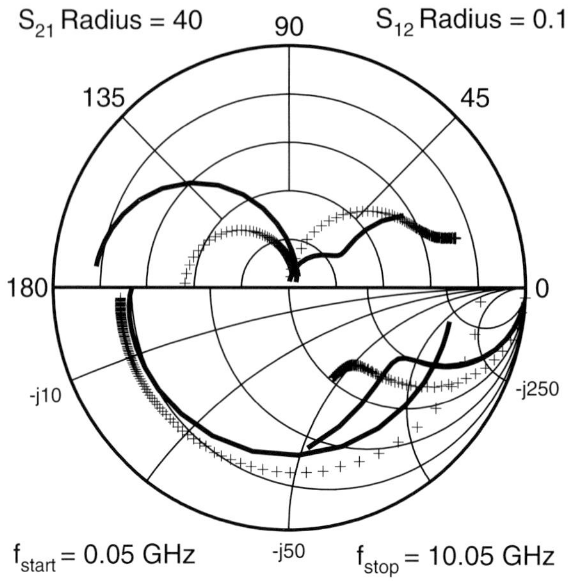

Fig. 5.40. Simulated (–) and measured (+) S-parameters in a Smith-polar chart from 50 MHz to 10 GHz at $V_{\mathrm{CE}} = 3$ V and $\mathrm{J_C} = 15$ kA/cm^2.

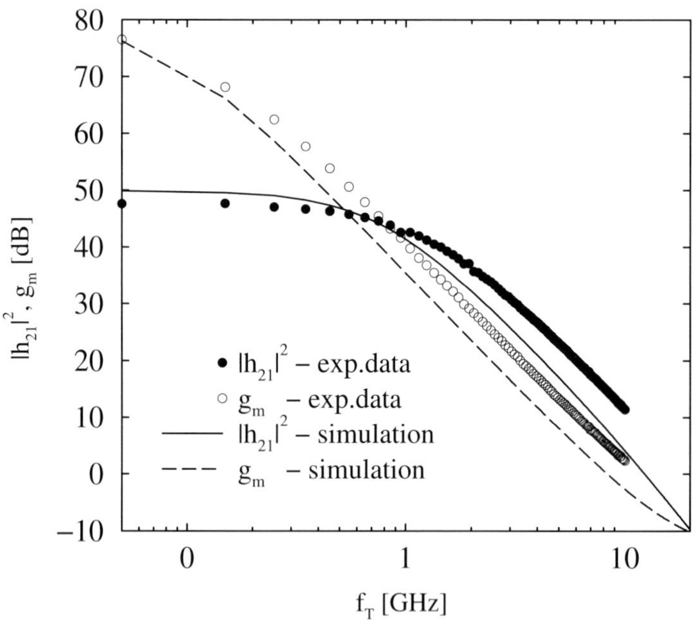

Fig. 5.41. Simulated (lines) and measured (circles) short-circuit current gain ratio $|h_{21}|^2$ and matched gain as a function of frequency at $\mathrm{J_C} = 15$ kA/cm^2.

5.3.4 Analysis of HBT Behavior After Electrothermal Stress

This section presents simulations of one-finger power InGaP/GaAs HBTs before and after both electrical and thermal stress aging. The influence of the ledge thickness (see Fig. 5.42) and of the surface charges on the device performance and their impact on reliability is analyzed. Possible explanations of device degradation mechanisms, e.g. the effect of vanishing negative surface charges on the ledge/SiN interface, are proposed.

Importance of the InGaP Ledge. It is commonly accepted that GaAs HBTs with an InGaP ledge have an improved reliability [310]. In these devices, the emitter material completely covers the base layer, thus forming the so-called ledge between the extrinsic base and the nitride passivation. This allows the high surface recombination, which is typically observed in the extrinsic base of AlGaAs/GaAs HBTs, to be avoided.
The surface charges, which exist at the ledge/nitride (InGaP/SiN) interface, play an important role in the device. They have a large impact on the Fermi-level pinning at the surface, which results in surface depletion. In InGaP/InGaAs FETs this effect is not as strong for InGaP/SiN as for AlGaAs/SiN surfaces and results in an enhanced shift of the threshold voltage [399] depending on the process parameters.
The influence of the ledge thickness d and negative surface charges σ on the device performance is investigated using MINIMOS-NT. A schematic drawing of the simulated device structure is shown in Fig. 5.27.

InGaP Ledge Thickness. Fig. 5.42 shows measured and simulated collector and base currents of InGaP/GaAs HBTs with different ledge thicknesses operating under forward Gummel plot conditions with $V_{\mathrm{BC}} = 0$ V. The experimental data are taken from a device with a 40 nm ledge thickness. Surface charges at any of the device interfaces are not yet considered in the simulation. Note the strong increase of the base current at low bias with increasing ledge thickness.
As can be seen in Fig. 5.42, simulated and measured base currents differ significantly in the case of 40 nm ledge thickness. Only simulation with a ledge thickness of less than 20 nm delivers a good match. The reason is that at the semiconductor/insulator surface, the Fermi-level pinning is not taken into account if surface charges are not considered in the simulation. Therefore a non-physical electron current path occurs in the upper part of the ledge, as shown in Fig. 5.43. Fig. 5.44 shows the corresponding electron distributions in the ledge at x = 1.6 μm, 2.0 μm, and 2.4 μm using vertical cross-sections. The hole distribution in the middle of the ledge (x = 2.0 μm) is also included. In the next subsection, these concentrations are compared to those when surface charges are present.

Negative Surface Charges. As can be seen in Fig. 5.45, the base current decreases if more negative surface charges are introduced. The symbols represent experimental data for the collector current I_{C} and the base current I_{B}, and the simulation results are from a device with 40 nm ledge. The lower part of the ledge

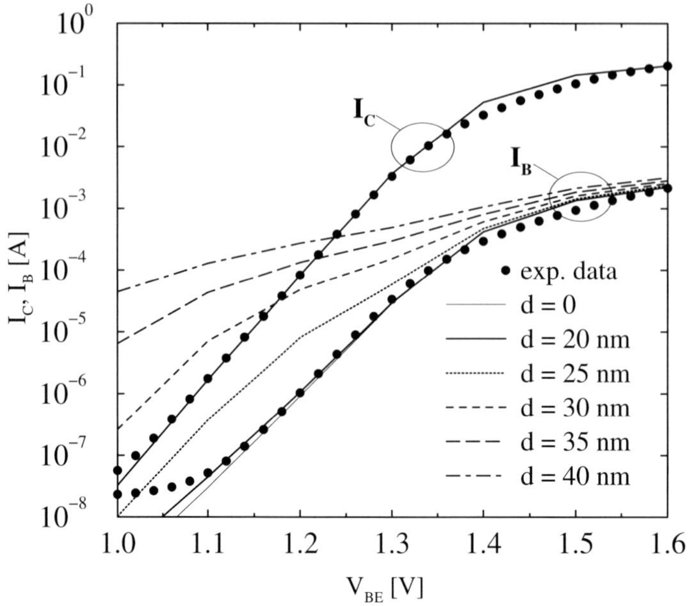

Fig. 5.42. Forward Gummel plots at $V_{\mathrm{BC}} = 0$ V: Dependence of I_{B} on the InGaP ledge thickness and comparison with measurement.

is depleted, due to the emitter-base space charge region. The negative charges deplete the upper part of the ledge as well [317], and the leakage is reduced (see Fig. 5.46). Fig. 5.47 presents the corresponding electron distribution in the ledge at x = 1.6 μm, 2.0 μm, and 2.4 μm, and the hole distribution at x = 2.0 μm. Note that even in this case the ledge is not completely depleted. However, the electron concentrations near the InGaP/SiN interface are significantly lower in comparison to the ones shown in Fig. 5.44. Thus with a surface charge density $\sigma = 10^{12}$ cm^{-2} the measured base current can be simulated very well. Note that in the presence of negative surface charges, the hole concentration in the ledge increases. Thus a path for the hole current can occur if the charge densities are sufficiently high.

Device Reliability. Based on these investigations, it is now possible to explain the base current degradation of an InGaP/GaAs HBT, which is strongly stressed under conditions far from normal operating conditions. For this device, the base current degradation in the medium voltage range can be explained by a decreasing surface charge density from 10^{12} cm^{-2} to $4{\times}10^{11}$ cm^{-2} along the interface between the ledge and the passivation. This might be due to compensation of the negative surface charges by H^+ ions, which are known to be present in the device due to the epitaxial MOCVD growth processes [57, 62].

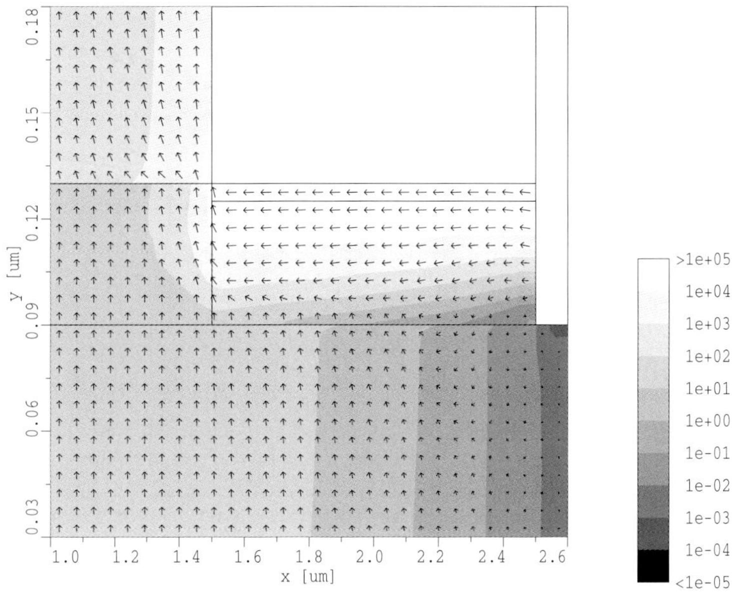

Fig. 5.43. Simulated electron current density [A/cm^2] at $V_{\mathrm{BE}} = 1.2$ V without surface charges.

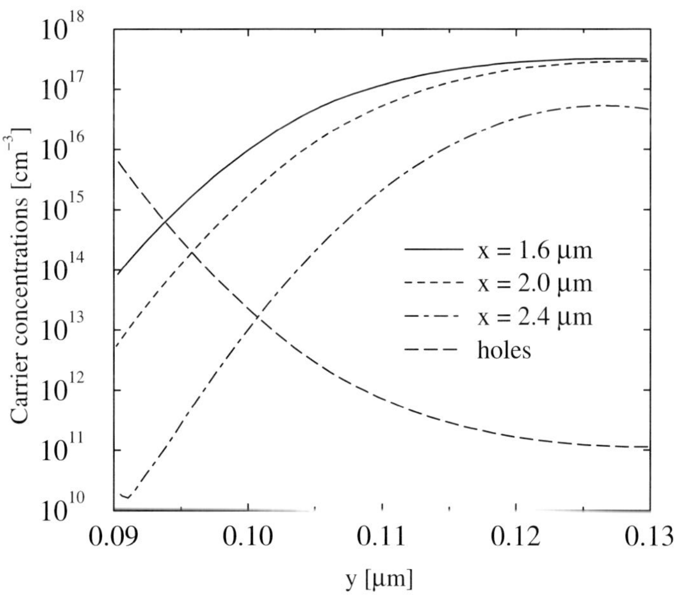

Fig. 5.44. Simulated electron and hole distribution in the ledge without surface charges.

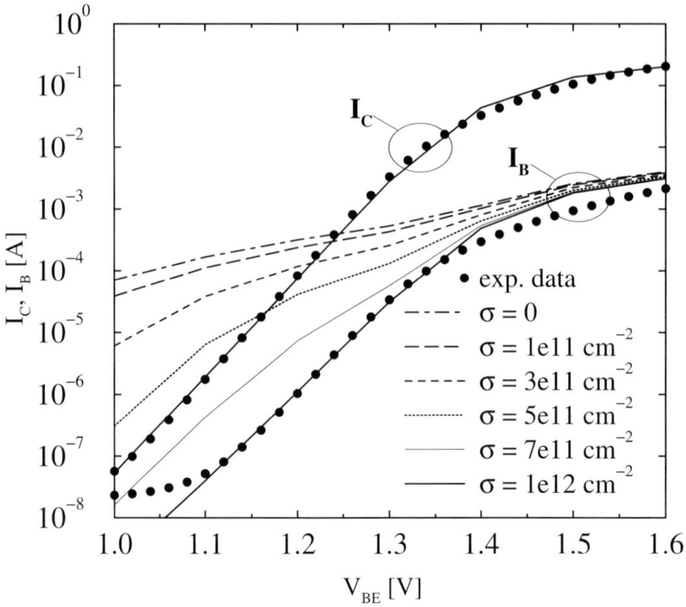

Fig. 5.45. Forward Gummel plots at $V_{BC} = 0$ V: Dependence of I_B on the negative charge density at the ledge/nitride interface, d= 40 nm.

Fig. 5.48 presents the electron current density for a surface charge density of $4{\times}10^{11}$ cm^{-2}. Fig. 5.49 shows the corresponding electron distribution in the ledge at x = 1.6 μm, 2.0 μm, and 2.4 μm, and the hole distribution at x = 2.0 μm. Note that the upper part of the ledge is now not completely depleted, thus allowing for a base leakage current.

Fig. 5.50 compares measured and simulated forward Gummel plots at $V_{BC} = 0$ V. Filled and open symbols denote measured characteristics of the non-degraded and degraded device respectively. The lines show the corresponding simulation results. The good agreement, which could also be obtained for stressed devices, demonstrates the applicability of physically-based device simulation to device reliability issues.

Several other effects are suggested in the literature (see for example [104, 344] and references therein) as a possible reason for the strong increase in the base leakage current, e.g. spread-out of the base contact at the metal/GaAs interface, increased recombination-generation in the InGaP layer, or degradation of the SiN/GaAs interface. The simulation analysis shows that such effects cannot be the dominant reason for degradation of the current gain of the device under investigation.

The decrease in the collector current at high-level injection is suggested in the literature (e.g. [344]) to be due to increased emitter resistance which could occur

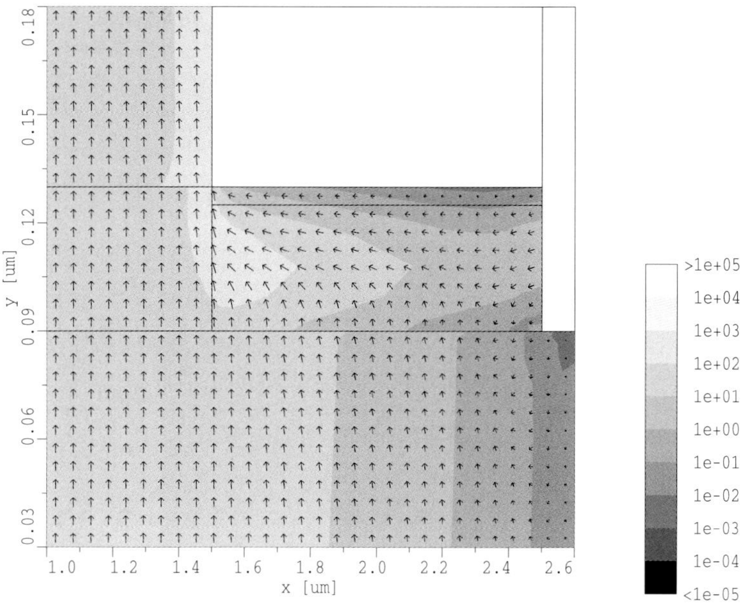

Fig. 5.46. Simulated electron current density [A/cm^2] at $V_{\mathrm{BE}} = 1.2$ V with a surface charge density of 10^{12} cm^{-2}.

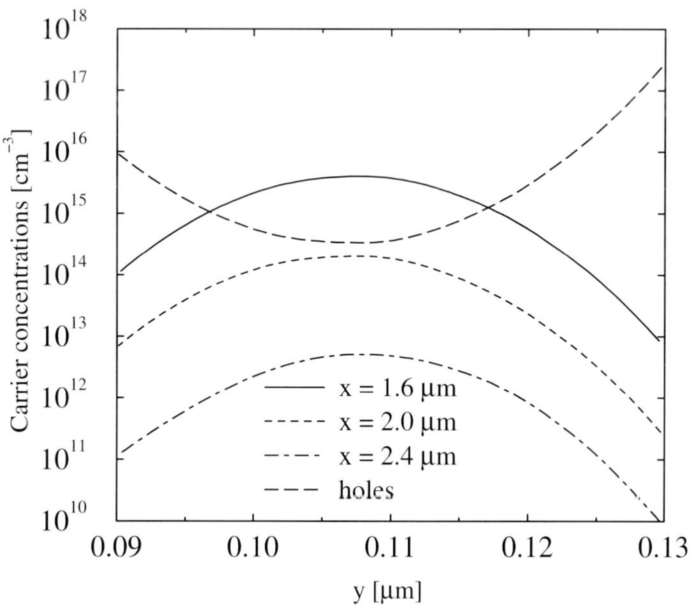

Fig. 5.47. Simulated electron and hole distribution in the ledge with a surface charge density of 10^{12} cm^{-2}.

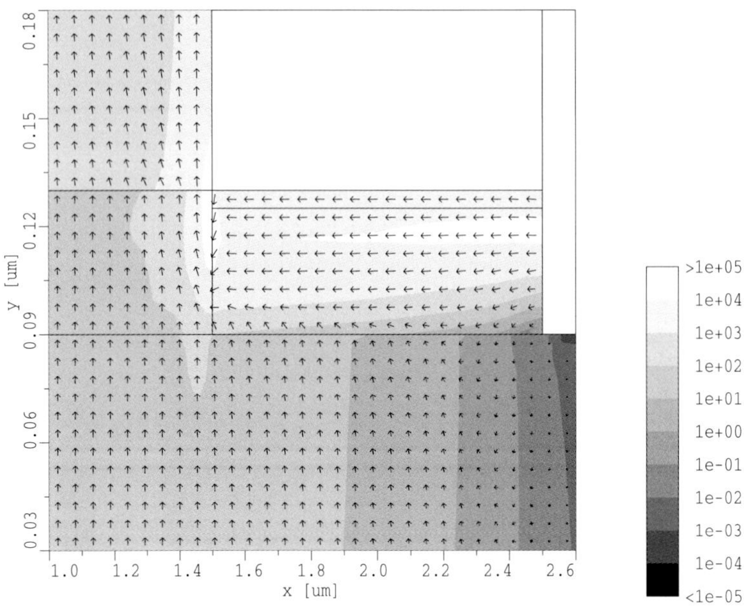

Fig. 5.48. Simulated electron current density [A/cm^2] at $V_{\mathrm{BE}} = 1.2$ V with a surface charge density of 4×10^{11} cm^{-2}.

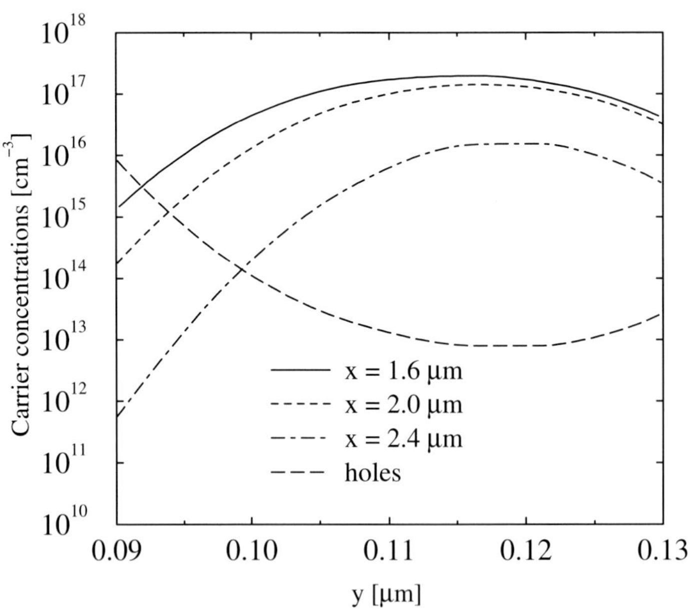

Fig. 5.49. Simulated electron and hole distribution in the ledge with a surface charge density of 4×10^{11} cm^{-2}.

due to emitter contact detachment, indium segregation in the metal layer, or dislocations at the InGaAs/GaAs interface. The simulation analysis shows that contact detachment leads to an electron current crowding in the remaining contact area, which leads to insignificant changes in the collector current. Only an emitter contact detachment of more than 80%, which is not very probable, could explain the measured values (see Fig. 5.51). Indium segregation in the contact metal could be a possible reason for degradation, since the emitter contact resistance increases. The corresponding decrease of the indium content in the cap has no significant influence on the emitter resistance.

Simulation for Improvement of Device Reliability. Once the possible reason for device degradation is known, it is natural to search for solutions. Some ways to avoid degradation are analyzed hereafter, assuming the worst case when the negative surface charges at the ledge/nitride interface of a device with 40 nm ledge thickness are completely compensated due to stress.
A possible solution is to avoid the electron leakage path in the ledge by means of electrically isolated base contacts, e.g. by introducing a nitride spacer between the ledge and the base metals.
The simulation analysis shows that the depth of such a spacer is of importance. On the one hand, there is the constraint not to exceed the ledge thickness in order to avoid surface recombination in the base. On the other hand, the spacer has to have a sufficient depth to prevent the electron current. Fig. 5.52 shows the electron current density at $V_{\mathrm{BE}} = V_{\mathrm{CE}} = 1.2$ V in a device with a distance of 10 nm between the spacer and the base layer. As can be seen in Fig. 5.52, a current path still exists under the spacer if the surface charges are compensated. Unfortunately it is technologically challenging to control the exact spacer depth, which has to be about 95% of the ledge thickness in order to solve the reliability problem.
Another possible solution is to isolate the ledge from the active emitter area. This approach is more efficient for lower biases. However, an increased surface recombination at the interface between the spacer and the emitter is observed leading to a lower current gain at all biases in comparison to the non-stressed device. In addition, a current path again exists under the spacer. Fig. 5.53 shows the electron current density at $V_{\mathrm{BE}} = V_{\mathrm{CE}} = 1.2$ V in a device with a spacer placed between the ledge and the active emitter. The distance from the bottom of the spacer to the base is 10 nm.
Fig. 5.54 shows the simulated forward Gummel plots for 40 nm ledge devices with 35 nm deep spacers on the base contact side or on the emitter side of the ledge, respectively. In addition, simulation results for a device with 20 nm ledge thickness are included for comparison in Fig. 5.54. Measured data for a non-stressed device (symbols) are included as a reference. Note that the 20 nm ledge device not only has better device characteristics, but is also easier to manufacture. Therefore, the simulation analysis leads to the conclusion that the particular degradation problem should be solved by partially removing the ledge layer.

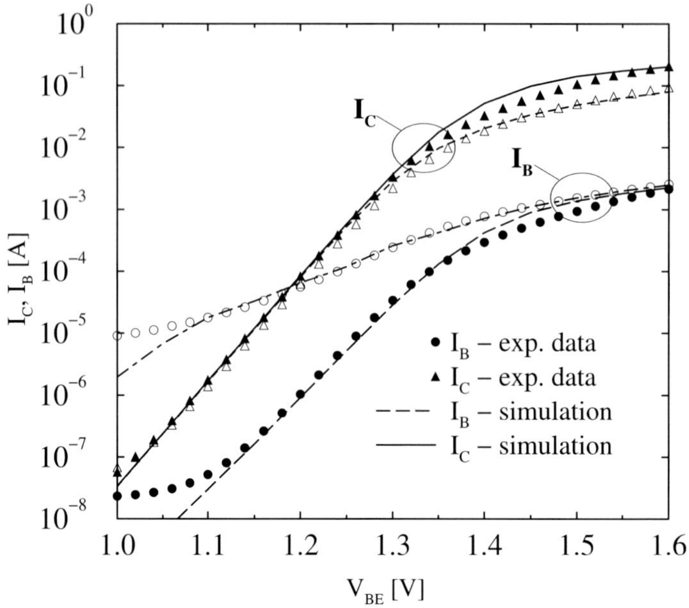

Fig. 5.50. Forward Gummel plots at $V_{BC} = 0$ V: Comparison of measurement (symbols) and simulation (lines) before (filled) and after (open) HBT aging.

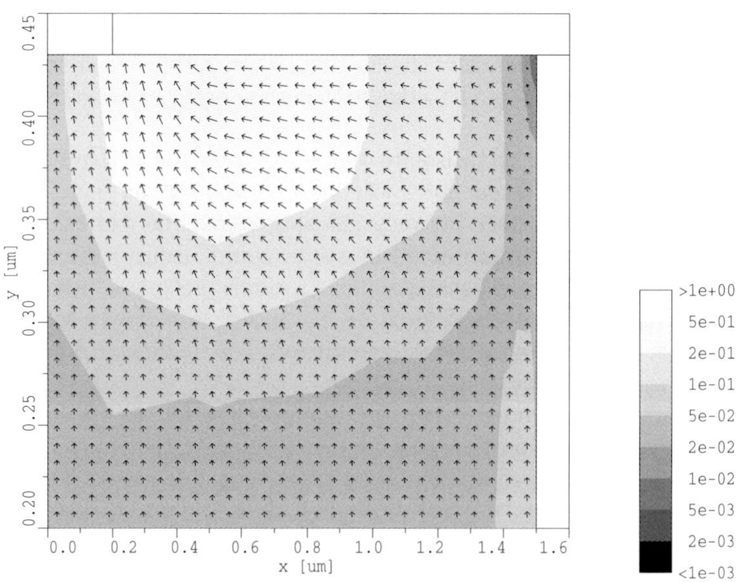

Fig. 5.51. Electron current density [A/cm^2]: Simulation of emitter contact detachment.

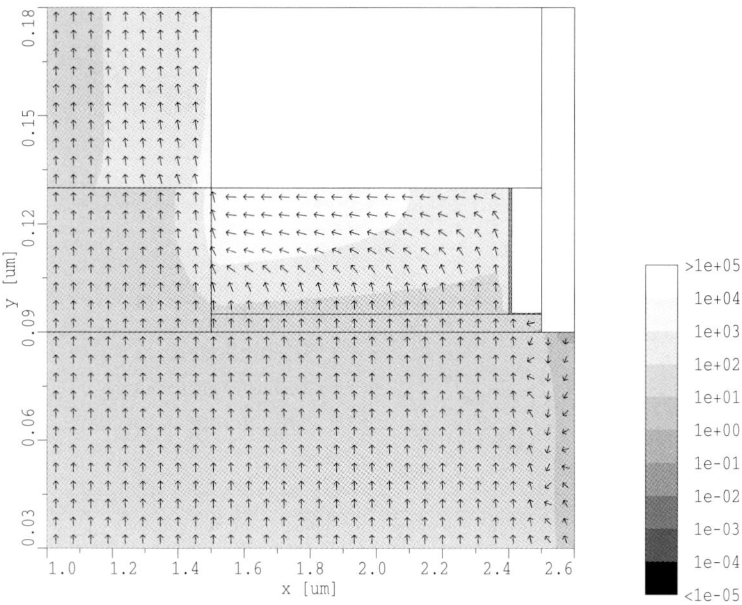

Fig. 5.52. Simulated electron current density [A/cm^2] at V_{BE} = 1.2 V with a spacer between ledge and base contact.

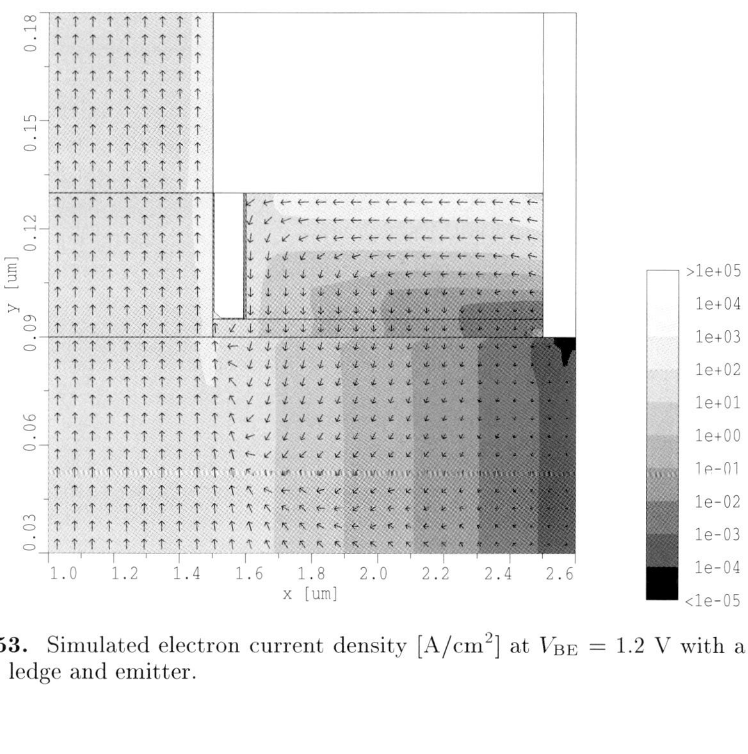

Fig. 5.53. Simulated electron current density [A/cm^2] at V_{BE} = 1.2 V with a spacer between ledge and emitter.

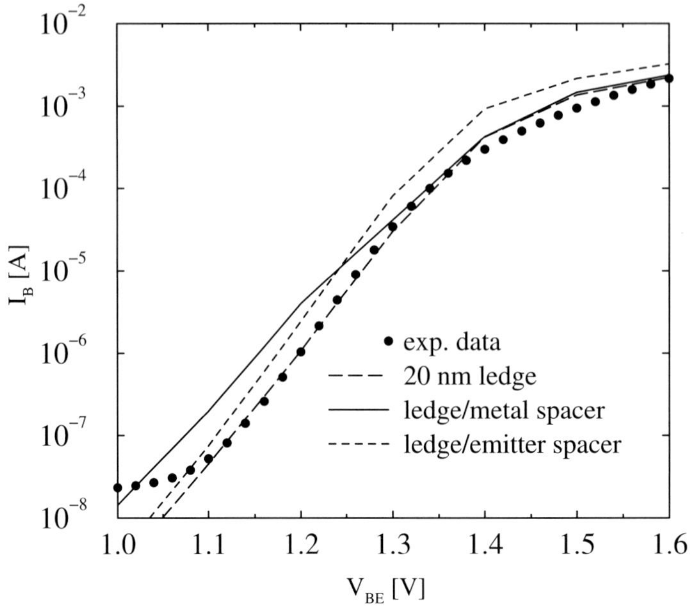

Fig. 5.54. Base current I_B in stressed devices using a spacer between the ledge and the base contact, between the ledge and the emitter, and a device with a thinner ledge. Comparison with measured data for a non-stressed device.

5.3.5 Three-Dimensional Analysis of Leakage Currents

Two-dimensional device simulation proved to be valuable for understanding the underlying device physics and for investigating device reliability, as shown in the previous sections. Three-dimensional simulation allows investigation of leakage currents in III-V HBTs. Good agreement between two-dimensional simulation and measurement can be achieved for medium and high bias [433], but not for low bias where high leakage currents exist (≈nA). The base-emitter leakage current can be reproduced by including certain generation/recombination mechanisms [303]. However, the collector-emitter leakage current cannot be explained with two-dimensional analysis, since there is no direct path between emitter and collector.

Fig. 5.55 shows the top view of a GaAs HBT. Conventional two-dimensional simulations of HBTs employ a cut in the x-direction (cut A-A in Fig. 5.55). Thus three-dimensional effects which typically occur at the edges of a real structure are neglected. In addition, non-homogeneities of the materials in the z-direction (cut B-B in Fig. 5.55) may result in non-homogeneous electrical performance and operating temperature of the device. Note that a direct path between the emitter and the collector exists only in the third dimension (see Region A in Fig. 5.55 where the emitter is not surrounded by the C-shaped base contact). This region is identified to be very sensitive to variations of the doping profile, surface charges, and traps, and the simulation results prove that collector-emitter leakage can occur there. Fig. 5.56 shows results from a three-dimensional simulation at $V_{\mathrm{BE}} = V_{\mathrm{CE}} = 0.5$ V using a two-dimensional cut in the B-B direction. Even for low voltages a high emitter current flows from the collector to the emitter. Fig. 5.57 shows the potential distribution in Region A.

Fig. 5.58 shows results from two-dimensional and fully three-dimensional simulations compared to measured forward Gummel plots for an InGaP/GaAs HBT both before and after aging. It has been shown in Section 5.3.4 that the electrical performance of the unstressed device in its active region (i.e. $V_{\mathrm{BE}} > 1$ V) can be modeled with high accuracy solely by means of two-dimensional simulations. By incorporating surface charges in the InGaP ledge, the two-dimensional approach can describe the high base leakage current (see Fig. 5.58) which happens at $V_{\mathrm{BE}} \leq 1$ V. However, as can be seen from the two-dimensional result for the collector current of Fig. 5.58, I_{C}(2D), this approach does not give any collector leakage. This is in discrepancy with the measurement data, which show collector leakage currents in the range of 10 nA.

The search for this leakage in Region A (Fig. 5.55) requires full three-dimensional simulation of the device. As can be seen from the I_{C}(3D)-curve in Fig. 5.58, a considerable collector leakage in the order of nA is possible. This leakage is due to the surface current flowing from the collector contact to the emitter contact, as shown in Fig. 5.56. However, the simulated leakage I_{C} is lower than the measured I_{C}. The difference has been attributed to a leakage current contribution due to the non-ideal insulation of the pads, which is not included in the simulation. The technology under consideration is unable to provide a perfect insulation, since the pads are located on the sub-collector layer with no dielectric insulation in between.

The intrinsic surface leakage and the extrinsic leakage, which occurs between the pads, have to be distinguished. A straightforward experimental solution is to measure the leakage of a so-called "open-test-structure" [445]. This test-structure has a setup absolutely identical to the active transistor structure, except the electrical contacts directly on the semiconductor material are missing. Without such a test-structure available, it is not possible to separate intrinsic and extrinsic leakage current contributions exactly. Unfortunately, the pads cannot be incorporated in a three-dimensional simulation due to their huge size and their great distance to the device in comparison to the tiny active device geometry. This is an unacceptable computational task.

For the analysis of the intrinsic leakage current, both two-dimensional and three-dimensional simulations are performed with MINIMOS-NT. The simulation grid contains more than 150 000 points for the three-dimensional simulation and about 3000 points for the two-dimensional simulation. The three-dimensional simulation requires about one hour of CPU time on a 1 GHz Pentium Linux computer for a given bias point.

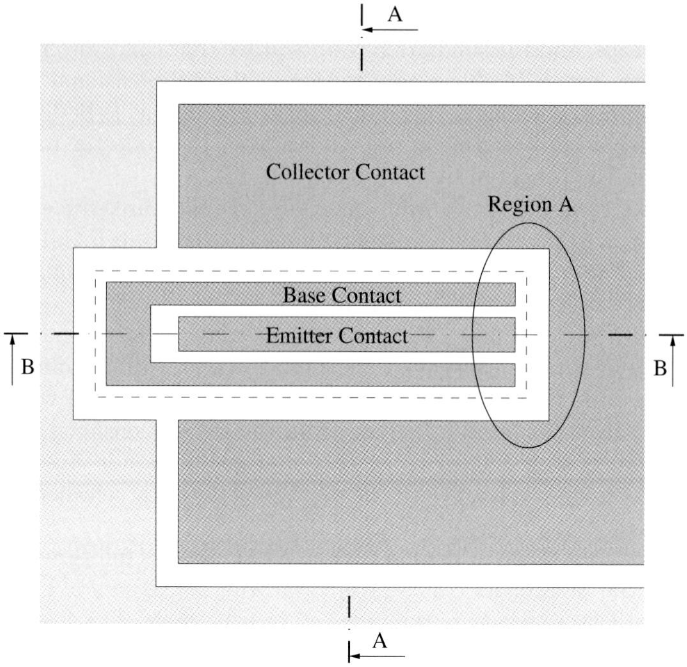

Fig. 5.55. Top view of the three-dimensional HBT structure.

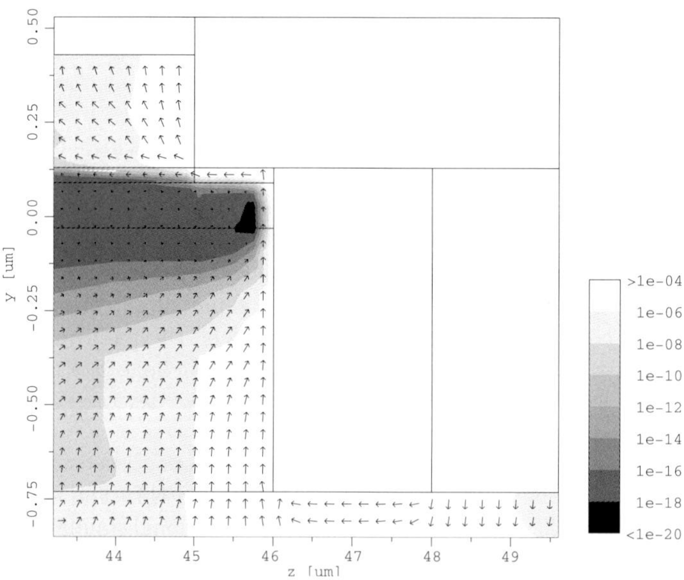

Fig. 5.56. Simulated electron current density [A/cm^2] in Region A of the HBT using a cut in B-B direction.

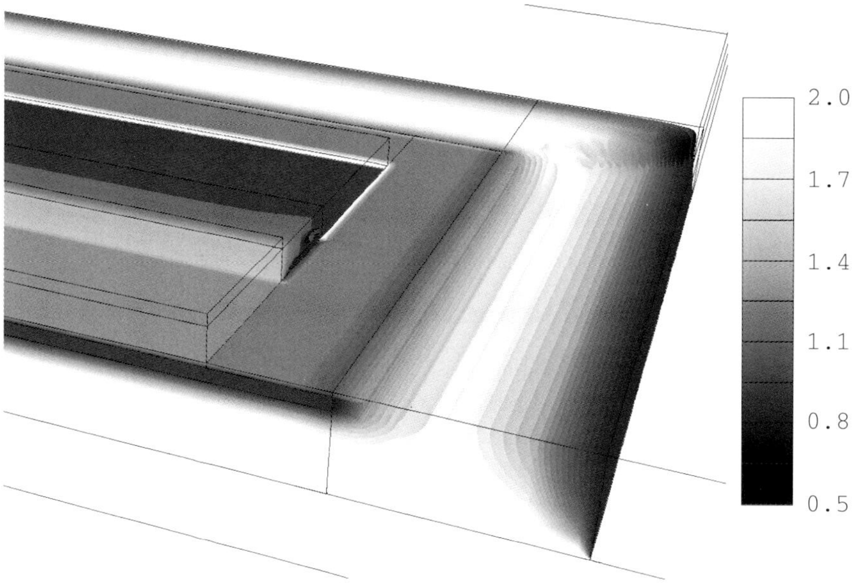

Fig. 5.57. Iso-surfaces of the potential [V] in Region A at $V_{\mathrm{BE}}=V_{\mathrm{CE}}= 0.5$ V.

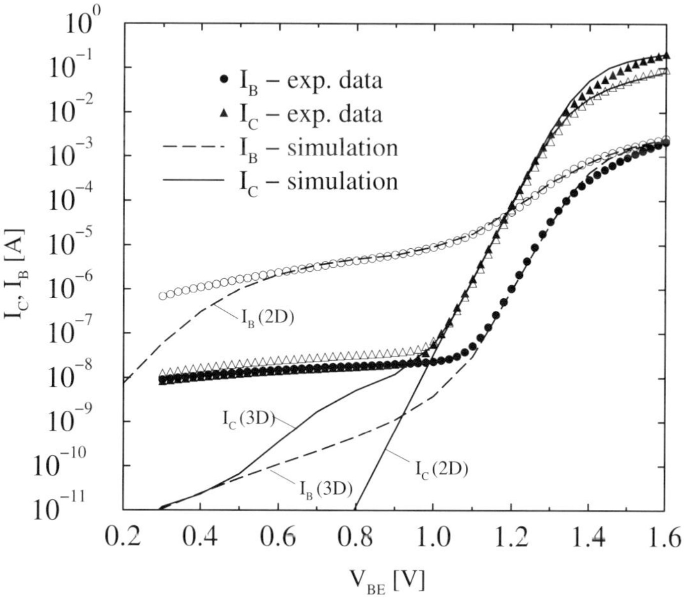

Fig. 5.58. Forward Gummel plots at $V_{\mathrm{BC}} = 0$ V: Comparison of measurements (symbols) and three-dimensional and two-dimensional simulations (lines) before (filled) and after (open) HBT aging.

5.4 High-Speed InP HBTs

InP HBTs are very promising III-V devices for low-power, high-speed applications. As stated in Chapter 2, these devices can be optimized to supply high cut-off frequencies and low gate-delay times at relatively low current levels. Double heterojunction bipolar transistors (DHBTs) can provide both breakdown voltages beyond 6 V and high-speed capabilities with cut-off frequencies beyond 250 GHz, as shown in Section 2.2.

5.4.1 Device Design and Manufacturing Issues

Epitaxy. Fig. 5.59 gives a typical layer structure of a single heterojunction bipolar transistor (SHBT) realized in the InAlAs/InGaAs material system. The device contains an InAlAs emitter, heavily doped to reduce the emitter access resistance at the metal/semiconductor interface and lightly doped to take advantage of the low-field mobility in the access region. Both MBE and MOCVD grown InP SHBT devices are found. The $In_{0.53}Ga_{0.47}As$ base is typically p-doped at $1-3\times10^{19}$ cm^{-3} and is followed by a lightly n-doped $In_{0.53}Ga_{0.47}As$ collector. The base-collector design is crucial for the speed/breakdown performance. Higher doping leads to increased speed; lower doping leads to increased breakdown voltage.
Fig. 5.60 shows a typical InP DHBT layer structure. The emitter design in an InP DHBT depends on the availability of aluminum in phosphorus-based MBE or MOCVD epitaxial growth reactors for use in InP DHBTs. Typically, either aluminum or phosphorus are available, which results in InAlAs or InP as an emitter material. Generally for any HBT, base thickness and base transit time need to be reduced to obtain high-speed devices. However, a minimum base thickness is required to allow diffusive base transport, since the emitter-base injection can be a hot-electron process. InAlAs as an emitter material delivers a partially hot-electron injection into the base, since increased bandgap and bandgap offset prevail for InAlAs relative to InP.

Processing. General InP HBT processing considerations further involve the definition of the base contact, either in self-aligned or non-self-aligned mode. Self-aligned optical definition is typically used for submicron base contact geometries suitable for low C_{jc} and thus high-speed applications [159]. High-speed devices also typically employ underetched base and emitter contacts. The critical difference between Fig. 5.59 and Fig. 5.60 is the transition of the second heterojunction. A linear or staggered bandgap alignment from the $In_{0.53}Ga_{0.47}As$ base to the collector material by means of quaternary material is used to smoothen the bandgap discontinuity at the $In_{0.53}Ga_{0.47}/In_{0.52}Al_{0.48}As$ or $In_{0.53}Ga_{0.47}$/InP interfaces. An alternative is to use an n-doped launcher structure made of $In_{0.53}Ga_{0.47}As$ [159] and thus overcome the second heterojunction barrier by nearly ballistic transport.

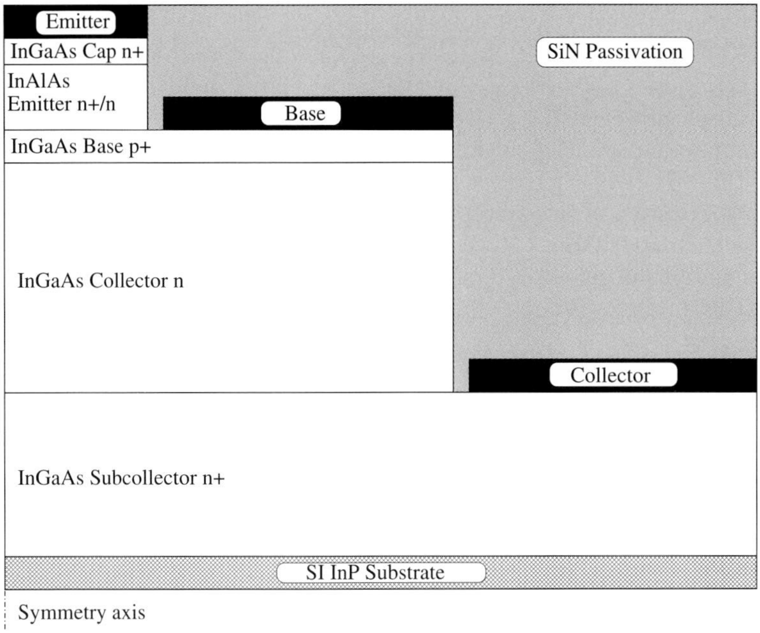

Fig. 5.59. Layer structure of an InP single heterojunction bipolar transistor.

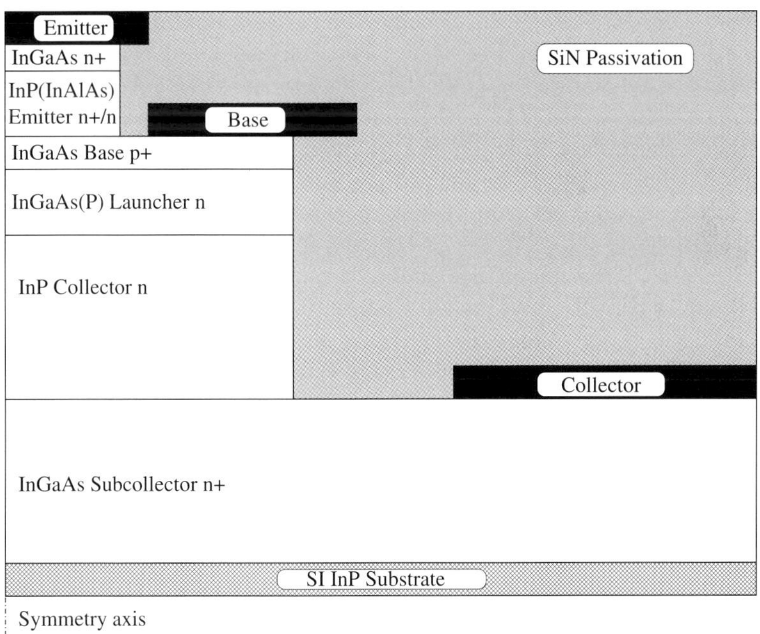

Fig. 5.60. Layer structure of an InGaAs/InP double heterojunction bipolar transistor.

InP is preferred as a typical collector material, due to its superior transport properties in comparison with $In_{0.52}Al_{0.48}As$. Similarly to the SHBT, the collector doping determines both BV_{CE0} (decreasing at higher doping) and f_T (increasing with doping). However, BV_{CE0} for the DHBT is enhanced by a factor of at least three compared to some of the best submicron high-speed SHBTs [477]. Another trade-off must be considered for the collector layer thickness. Smaller thickness leads to a reduction of the collector transit time τ_C, but also to an increase of the base-collector capacitance C_{jc}. Device simulation offers the potential of numerical analysis and of the given trade-offs and design optimization.

5.4.2 InP SHBTs: Calibration for Large Area Devices

Similar to the calibration methodology used for SiGe and GaAs HBTs, the simulation parameters are first calibrated for large-area devices to minimize the influence of surface effects in the mesa structure. Fig. 5.61 shows the simulated and measured Gummel plot of two large area $100 \times 100\ \mu m^2$ $In_{0.52}Al_{0.48}As$/ $In_{0.53}Ga_{0.47}As$ SHBTs. The devices are identical, except for the doping concentrations in the emitter, which amount to $N_D = 3{\times}10^{17}\ cm^{-3}$ and $N_D = 10^{15}\ cm^{-3}$ (nid - non-intentionally doped). For both devices, good agreement between simulation and measurement is achieved using the same set of simulation models and parameters and changing only the doping concentration in the emitter.

Simulation Details. The effect of minority carrier mobility is considered, as given for $In_{0.53}Ga_{0.47}As$ in Fig. 3.55. The calibration of the mobility is obtained from the saturation region of the collector current I_C. At the emitter-base $In_{0.48}Al_{0.52}As$/ $In_{0.53}Ga_{0.47}As$ heterojunction, the simulation can be performed using either a thermionic emission or a tunneling assisted interface model. Due to the larger bandgap offset of the $In_{0.48}Al_{0.52}As/In_{0.53}Ga_{0.47}As$, transition tunneling and subsequent hot electron injection has a significant impact on the forward Gummel plot.

Thus in this case, a thermionic field emission model must be used as in the case of InAlAs/InGaAs HEMTs described in Chapter 6. The importance of bandgap narrowing for highly p-doped base materials has already been stressed for SiGe HBTs and GaAs HBTs. This effect is significant also for InP HBTs with p-doping levels of $\geq 10^{19}\ cm^{-3}$ in $In_{0.53}Ga_{0.47}As$, as shown in Chapter 3. Further recombination is described by the SRH model (3.155) using a trap density of $3 \times 10^{12}\ cm^{-3}$. A surface charge density of $10^{12}\ cm^{-2}$ is used to model the interface between the SiN emitter mesa-passivation and the emitter semiconductor materials InAlAs and InGaAs. With the parameters found in the calibration methodology, more complicated devices, the output characteristics, and RF quantities can be addressed in the next section, using the calibration methodology proposed in Section 5.2 and Section 5.3.

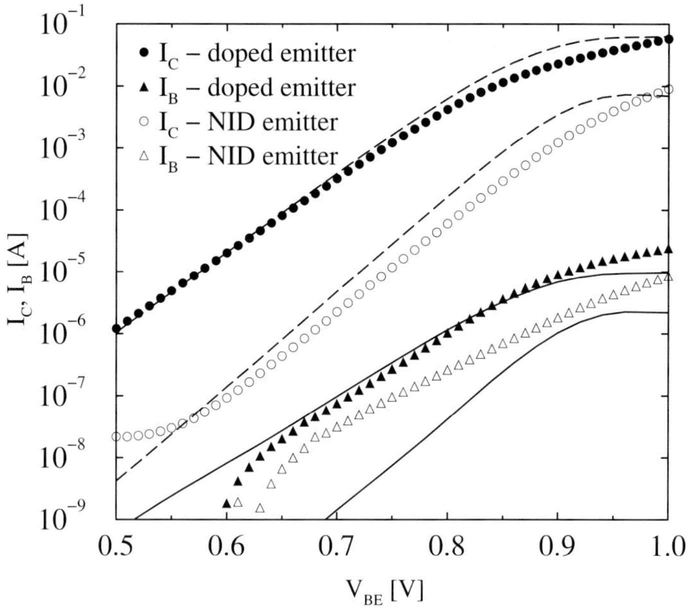

Fig. 5.61. Forward Gummel plots at $V_{\mathrm{BC}} = 0$ V: Simulation of two $100\times100\mu m^2$ InAlAs/InGaAs/InP SHBTs compared to measurements.

5.4.3 InP DHBT: A High-Speed Device with a High Breakdown Voltage

Contrary to SHBTs, DHBTs do not suffer from breakdown problems and base push-out effects due to the energy barrier at the base-collector heterojunction. However, other effects, such as charge accumulation in the base and barrier transport, occur in DHBTs and need to be accounted for.

In this simulation study, a nominally $1 \times 8\ \mu m^2$ InP/InGaAs/InGaAsP/InP self-aligned DHBT is investigated. The layers are grown by MOCVD on three-inch InP substrates. An InGaAs emitter cap is followed by an InP emitter and an $In_{0.53}Ga_{0.47}As$ carbon-doped base at 2×10^{19} cm^{-3}. A quaternary graded spacer is used to connect the base and the collector. The collector contact is placed on the subcollector. A low-k dielectric is used for the passivation. Typical characteristics observed experimentally are $f_{\mathrm{T}} \approx 130$ GHz and a breakdown voltage of $BV_{\mathrm{CE0}} = 5.5$ V.

Simulation Details. The DC calibration is performed based on the results for forward Gummel plots of the SHBT in the previous section. Similar to the SHBT, a TFE interface model is used at the emitter-base heterojunction. At the base-collector junction, the TFE model is also used to describe the transport over the blocked heterojunction. The use of the TFE interface model including tunneling

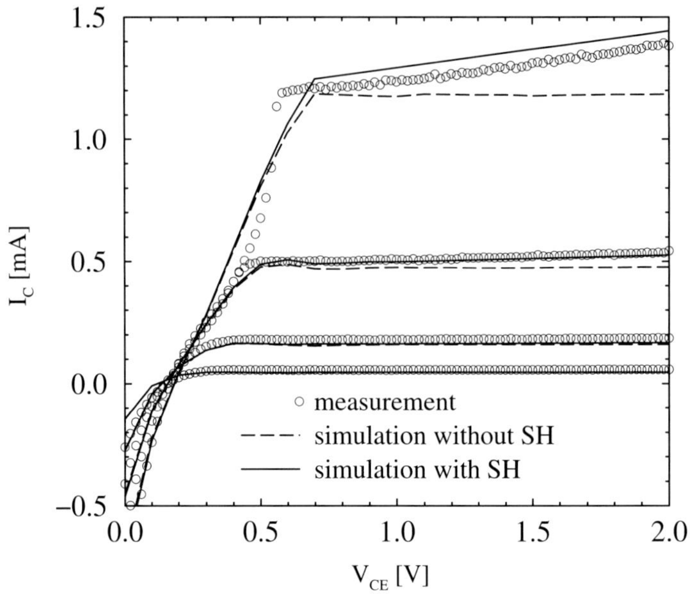

Fig. 5.62. Output characteristics InP/InGaAs DHBT: Comparison of simulation with and without self-heating and measurements for V_{BE} stepped by 0.05 V from 0.65 V to 0.8 V.

is especially important for simulation of DHBTs, as the high-field, high-energy electron transport in vertical direction is typically used to surmount the energy barrier at the interface. The material parameters for the spacer material InGaAsP are interpolated between $In_{0.53}Ga_{0.47}As$ and InP, as given in Chapter 3.
The simulation of the output characteristics is obtained using a methodology similar to that used for GaAs HBTs. Fig. 5.62 shows the results for the simulated output characteristics for V_{BE} stepped by 0.05 V from 0.65 V to 0.80 V in comparison to measurements. Despite the good thermal conductivity of InP, the output characteristic is critically influenced by self-heating, due to the small feature size of the device and the resulting high current densities. Therefore, self-heating must be included in the simulation. The deviation of the simulation with SH relative to the simulation without SH is especially significant for $V_{CE} \geq 1$ V, as can be observed in Fig. 5.62.

5.4.4 High-Speed Optimization by RF Simulation

With the figures-of-merit given in the previous section, the simulation example represents a high-speed device. For the optimization of high-frequency performance, small-signal simulation results can be obtained with an AC solver. Fig. 5.63 compares simulated and measured S-parameters at $V_{CE} = 1.1$ V in a

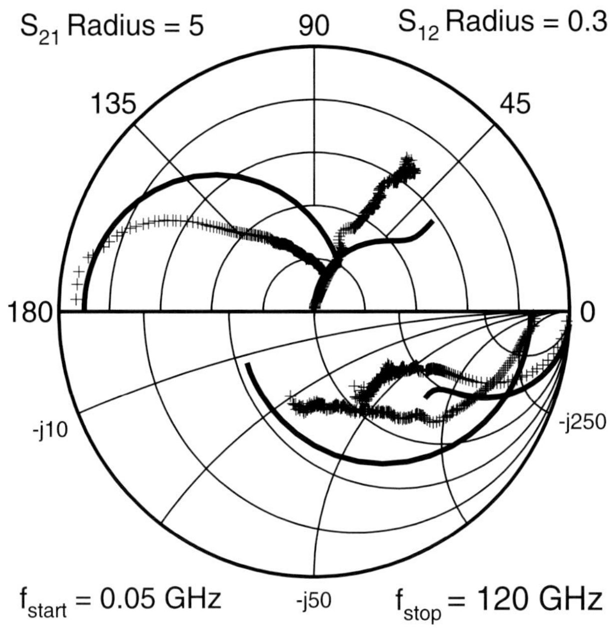

Fig. 5.63. Simulated (–) and measured (+) S-parameters of an InP/InGaAs DHBT in a Smith-polar chart from 0.25 to 120 GHz at $V_{CE} = 1.1$ V.

frequency range between 250 MHz and 120 GHz. The comparison between simulation and measurement reveals two different aspects. First, for lower frequencies the agreement is good for all four parameters and confirms the intrinsic DC modeling of the device with similar deviations, as shown before for SiGe and GaAs HBTs. The parameter S_{21} representing the gain and the reverse isolation S_{12} are modeled correctly. Further, input and output reflection S_{11} and S_{22} match the measurement properly. However, there is a growing discrepancy between simulation and measurement as a function of frequency, especially beyond 40 GHz. The reason for the deviation is found in the complicated passive environment of such a small high-speed device, similar to that realized experimentally by NTT [224]. Thus the description of the extrinsic passive small-signal equivalent circuit elements requires more attention, and the topology for the passive elements is more complicated than the one-shell approach given in Fig. 4.3. However, with the confirmation of the modeling parameters at frequencies below 40 GHz, further analysis is possible with respect to the high-speed design trade-offs described above for the intrinsic and extrinsic capacitance C_{jc}, the design of the base layer, and the collector region.

6. High Electron Mobility Transistors

6.1 General Considerations

High Electron Mobility Transistors (HEMTs) [340] are an advanced modification of the simple bulk FET, such as the MEtal Semiconductors Field Effect Transistor (MESFET). Typically, a semiconductor material (barrier) with a comparably wider bandgap is grown on top of a semiconductor material with a higher mobility and comparably lower bandgap. If the bandgap alignment of the two materials is appropriately chosen, a channel forms due solely to the alignment of the band edges and not, as for silicon MOSFETs or III-V MESFETs [460], due to an oxide/semiconductor interface or doping profiles. The channel material is not intentionally doped.
Growing only two materials on top of each other results in a single heterostructure device. When a material similar or identical to the barrier material below the channel is introduced, a second heterostructure forms, resulting in a double heterostructure HEMT.
As is typical for FETs, the channel region is controlled by a gate Schottky contact positioned above the barrier layer to control the flow of carriers in the channel. Two ohmic contacts, which are positioned on top of highly conductive and highly doped cap layers, lead to the control of the current flow from the source ohmic contact to the drain ohmic contact. Ideally the control of the channel region is purely electrostatic without current flow through the gate. In reality a low gate current is visible; the current amplification of HEMT, however, amounts to at least four orders of magnitude.

6.2 High-Speed and High-Power AlGaAs/InGaAs PHEMTs

AlGaAs/InGaAs HEMTs on GaAs substrates are standard devices in modern microwave and millimeter-wave industrial applications. HEMTs provide a combination of high-speed performance and high breakdown voltages enabling operation up to 100 GHz. This allows extremely low-noise performance and very good high-power capabilities. This section presents simulation examples of state-of-the-art industrial HEMT technology development using MINIMOS-NT.

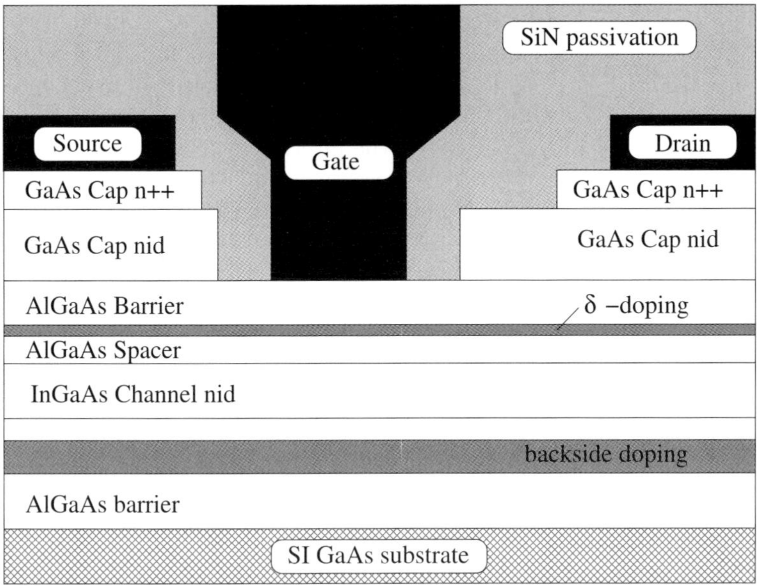

Fig. 6.1. Layer structure of a double-recess double-heterojunction AlGaAs/InGaAs pseudomorphic HEMT.

6.2.1 Device Fabrication

A typical modern HEMT structure is given in Fig. 6.1. An InGaAs channel, usually with 25% In content, is sandwiched between $Al_{0.2}Ga_{0.8}As$ layers using pseudomorphic growth on semi-insulating (s.i.) GaAs substrate. For the ohmic contact formation, highly n-doped cap layers are grown on top of the AlGaAs layers. HEMTs are unipolar devices and n-doping is achieved by so-called δ-doping or bulk layer-doping in the two AlGaAs layers by introducing silicon with a thickness of 1-5 nm and concentrations of up to 6×10^{12} cm^{-2}.
High-performance devices contain additional layers between the barriers and the cap layers. These layers vary in doping and material composition. The design of the gate-drain region, with single or double recesses, represents the key to ultimate performance, providing at the same time manufacturability, in terms of the etch processes involved, and high breakdown voltages. The key features to achieve are carrier-depleted barrier layers, to increase the robustness of the Schottky contact, and reduced gate-to-drain capacitances.
The gate metalization is realized in a T-like shape to reduce the fringing contribution to the capacitances. A slim stem of the gate foot is used, broadening towards the gate contact far away from the semiconductors.
Crucial steps for modern high-power, high-speed performance are the design and manufacturing of the source-side HEMT access region, which provides source resistances of $\leq 0.3\Omega$·mm.

With the design issues mentioned so far, the passivation/semiconductor interfaces at drain and source gain an importance similar to that of the SiO_2/silicon interface in MOSFETs. The interface is typically realized by using SiN in different modifications and thin GaAs or $Al_{0.2}Ga_{0.8}As$. This material combination leads to a strong Fermi-level pinning at the interface near the midgap of GaAs or AlGaAs. This pinning makes the recess design as described in Section 6.2.4 possible. Single recess structures are typical for high-gain, high-speed devices, due to the scaled source region. Double-recess structures are typical for AlGaAs/InGaAs power HEMTs, as shown in Fig. 6.1.

6.2.2 Calibration Methodology and DC Analysis

Modern III-V TCAD can provide excellent control of heterostructure device properties. During the calibration of a HEMT TCAD model for a specific process, the derivation of material properties has to be in agreement with both DC currents and the charge control, i.e. the simulation and extraction of capacitances and other DC and RF quantities, as provided in the next section. In addition to the layers added for the design of the gate-drain region, thin etch-stop layers provide etch depth control for high-yield and high-uniformity processes, making use of the high chemical selectivity of GaAs and $Al_{0.2}Ga_{0.8}As$ etch rates. Etch-stop layers with thicknesses smaller than 2 nm can be neglected in the simulation for numerical efficiency, if a possible effect was properly evaluated by simulation including the etch-stop layer.
Even the simulation of simple transfer characteristics of a single-recess device contains a detailed set of information which can be matched by simulation with high precision for a specific process. The transfer characteristics at low V_{DS}, e.g. at V_{DS}= 1 V, can be divided into several bias regions, as shown in Fig. 6.2.
The different bias regions can be used to isolate different quantities, as given in the following enumeration:

1. The device pinch-off region is determined by the surface and buffer isolation.
2. The maximum transconductance is affected by channel mobility and channel charge density.
3. The parasitic MESFET behavior is influenced by the AlGaAs/InGaAs heterostructure bandgap discontinuity (real-space-transfer) and by the contacting of the channel (ohmic contact situation).
4. The opening of the gate diode at $V_{GS} \geq V_{bar}$ is characterized by a significant forward current.
5. The maximum drain current I_D is determined by overall δ-doping, indium content in the channel, access resistances R_S and R_D, and self-heating.

Even within very advanced HEMT processes, several parameters vary within certain margins. The material parameters to be calibrated include:

- The exact carrier mobility in the channel

- The activation of the δ-doping: high-performance HEMTs are typically doped with δ-doping, which corresponds to a degenerate bulk doping concentration
- The doping activation, which is further influenced by surface and buffer compensation effects.

The variations further include geometrical quantities, such as:

- The exact gate-to-channel separation (vertical depth)
- The exact gate recess geometry within the lateral accuracy of the definition of a submicron gate process.

A non-ambiguous calibration method is necessary with respect to such variations. The calibration can be achieved by a combination of bias scans, temperature dependent measurements, RF measurements, and, if available, pulsed measurements. Fig. 6.3 and Fig. 6.4 give the simulation of the maximum drain current I_{Dmax} and the threshold voltage V_{thr} as a function of the active doping concentration for the doping activation.
In Fig. 6.3 and Fig. 6.4, the expected increase of the threshold voltage V_{thr} with reduced doping and the increase of I_{Dmax} with increasing doping can be observed. Performing the same analysis of the gate-to-channel separation for a given doping, one can obtain the variation of the threshold voltage by comparing it to different actual wafer runs.

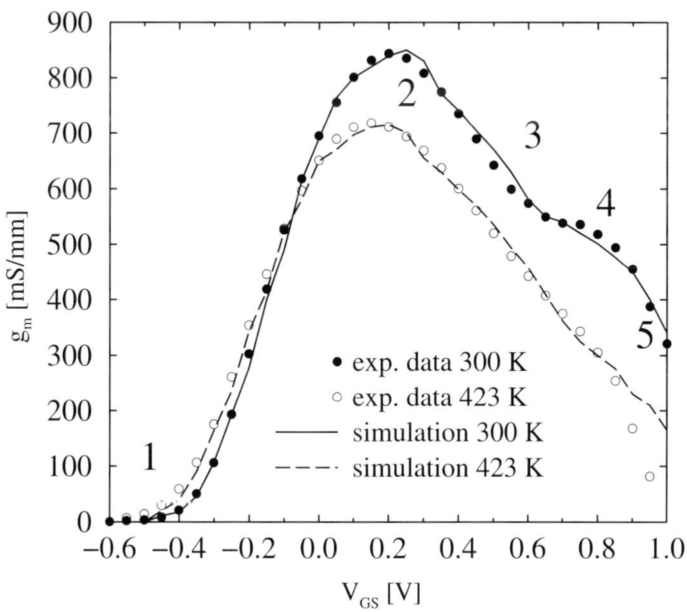

Fig. 6.2. Simulated and measured transconductance g_{m} as a function of lattice temperature including self-heating at V_{DS}= 1 V.

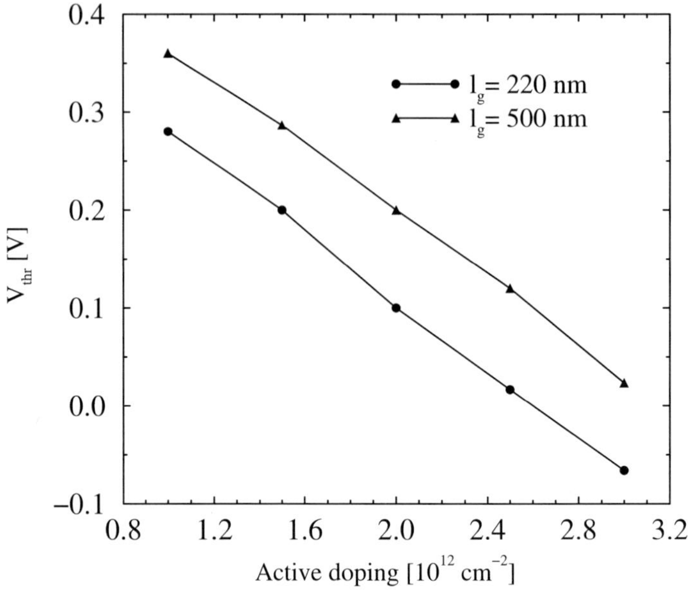

Fig. 6.3. Simulated threshold voltage V_{thr} as a function of active doping level N_{D} for two different gate lengths l_{g}.

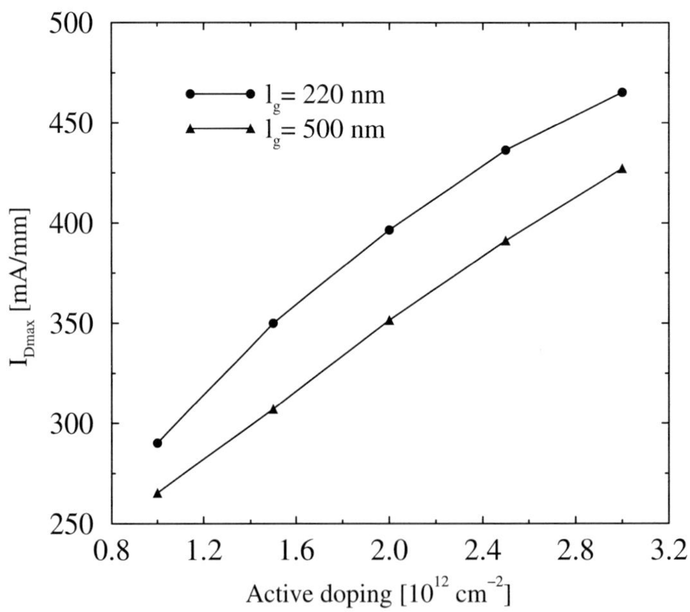

Fig. 6.4. Simulated maximum drain current I_{Dmax} as a function of active doping level N_{D} for two different gate lengths l_{g}.

In Fig. 6.2, the measured dependence of the transconductance on temperature is shown between T_{sub}= 300 K and T_{sub}= 423 K. The decrease of $\Delta g_m/\Delta T_{sub}$ for this on-wafer measurement amounts to 100 mS/mm/100 K. The overall reduction for the observed temperature range amounts to $\approx$ 22%. The V_{GS} voltage, for which $g_{m,max}$ is obtained, changes to more negative values. This effect can be correlated to a balance of the effective reduction of the Schottky barrier height and the bias shift due to the increase of the parasitic ohmic resistances. The temperature-dependent simulation can precisely reproduce the g_m curve for both temperatures, except for the decrease at $V_{GS} \geq 1$ V, where the simulation starts to overestimate the g_m for both temperatures. This overestimation stems from the fact that self-heating is only modeled with fitted boundary conditions. To match the g_m curve with such high precision, the activation level of the upper δ-doping was adjusted to fit the drain current I_D, the transconductance g_m, and the threshold voltage V_{thr}.
Once the simulation of the transfer characteristics is accomplished with the methodology described, a more complete extraction of parameters is required. The following steps can be applied to understand the output characteristics:

- The comparison to the simulation of both continuous-wave (cw) DC and pulsed DC measurements with a very low-duty cycle
- The separation of the influence of self-heating and of dispersion in pulsed measurements
- The comparison of the linear part of the output characteristics between V_{DS}= 0 V and the knee voltage $V_{knee} \approx 1$ V.
- The inclusion of impact ionization in the simulation for the open channel situation, i.e. $V_{GS} \geq V_{thr}$
- The inclusion of reverse diode effects in the simulation of the output characteristics for $V_{DS} \approx V_{breakdown}$.

Fig. 6.5 compares a simulation without self-heating to the measured output characteristics with a very short pulse width of 200 ns with 1 ms pulse repetition. With pulsed operation mode available, the decreasing effect of self-heating can be visualized and separated to control the calibration of the isothermal simulation. Effects of dispersion can be neglected in these measurements due to the well-developed process, so that for the branches in the output characteristic, the changes can be attributed to self-heating only. Fig. 6.6 then gives the full comparison with self-heating accounted for in the simulations. Summarizing the calibration methodology, one can obtain the following information in parallel calibration of:

- the exact channel and spacer mobility from saturation current obtained from cw and pulsed measurements,
- the doping activation of degenerate doping densities from saturation current and pinch-off voltage, and

- the gate recess geometry from the analysis of pinch-off voltage and gate capacitances.

If this methodology includes several device characteristics and different thermal and RF dispersive conditions, a detailed set of information is available and the amount of free parameters within the calibration is reduced significantly. As can be seen in Fig. 6.6, very good agreement in a wide bias range is possible. As PHEMTs are currently in mass production in industry, the amount of process statistics available is considerable. This enables the inverse modeling of the processes in a more statistical sense: with the knowledge of mean values, statistical and systematic variations, statistical methodologies can be applied, as is done, for example, in [279].

6.2.3 RF Analysis of AlGaAs/InGaAs PHEMTs

As was stated in the previous section, RF quantities are a valuable means for further analysis of charge control in HEMTs. Fig. 6.7 compares simulated and measured S-parameters of a 2×60 μm $Al_{0.2}Ga_{0.8}As$/ $In_{0.25}Ga_{0.75}As$ HEMT at an elevated substrate temperature of T_{sub}= 373 K.
A crucial requirement for the comparison of measured and simulated RF quantities is the exact definition of a microwave reference plane to separate the microwave effect within and beyond the terminals of the simulation domain, as outlined in Chapter 4.

RF Parasitic Extraction. For a simulation example of a PHEMT with a gate length of l_g = 140 nm the resistive, capacitive, and inductive parasitic elements are extracted. The values are given in Table 6.1. R_{SM} and R_{DM} describe the

Table 6.1. Parasitic elements applied for PHEMT simulation

Element	unit		Element	unit	
R_S	[Ω mm]	0.3	L_S	[pH]	0.5
R_G	[Ω]	1.9	L_G	[pH]	22
R_D	[Ω mm]	0.35	L_D	[pH]	23
R_{SM}	[Ω mm]	0.1	C_{pgs}	[fF]	9
R_{DM}	[Ω mm]	0.1	C_{pgd}	[fF]	3
port extension	[ps]	1.45	C_{OUT}	[fF]	4.5

resistances of the metalization at source and drain. For this HEMT technology, the inductive and capacitive parasitic elements are extracted for a gate-width with W_g = 2×60 μm. Table 6.2 gives the values for the intrinsic device parameters (see Fig. 4.1), as obtained from simulation and by a consequent extraction procedure for the pseudomorphic HEMT. The extracted intrinsic elements are given for

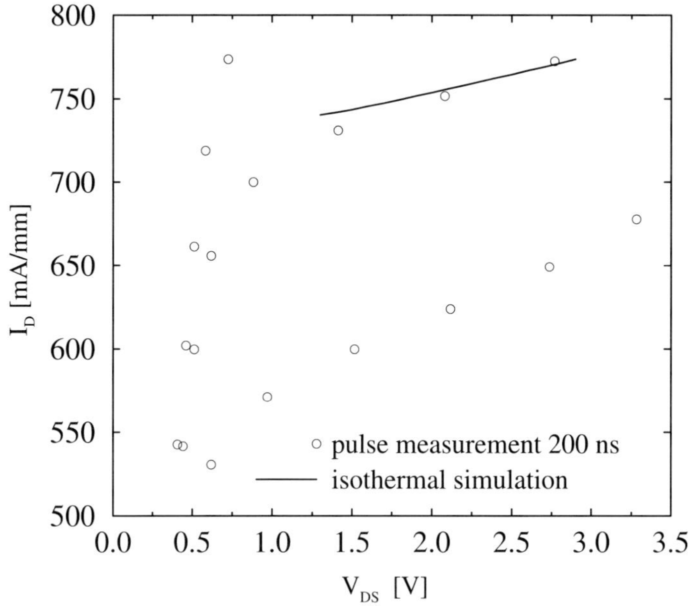

Fig. 6.5. Pulsed and cw measurements of the output characteristics compared to simulation at 300 K.

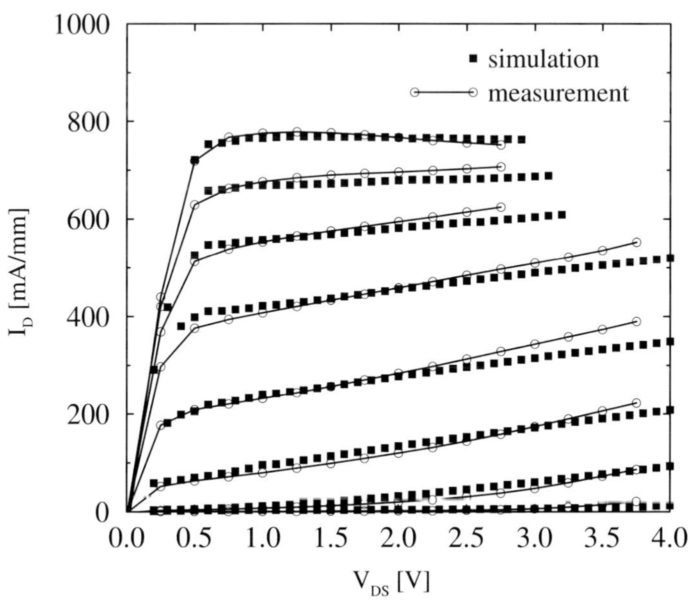

Fig. 6.6. Simulated output characteristics compared to cw measurements at $T_{\mathrm{sub}} =$ 300 K.

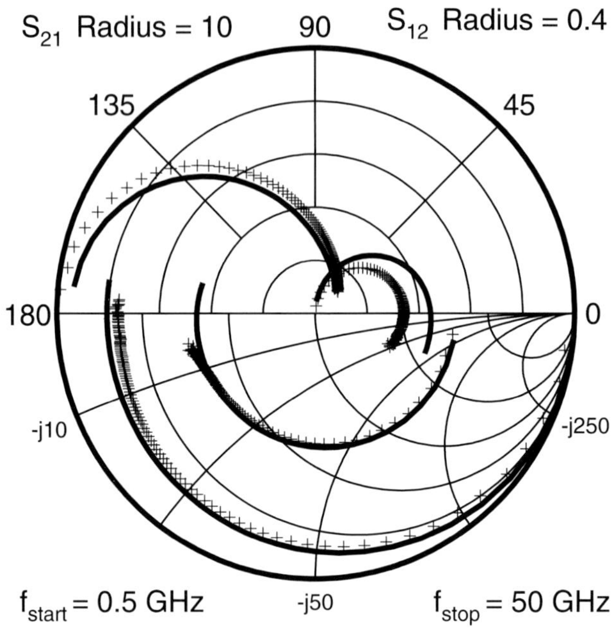

Fig. 6.7. Simulated (–) and measured (+) S-parameters $S_{\rm ij}$ as a function of frequency between 0.5 GHz and 50 GHz at $T_{\rm sub}$ = 373 K.

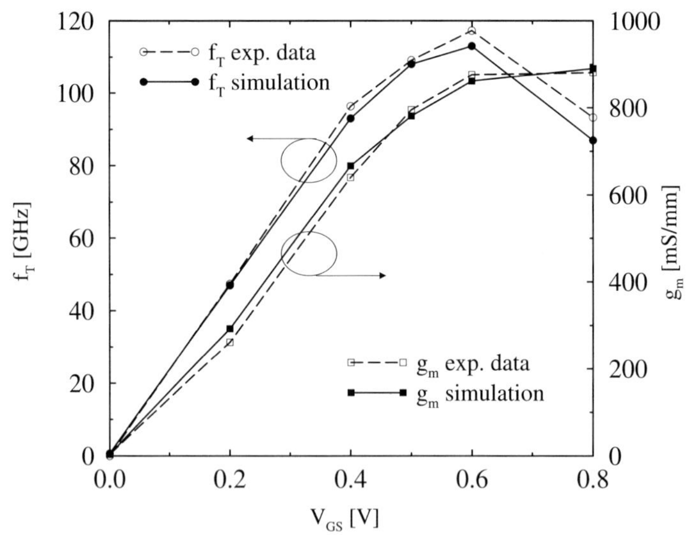

Fig. 6.8. Simulated and measured transconductance $g_{\rm m}$ and current gain cut-off frequency $f_{\rm T}$ as a function of the gate-source bias $V_{\rm GS}$ at $V_{\rm DS}$= 1 V.

Table 6.2. RF simulation results extracted for a PHEMT

$g_{\rm m}$ [mS/mm]	$C_{\rm gs}$ [fF/mm]	$C_{\rm gd}$ [fF/mm]	$C_{\rm ds}$ [fF/mm]	$g_{\rm ds}$ [mS/mm]	$R_{\rm gs}$ [Ω mm]	$R_{\rm gd}$ [Ω mm]	τ [ps]
1000	1060	280	150	50	0.11	0.31	0.5

$V_{\rm DS}$= 1 V in good agreement with measurements. The bias dependence of the intrinsic elements is discussed in the next section for a similar device using the same technology.

Cut-off frequency $f_{\rm T}$. The most important figures-of-merit, the cut-off frequencies $f_{\rm T}$ and $f_{\rm max}$, are extracted. Fig. 6.8 compares simulated and measured current gain cut-off frequency $f_{\rm T}$ as a function of $V_{\rm GS}$ at $V_{\rm DS}$= 1 V. It can be observed that $f_{\rm T}$ follows the behavior of the transconductance $g_{\rm m}$ extracted from RF simulations for this transistor. A maximum value of $f_{\rm T}$= 120 GHz is obtained in both simulation and measurement. Understanding the bias dependence of the small-signal elements constituting $f_{\rm T}$ is a crucial point for understanding the device.

Bias Dependence of Small-Signal Equivalent Circuit Elements. The dependence of $g_{\rm m}$ on $V_{\rm GS}$ at $T_{\rm L}$= 300 K is shown in Fig. 6.8. Similarly to Fig. 6.2, the channel is depleted at low $V_{\rm GS}$ voltages. With increasing $V_{\rm GS}$, the size of the depletion region diminishes and the channel fills with carriers. The maximum $g_{\rm m}$ is reached when the carriers pass the channel without conduction in the AlGaAs barrier and buffer. The limit for higher $V_{\rm GS}$ voltages is given by the opening of the gate diode in forward direction.

Fig. 6.9 compares $C_{\rm gs}$ as a function of $V_{\rm GS}$ as extracted from simulations and from measurements. The good agreement shows the precise charge control achievable in the simulation. The charge control is necessary to separate the transport quantities, such as velocities and mobilities, and the charge carrier densities.

Fig. 6.10 compares the simulated $C_{\rm gd}$ with measurements. The decrease of $C_{\rm gd}$ with $V_{\rm DS}$ is due to the approach of the mean channel carrier concentration towards the gate contact on the drain side of the gate. The agreement for $C_{\rm gd}$ demonstrates the precise charge control achievable at the drain side of the gate region, which is especially important due to its impact on high-power performance. Most important for the $C_{\rm gd}$ are the fringe contributions due to the coupling of gate metalization and to the drain cap carrier concentration.

The dependence of the cut-off frequency $f_{\rm T}$ on $V_{\rm DS}$ is investigated, as $C_{\rm gd}$ decreases as a function of $V_{\rm DS}$. For pseudomorphic GaAs HEMTs, the decrease of $C_{\rm gd}$ [415] is overcompensated by the increase of $C_{\rm gs}$ with $V_{\rm DS}$ and by the decrease of $g_{\rm m}$. Thus for $V_{\rm DS}$ voltages $\geq$ 1 V, $f_{\rm T}$ and $f_{\rm max}$ decrease as functions of $V_{\rm DS}$ in AlGaAs/InGaAs/GaAs HEMTs. On the contrary, in the newly developed AlGaN/GaN HEMTs of the same gate lengths, $f_{\rm T}$ and especially $f_{\rm max}$ rise for the greater part of $V_{\rm DS}$ values. The basic reason for this difference is the charge control in the channel. HEMT performance is improved due to the superior

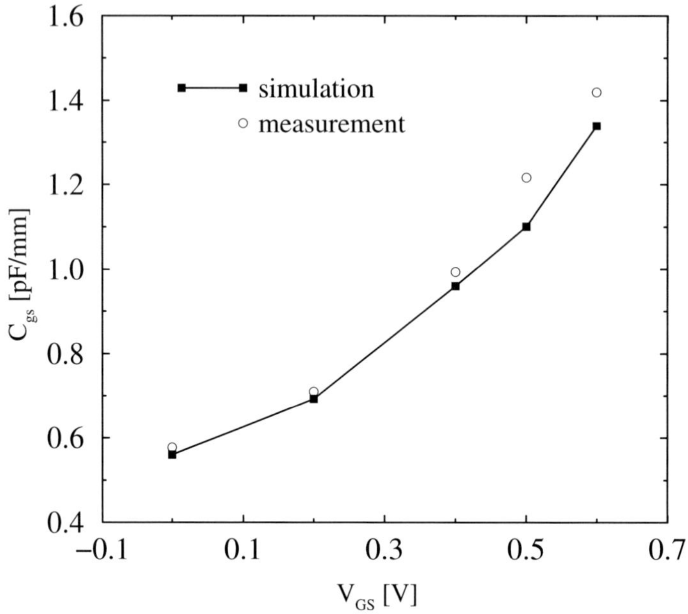

Fig. 6.9. Simulated gate-source capacitance C_{gs} as a function of V_{GS} compared to measurements at V_{DS}= 1 V.

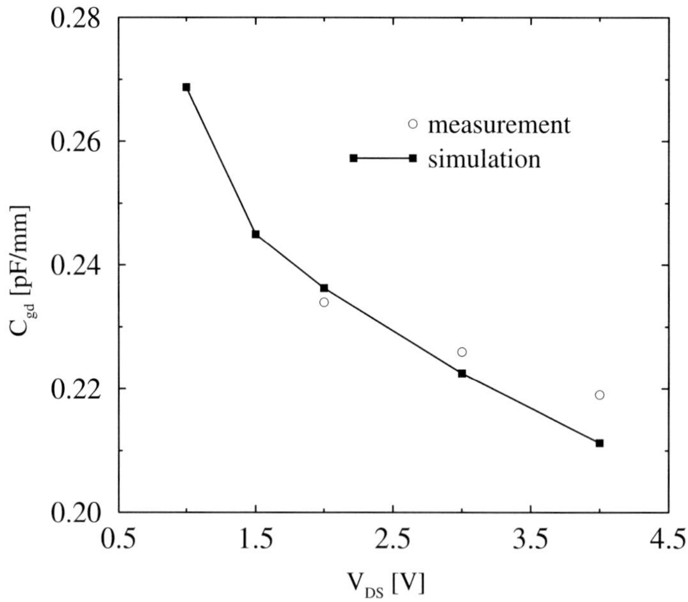

Fig. 6.10. Simulated gate-drain capacitance C_{gd} at constant V_{GS} as a function of V_{DS} compared to measurements.

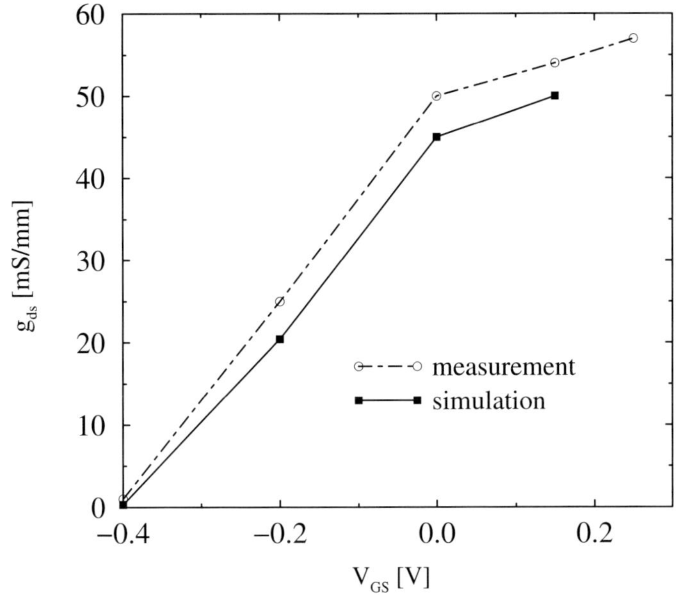

Fig. 6.11. Simulated and measured RF output conductance g_{ds} as a function of V_{GS} for V_{DS}= 2 V.

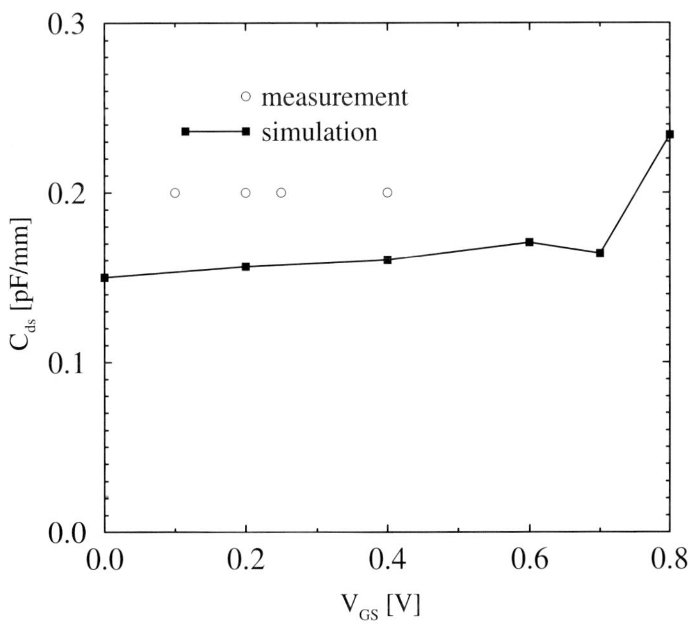

Fig. 6.12. Simulated and measured drain-source capacitance C_{ds} as a function of V_{GS} for V_{DS}= 2 V.

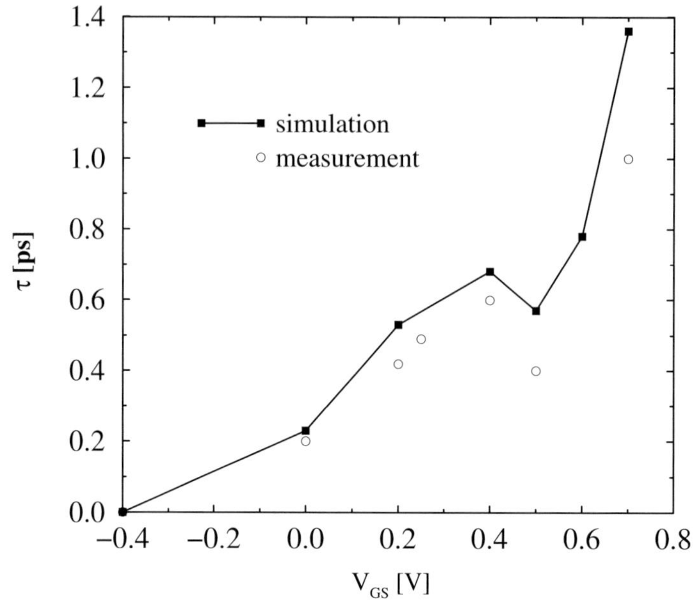

Fig. 6.13. Simulated phase term τ as a function of V_{GS} bias for V_{DS}= 2 V compared to measurements.

transport properties of the channel material. However, the device characteristics are not as good, as could be estimated from ballistic transport considerations. A similar statement is made in [461]. One reason for this is a transfer in k-space, i.e. electrons surmount the separation of the first and second valley in the InGaAs channel material and are effectively slowed down, due to the increase in the effective mass in the second valley. This is accompanied by a reduction of the effective carrier velocity, which results in a reduction in g_{m} with V_{DS}. The main reason for the performance degradation, however, is the real space transfer, as stated in [411]. The capacitance C_{gs} increases as a function of V_{DS} due to the spread of carriers into the drain-side barrier layer and buffer when passing the gate. This spread is due to scattering events.

The output conductance g_{ds} as well as the output capacitance C_{ds} determine the output match of the active transistor towards the environment in a circuit. The simulated output conductance g_{ds} is given in Fig. 6.11. The extracted RF g_{ds} is always positive. The qualitative and the quantitative behavior of g_{ds} is modeled correctly. A typical value of g_{ds}= 50 mS/mm is observed for a class A bias, i.e. $I_{\mathrm{D}} = 50\%\ I_{\mathrm{Dmax}}$, for this pseudomorphic device. Fig. 6.12 shows the bias dependence of the measured C_{ds} as a function of V_{GS} for V_{DS}= 2 V. A nearly constant behavior is also observed in the simulation which underestimates the available experimental data. An explanation for the underestimation can be

found in (4.15) in Chapter 4. C_{ds} is very sensitive to additional contributions from the buffer layer, which are not properly modeled with the trap models available. Fig. 6.13 compares the phase term τ as a function of V_{GS} as extracted from simulations and measurements. Both the qualitative and the quantitative behavior are reproduced correctly by the simulations. τ rises as a function of V_{GS} for V_{DS}= 2.0 V. A maximum value can be observed for the maximum g_m. τ slightly drops with higher V_{GS}. When the diode opens at $V_{GS}\approx$ 0.7 V, τ rises sharply in the simulation in agreement with the measurements.
Quantization effects occur in a HEMT channel, due to channel layer thicknesses below the de Broglie wave-length of the electron. The quantization effects influence the transport properties such as the carrier mobility relatively modestly. They are accounted for by calibration of the mobility in AlGaAs/InGaAs and InAlAs/InGaAs heterostructures. The effects are more significant with respect to the charge control, as the tunneling of the carriers through the heterobarriers determines the capacitances. The interface models given in Chapter 3 are therefore of utmost importance. As shown in the previous examples, the interface models can effectively account for the heterojunction transport effects occurring in the simulation as a function of bias.

6.2.4 Breakdown Analysis in AlGaAs/InGaAs PHEMTs

The optimization of FETs towards higher operational voltages is a very typical TCAD task. Primary analysis of the device structure concentrates on the calculation of the carrier distribution and the field distribution for several biases of operation. The optimization strategy of AlGaAs/InGaAs HEMTs towards increased breakdown voltage includes:

- The evaluation of doping concentrations for cap layers and δ-doping to deplete the high-field region to carrier concentrations below 10^{13} cm^{-3}
- The expansion of the depletion region to reduce and avoid electric field peaking beyond a critical field for all V_{DS} voltages statically and dynamically applied to the device terminals
- The evaluation of the impact of the breakdown optimization on device speed and gain.

Single-Recess Devices versus Double-Recess Devices. The strategy for a single-recess device is to find doping concentrations and recess geometries suitable for the gain and the breakdown requirements, which are traded off against each other. Devices with smaller gate recesses reach more gain; those with larger recesses deliver higher breakdown voltages. For the double-recess approach, the optimization concept is based on a carrier-depleted inner and outer recesses. As shown in Fig. 6.16, the electric field extends between the high electron concentration at the lateral edge of the outer recess and the gate metal. In the case of the depleted recess, the cap layer thickness is limited and the inner recess is

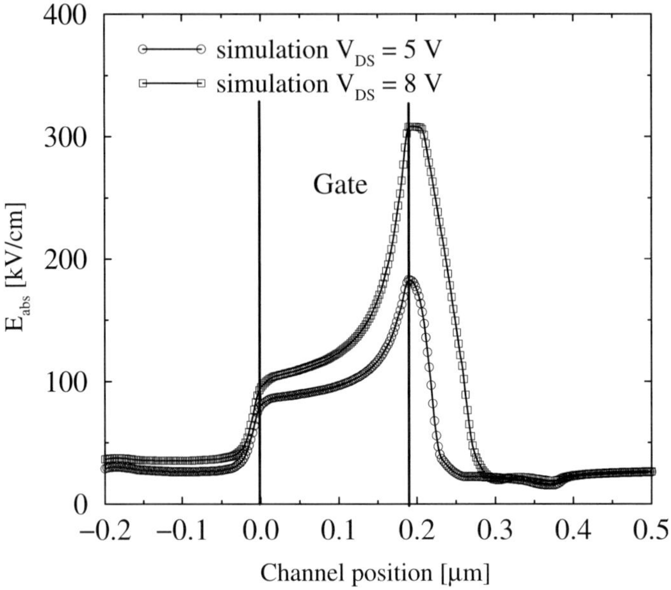

Fig. 6.14. Simulated electric field along the channel for V_{DS}= 5 V and V_{DS}= 8 V at $I_D\approx$ 250 mA/mm for a l_g= 190 nm PHEMT with a depleted recess.

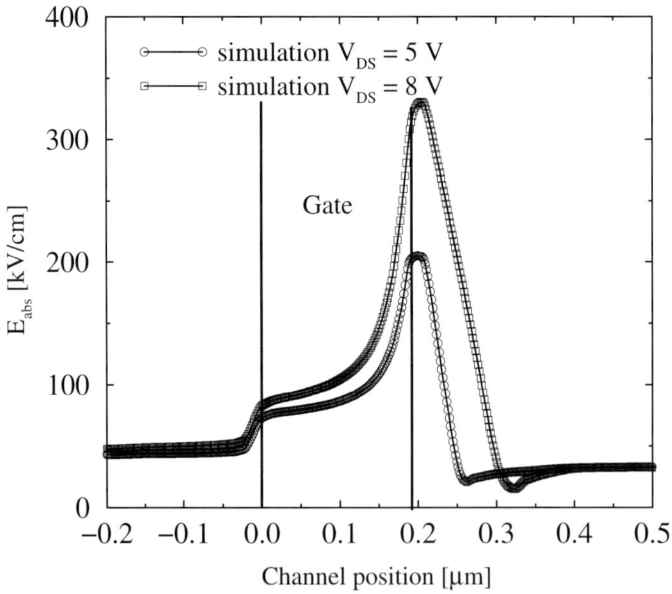

Fig. 6.15. Simulated electric field along the channel for V_{DS}= 5 V and V_{DS}= 8 V at $I_D\approx$ 250 mA/mm for a l_g= 190 nm PHEMT with a non-depleted outer recess.

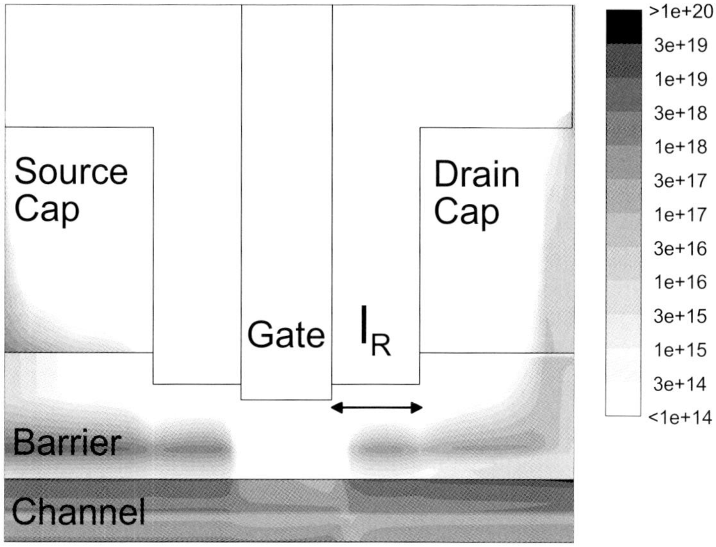

Fig. 6.16. Simulated electron concentration (given in [cm^{-3}]) in a PHEMT for V_{DS}= 5 V at I_D= 150 mA/mm.

controlled by the surface potential. The carrier concentration in the cap layer needs to be limited enough to enable the surface potential to be able to deplete the inner recess and the cap. The doping concentrations in the cap and in the δ-doping layer and the recess lengths have to be chosen accordingly, as given in Fig. 6.16. Increased length of the second recess relaxes the maximum fields in the spacer and thus in the channel at high operational V_{DS} bias, typically $\geq$ 5 V for devices with a gate length of $\leq$200 nm. The exact geometry determines for which bias conditions the underlying barriers and channel are depleted and to what extent. Using this concept, extremely high on- and off-state breakdown voltages can be achieved for pseudomorphic AlGaAs/InGaAs/GaAs high-power HEMTs. If the second recess is chosen not to be depleted, as shown Fig. 6.15, then the length of the inner recess is of primary importance and determines the fields in the barrier and the channel. A sharp drop occurs between the drain side of the gate and the lateral edge of the first recess similar to the single-recess situation. The sum of the inner and outer recess lengths leads to a marginal additional relaxation. The depth of the inner recess is of no importance once it exceeds a certain value. For the concept of a non depleted recess, the first recess protects the diode especially for the off-state. It is found experimentally that, as a rule of thumb, when such a concept for a l_g= 200 nm device is used, a maximum breakdown voltage of BV_{GD}= 11 V evolves.
Comparing Fig. 6.14 and Fig. 6.15, the advantage of the depleted cap can be seen: there is a reduction in the electric fields in the channel. Breakdown voltages as

high as 25 V can be obtained for the same gate length for devices with a depleted recess.

Inclusion of Breakdown Effects. For more advanced simulations, the breakdown mechanisms can actually be included in the model itself. The breakdown effects occurring in PHEMTs at both high V_{GS} and V_{DS} voltages have been thoroughly discussed in the literature [25, 120]. Two things are central for understanding the device breakdown: first, the hardness of the gate diode and, second, impact ionization in the channel and in the barrier layers. Optimization of a device towards high voltage operation therefore requires the implementation of the Schottky contact model, including tunneling and calibration of the free parameters involved in (3.75) and (3.78). The following methodology is proposed to calibrate the breakdown behavior in a given HEMT process:

- Modeling of the output characteristic at low V_{GS} and high V_{DS} voltages (off-state situation) to calibrate tunneling parameters, such as tunneling length or tunneling mass
- Calibration of impact ionization behavior from a high-current V_{GS} bias at medium V_{DS} (on-state situation for opening gate diode)
- Comparison of the model to MC simulations for the same potential profile

The physical background for this methodology can be found in (3.193), for example, in Chapter 3. In a very general view, the local impact ionization rate is proportional to the local current density. For the off-state behavior, the zero-current situation is approximated so that the breakdown is determined by the diode breakdown.
Generally, one could argue that, with sufficient modeling, calibration for a specific process technology is not necessary and that MC simulation alone could provide sufficient information. However, some factors are process-dependent and strongly suggest a calibration method:

- The high density of states at the SiN/barrier interface, which varies to a certain extent
- The exact shape of the silicon profile in the AlGaAs barrier, which varies within certain margins

A similar methodology was used to investigate breakdown issues in PHEMTs with the device simulator DESSIS [227].

6.3 High-Speed InAlAs/InGaAs HEMTs on InP and GaAs

HEMTs based on the InAlAs/InGaAs material system are usually used for the highest frequencies of interest. Currently, frequencies used in applications range from 60 GHz to 220 GHz in analog MMICs and digital circuits.

6.3.1 Device Manufacturing

Fig. 6.17 gives a layer structure of an InAlAs/InGaAs HEMT on either GaAs or InP substrates. In the case of metamorphic growth on GaAs, an additional quaternary buffer is placed between the semi-insulating GaAs substrate and the InAlAs buffer layer to compensate the lattice mismatch of the channel material $In_{0.53}Ga_{0.47}As$ and the GaAs substrate.

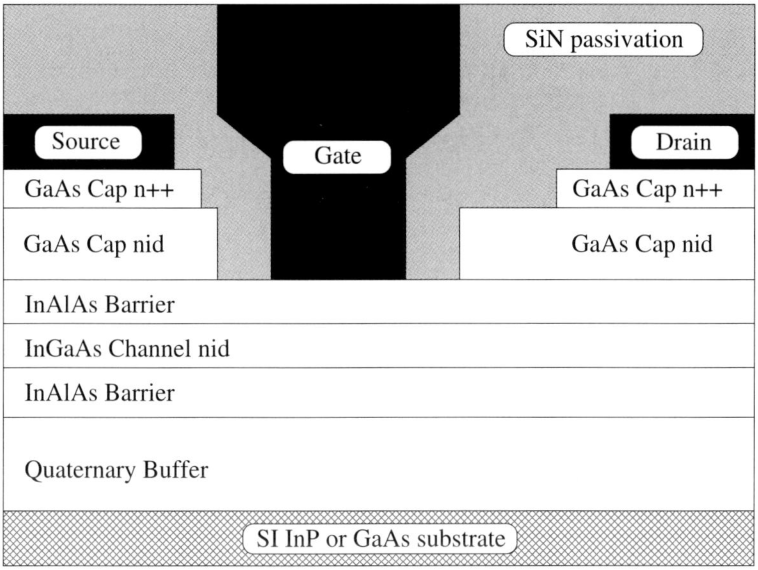

Fig. 6.17. Layer structure of a (metamorphic) double heterojunction InAlAs/InGaAs HEMTs on GaAs or InP substrate.

Device speed depends to a large extent on the choice of the indium content x of the $In_xGa_{1-x}As$ channel during the epitaxial growth. The material is either grown lattice-matched to InP with $x= 0.53$ or pseudomorphically with $x \geq 0.53$. Dual channel concepts are frequently used for carrier cooling to avoid high-field effects in the indium-rich material. For high-power operation, combined $In_xGa_{1-x}As$/InP dual-channel concepts are also applied. Typical overall channel layer thicknesses amount to 20 nm. Since InAlAs/InGaAs HEMTs always address the highest frequencies of application, gate definition modules suitable for smallest feature size definition are required. Currently the gate length ranges from 150 nm to about 70 nm for statistically reproducible processes. Devices with gate lengths as small as 30 nm have been demonstrated. Critical process manufacturing issues involve the control of the InAlAs/SiN interface and the exact definition of the gate recess to suppress breakdown effects. For this purpose, the InAlAs barrier can be capped by InP [554].

6.3.2 DC Simulation

InAlAs/InGaAs/InP HEMTs. The calibration methodology for the simulation of InAlAs/InGaAs HEMTs is similar to the methodology applied for PHEMTs [414]. However, since high-field effects occur at lower voltages than in GaAs PHEMTs, some additional steps are necessary, as given below.
Fig. 6.18 shows the transfer characteristics for a dual channel $In_{0.48}Al_{0.52}As$/ $In_{0.66}Ga_{0.34}As$/ InP device with l_g= 150 nm as a function of temperature T_{sub}. The device shows a DC transconductance g_m of nearly 1000 mS/mm at V_{DS}= 0.75 V. The transfer characteristic further shows a decrease of the transconductance g_m and the maximum drain current I_{Dmax} with rising temperature.

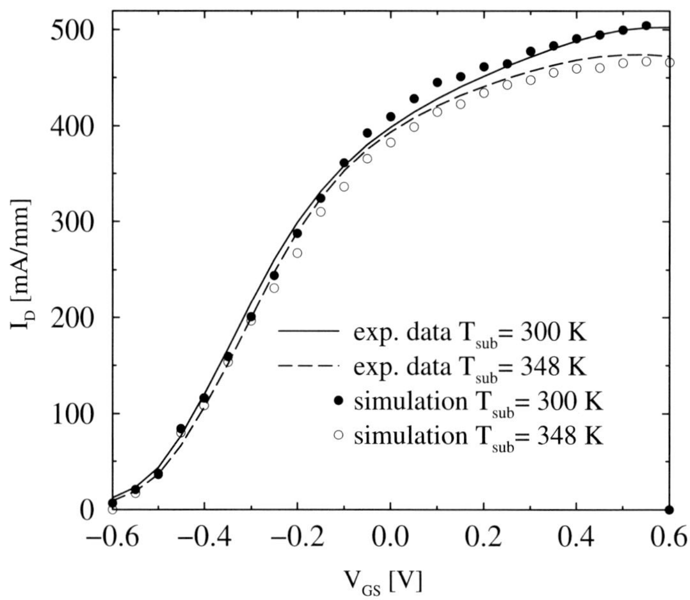

Fig. 6.18. Simulated and measured transfer characteristics as a function of temperature T_{sub} at V_{DS}= 0.75 V.

Fig. 6.19 shows the simulated output characteristics and comparison with measurements for T_{sub}= 300 K for the same device as in Fig. 6.18. Very good agreement is achieved between simulation and measurement using the model parameter values given in Chapter 3.
As a principal result, the output conductance in Fig. 6.19 cannot be simulated without taking both electrons and holes and their generation/recombination into account. Four factors of influence are found to be significant for this high-gain device and to be responsible for the output conductance in this simulation analysis.

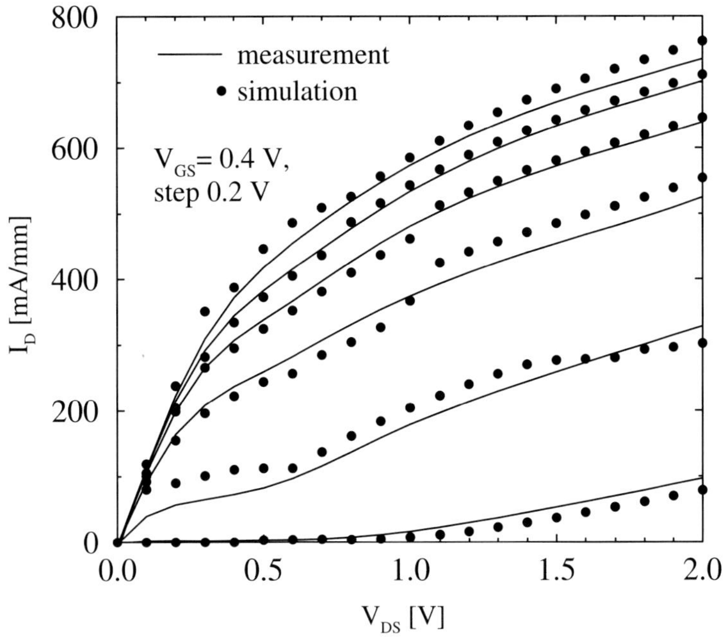

Fig. 6.19. Simulated and measured output characteristics of a pseudomorphic dual-channel InP HEMT with l_g= 150 nm.

The transient behavior of the generation/recombination is not included. These factors include:

- The surface potential/surface charges at the SiN/InAlAs interface
- The source recess geometry
- The SRH and Auger generation/recombination
- Impact ionization

Often a so-called "kink" is observed in the output conductance g_{ds} [490]. An analysis of Fig. 6.19 shows that the kink is only the result of a very special combination of the factors mentioned above [490]. The factors have an influence even if a "kink" is not visible at all. If a stable surface treatment and SiN passivation is provided, the surface potential for $In_{0.52}Al_{0.48}As$ is only subject to minor changes due to the pinning of the Fermi level. The overall geometry of the source side influences the current flow. First, for the pinned surface potential at the SiN/InAlAs interface, a depletion region prevails near the surface at both source and drain, which attracts holes from the material background concentration and the manifold generation processes (SRH, impact ionization, Auger). This results in a change of the source side potential near the gate and thus results in changes in the currents I_G and I_D. Second, and especially relevant for the RF properties,

the depth of the recessed cap d_R, in combination with the available cap doping, determines the current path in the barrier layer and thus the source resistance R_S. Since the current path is involved, the exact ohmic contacting at the source side is decisive.
The SRH generation/recombination increases significantly when increasing V_{DS} between 0.5 V$\leq V_{DS} \leq$2 V. This is due to the fact that the shape of the space charge region changes drastically in the bias range 0.5 V$\leq V_{DS} \leq$1.5 V for constant V_{GS}. The pronounced asymmetrical shape of the space charge region develops under the gate for normal operation, since normally 0$\approx V_{GS} \ll V_{DS}$.
Impact ionization in the InAlAs/InGaAs HEMT introduces changes in the output conductance due to the following factors:
An additional contribution is added to the drain current I_D due to increased electron concentration. However, this contribution is small for the typical region 1 V$\leq V_{DS} \leq$2 V. The generation of holes enhancing the gate-current I_G and causing a change in the effective potential near the gate is much more significant. This potential shift acts like a V_{DS}-dependent positive shift of the gate potential. Once the accumulation of holes at the source side gets higher, there are shifts of the output conductance g_{ds} resulting in "kinks." Higher impact ionization due to self-heating for channel materials with high indium contents is an additional effect at high V_{DS} bias. The changes in the output conductance can also be seen in other device characteristics. Fig. 6.20 shows the input characteristics, i.e. I_G as a function of V_{GS}, for the l_g= 150 nm device with V_{DS} as a parameter. A strong increase of the gate current I_G can be observed leading to the well-known bell-shaped curve. The gate currents I_G are influenced by direct/SRH generation/recombination and by impact ionization.
Aside from impact ionization, other generation and recombination processes do significantly influence the gate currents of the HEMT (due to the direct bandgap of InAlAs/InGaAs). Actually, due to the space charge region at the gate, all generation/recombination processes must be considered for a detailed analysis. In a HEMT biased at V_{GS} at $g_{m,max}$, carrier depleted regions (barrier) are adjacent to highly doped regions (semiconductor caps) and regions with high carrier concentrations and current flow. This carrier distribution is a prerequisite for the following effects: SRH and direct generation/recombination in the barrier, Auger effects in the caps, and Auger and impact ionization in the channel. Since AlGaAs, InAlAs, and InGaAs are direct semiconductors for typical material compositions used in HEMTs, direct processes are considered. Fig. 6.20 and Fig. 6.21 show that the gate currents can be reproduced well using the models presented in Chapter 3.
Fig. 6.21 shows the simulated and measured input characteristics of the previous device as a function of temperature. The gate current I_G increases with increasing substrate temperature T_{sub}, which is typical for an indium content $x \geq 0.53$. One reason for this is the decrease of the bandgap and the positive temperature coefficient of the impact ionization, which is discussed in Chapter 3.

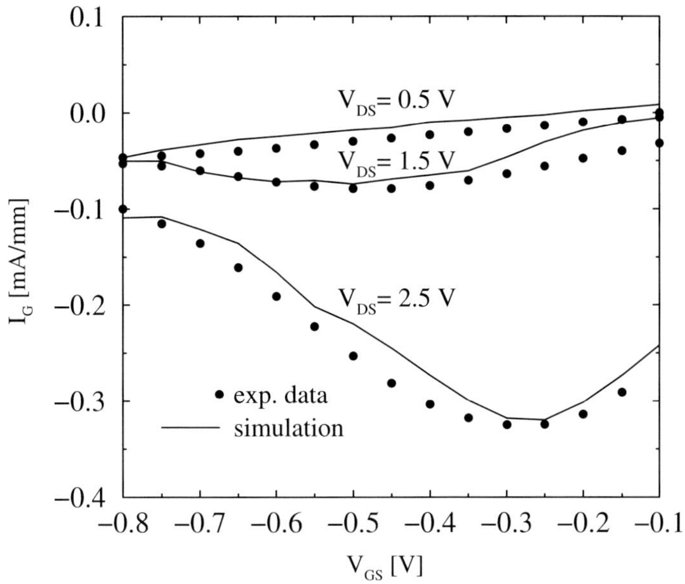

Fig. 6.20. Simulated and measured gate current I_{G} as a function of V_{GS}.

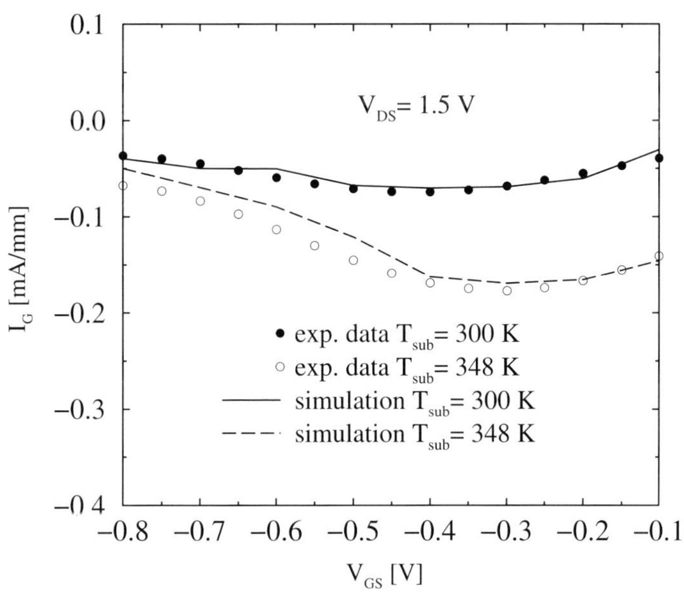

Fig. 6.21. Simulated and measured input gate current I_{G} as a function of V_{GS} with the substrate temperature T_{sub} as a parameter.

Metamorphic HEMTs. For digital and low-noise applications, metamorphic InAlAs/InGaAs HEMTs on GaAs are developed similar to those on InP for gate lengths between l_g= 150 nm and 70 nm. Fig. 6.22 shows the simulated and measured transfer characteristics of a 150 nm metamorphic HEMT on GaAs as a function of V_{GS} voltage in the low-power range for two V_{DS} voltages.
Modeling of the device in good agreement with measured data is possible with the calibration methodology described for the pseudomorphic GaAs HEMT above. The transfer characteristics can be precisely fit for the two V_{DS} voltages, as shown in Fig. 6.22.

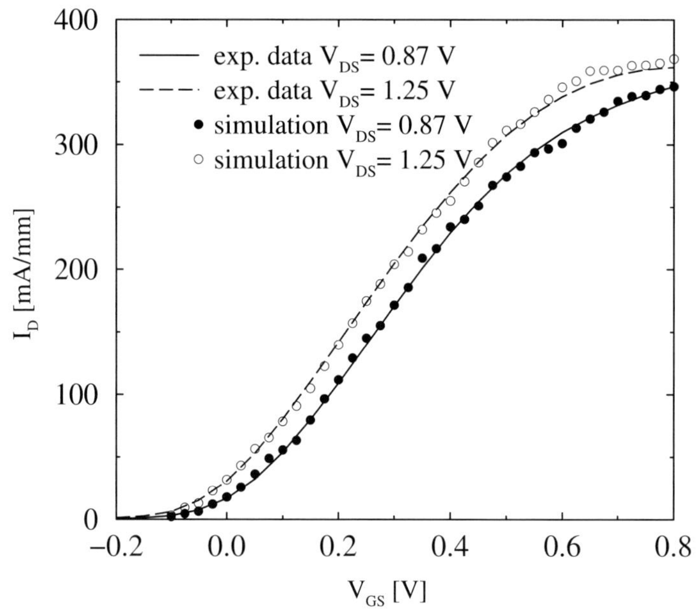

Fig. 6.22. Simulated transfer characteristics of an enhancement-type InAlAs/InGaAs HEMT at two V_{DS} bias for l_g= 150 nm and comparison with measurements.

The calibration methodology to obtain the agreement between simulation and measurement in Fig. 6.22 involves:

- The exact channel and spacer mobility and velocity saturation for the saturation current
- The exact definition barrier and recess geometry from various cross-comparisons with respect to other factors of influence, such as doping activation, similarly to PHEMTs.

The InAlAs/SiN interface is generally less stable than the AlGaAs/SiN interface. Thus inverse modeling based on various wafer runs is of great importance. Since

InAlAs/InGaAs HEMTs are typically designed for high-speed performance, the access source and drain regions of the gate are of particular importance for the reduction of the fringing capacitances. With the agreement obtained for the DC data, the modeling of charge control is addressed in the next section

6.3.3 RF Simulation

RF simulation allows the exact separation of the charge information from the transport parameters, such as mobility. Multibias S-parameter simulation is required for the speed evaluation in agreement with several material parameters. Fig. 6.23 gives a comparison of measured and simulated S-parameters up to 120 GHz for a l_g= 150 nm InAlAs/InGaAs HEMT on InP substrate.

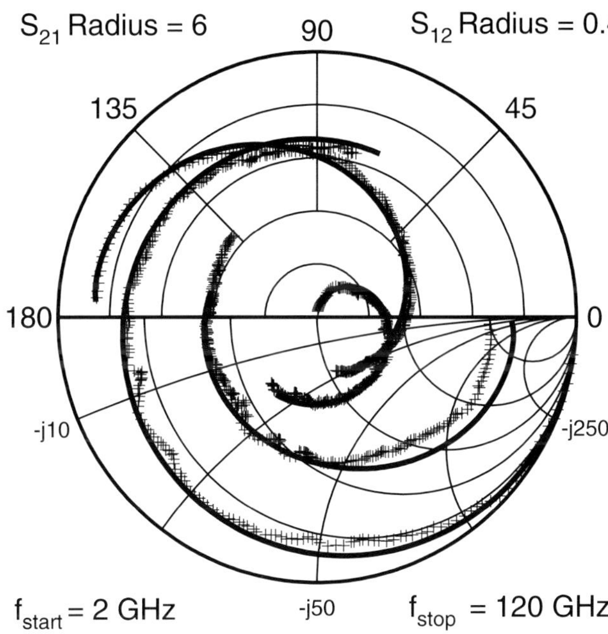

Fig. 6.23. Simulated (–) and measured (+) S-parameters S_{ij} in a Smith-polar chart between 2 GHz and 120 GHz for a InAlAs/InGaAs/InP HEMT with l_g= 150 nm.

The agreement obtained in Fig. 6.23 shows that both the reference level for the separation of the intrinsic and extrinsic device, as well as the material parameters, could be matched in the simulation. Fig. 6.24 further compares the simulated and measured current gain cut-off frequency f_T versus indium content in the channel for a l_g= 150 nm device at V_{DS}= 1 V. As expected, an increase of the indium content leads to an increase of the device speed, which is confirmed for several

devices experimentally in this work. Bollaert et al. [52] found a maximum f_T value at a channel indium content $x = 0.43$ for metamorphic HEMTs. This idea is not confirmed using the parameter set given in Chapter 3, as shown in Fig. 6.24. On the contrary, a nearly linear dependence is observed.
The experimental f_T data of four devices available for $l_g = 150$ nm and the simulation results are compared in Fig. 6.24.
The values for the composition range $x \leq 0.25$ are taken from typical pseudomorphic $Al_yGa_{1-y}As/In_xGa_{1-x}As/GaAs$ HEMTs. Both experimental and simulated values suggest a linear dependence of f_T. No devices are experimentally available between $x = 0.3$ and $x = 0.35$ for this work. The simulation results obtained for $x = 0.35$ and $x = 0.3$ support the measured values and the steadily decreasing behavior for decreasing x. Metamorphic devices with $x = 0.3$ were demonstrated by Bollaert et al. in [52] for power applications, where the further reduction of the indium content from $x = 0.43$ to $x = 0.3$ leads to a further decrease of f_T.
Thus the indium content in metamorphic devices serves mainly to adjust the device in the trade-off between speed and breakdown considerations. The InAlAs/-InGaAs HEMTs are limited by the on-state breakdown. In [426] no significant improvement of the device burnout was found using a reduced channel indium content of $x = 0.41$ in comparison with $x = 0.53$. This suggests that when using the $In_yAl_{1-y}As/In_xGa_{1-x}As$ in high-power applications, the indium content x needs to be reduced further than $x = 0.41$. However, such a reduction also decreases the improved speed performance possible in comparison with the PHEMT on GaAs.
Fig. 6.25 shows the simulated and measured f_T as a function of V_{DS} bias at constant V_{GS}. A strong decrease of f_T with increasing V_{DS} can be observed, which is based on three factors of influence. First, the increasing gate currents I_G shown in Fig. 6.20 decrease the current gain I_D/I_G. Second, C_{gs} increases due to the occurrence of RST, i.e. electrons leave the channel into the spacer and the barrier. Thus the center of the channel charge moves closer to the gate, which leads to an increase in C_{gs}. A third factor is the change of generation/recombination with rising V_{DS} due to the change of the extension of the space charge region. This enhances hole gate currents and leads to a potential shift at the gate. The holes lead to an effective positive shift of V_{GS} causing higher C_{gs}. Additional carriers are further generated by impact ionization. All these factors contribute to the observed decrease of f_T.
Scaling Considerations of InAlAs/InGaAs HEMTs. Similar to advanced CMOS devices HEMT scaling includes the reduction of several quantities:

- Of the nominal gate length from 250 nm to below 70 nm
- Of the operational V_{DS} voltage; most of all, reduction of the knee voltage
- Of the access resistances R_S and R_D
- Of the gate-to-channel separation to translate the nominal gate length reduction into effective channel length reduction

Technological and epitaxial means to accomplish the scaling are the gate length reduction, the increase of the indium content in the channel, the shrinking of the

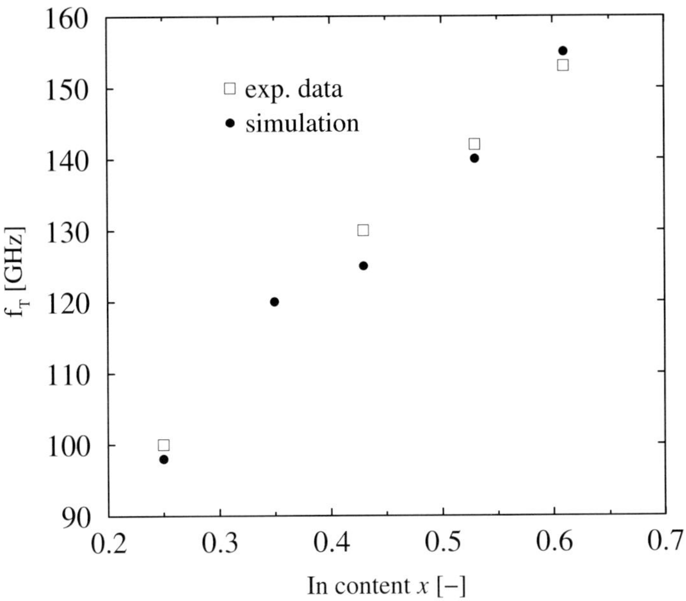

Fig. 6.24. Simulated current gain cut-off frequency f_T of a l_g= 150 nm InAlAs/InGaAs HEMT on GaAs and InP substrate as a function of indium content at V_{DS}= 1 V and comparison to measurements.

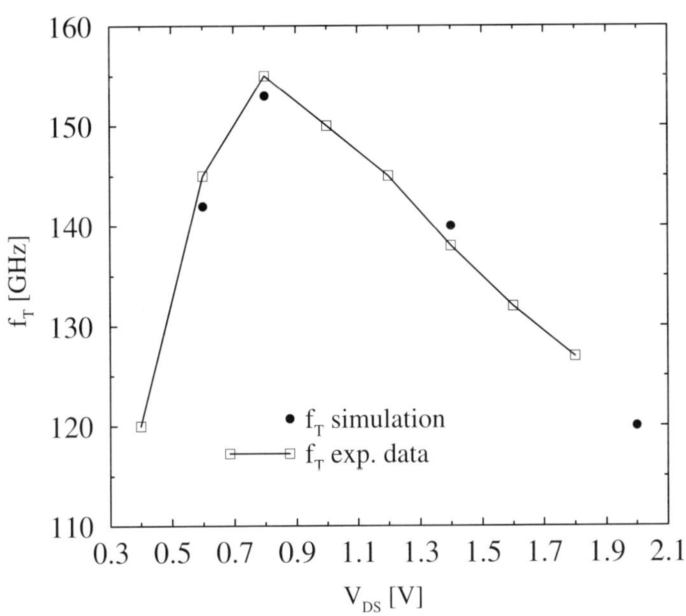

Fig. 6.25. Simulated and measured current gain cut-off frequency f_T as a function of V_{DS} bias at constant V_{GS} for the l_g= 150 nm InP HEMT.

layers thicknesses, and the introduction of new materials, such as InP as a channel material for high-power applications. For the HEMT scaling, the following engineering tasks have to be solved:

- Control gate leakage through the gate diode to maintain on-state breakdown voltages sufficient for the application.
- Control, i.e. limit, output conductance of operational voltages $V_{\mathrm{DS}} \geq V_{\mathrm{knee}}$.

A final solution of these engineering tasks will decide the "end of HEMT scaling" for conventional HEMT concepts, which currently is projected for sub-30 nm gate lengths.

6.4 High-Power High-Speed AlGaN/GaN HEMTs

AlGaN/GaN HEMTs are very competitive devices for high-power and high-power high-speed applications. GaN-based HEMTs have recently attracted much attention due to their outstanding properties already described in Section 2.2. In this section, the principle optimization issues are described and examples of the simulation of the terminal characteristics are provided.

6.4.1 Device Manufacturing

Substrates. AlGaN/GaN devices are currently grown typically on semi-insulating SiC and sapphire substrates. Another alternative substrate under discussion is highly resistive silicon, which has better thermal conductivity than sapphire but the highest lattice mismatch with respect to the GaN channel material. Genuine GaN substrates, although very desirable for optoelectronic applications, are not yet available in sufficient quantity, size, and quality. For all three materials, a buffer layer is required to compensate the lattice mismatch between GaN and SiC (3.4 %), GaN and sapphire (13 %), and GaN and silicon (17 %).

Polarization Effects. Pronounced spontaneous and piezoelectric polarization effects are a unique property of GaN materials which are not found in conventional silicon or III-V semiconductor devices. The polarization effects are advantageous but, at the same time, the most challenging material property of the AlGaN/GaN material system. They are advantageous, as intentional doping is not required to obtain sheet carrier concentrations in the channel of $\geq 10^{13}$ cm^{-2}, i.e. higher than obtained in any GaAs or InGaAs-based heterostructures. The polarization effects are challenging, as the amount of available carriers is controlled by interface quality and surface treatments, which are subject to epitaxial and process variations. The material composition at the heterointerface is the quantity which is directly accessible for the device designer and which controls the available sheet carrier density. Since the surface polarization can be partially compensated by surface treatment or passivation, charge control in AlGaN/GaN devices becomes a major task for TCAD simulation.

Epitaxy and Doping. GaN and $Al_xGa_{1-x}N$ are grown by either MBE or MOCVD and very similar results are obtained for the material quality. Fig. 6.26 gives an example of a typical fully planar AlGaN/GaN HEMT structure. A thick insulating GaN buffer of μm thickness is grown on top of the substrate of choice. An $Al_xGa_{1-x}N$/GaN heterointerface is grown on top of the buffer. The material composition x typically amounts to $0.1 \leq x \leq 0.35$. Bulk or δ-type silicon doping can be introduced to provide additional carriers and to improve the access resistances, as discussed below.

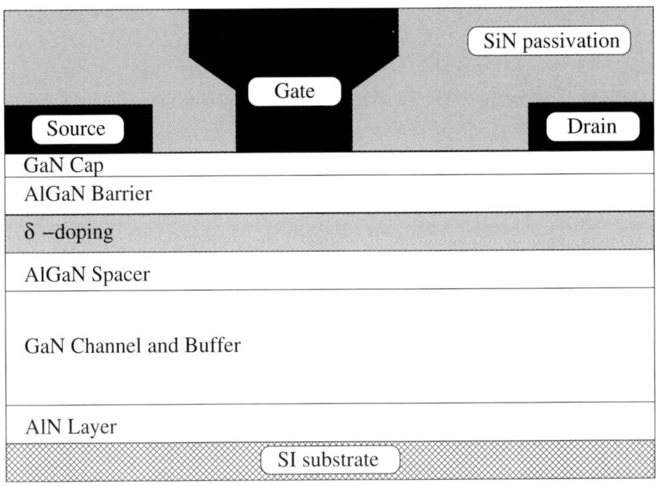

Fig. 6.26. Layer structure of a single heterojunction AlGaN/GaN HEMT.

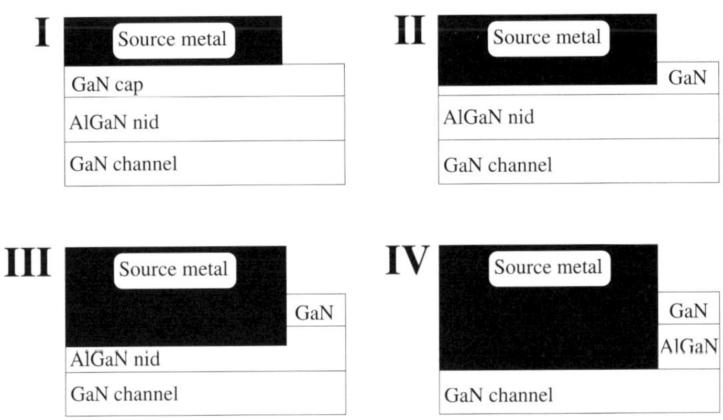

Fig. 6.27. Source side contact cases (I-IV) for AlGaN/GaN HEMTs.

Processing. AlGaN/GaN HEMTs today are non-recessed devices. The ohmic contacts are situated on the epitaxial layers in planar geometry. To obtain acceptable ohmic contact resistances for the wide bandgap devices, rapid thermal annealing with temperatures beyond 1000 K is typically used. Depending on the annealing procedures, different effective contact situations exist. Fig. 6.27 illustrates the limiting cases. There are four cases of planar contact geometries. In situation I, the metal remains on top of the GaN cap layer. The current flows through the GaN/AlGaN interface and through the full barrier layer into the channel. It was shown by Trew in [517] that for some bias conditions the access resistance can be significantly increased due to the band bending for wide bandgap devices. Situation II has a metal connection of the contact to the AlGaN barrier layer through the GaN layer. Thus the access resistance is reduced depending on the metal penetration depth. Situation III is similar to II; however, if a silicon-δ doping is incorporated, the access resistance is additionally reduced. Situation IV supplies direct access of the metal to the GaN channel, theoretically the most promising situation. Although the access resistance is significantly reduced, avoiding the poor transport characteristics of the AlGaN barrier, the processing of such a contact situation remains very critical. The AlGaN/GaN polarization charge at the interface is a major source of carrier supply. Metal diffusion or recess etching can significantly interfere with the charge supply under the contacts, so contact situation IV needs careful consideration with respect to manufacturability. An understanding of the resistances R_S and R_D in this section is especially desirable when these devices are used at frequencies of operation above 10 GHz.

6.4.2 DC Simulation Results

In the channel the high density of induced polarization channel charges accumulates very close to the interface with thicknesses of 2 nm and below. Thus interface roughness and epitaxial quality play a significant role on the effective carrier transport parameters. In the simulation, the effects of polarization are represented by surface and interface charges. A positive charge is introduced at the channel barrier interface. This positive charge is compensated by a negative surface charge at the barrier SiN interface and at the substrate contact, as given in Fig. 6.26 and as suggested in [481].

The calibration of the mobility model is performed based on Hall measurements from the epitaxial layers. The calibration methodology for AlGaN/GaN HEMTs is similar to the methodology used for the GaAs PHEMT. Fig. 6.28 shows the simulated transfer characteristic of a l_g= 200 nm GaN HEMT.

The contact situation II from Fig. 6.27, i.e. non-alloyed ohmic contacts, is assumed. Electrons have to pass the $Al_{0.25}Ga_{0.75}N$/GaN bandgap discontinuity by RST on the drain side. The transport physics of this HEMT cannot be described correctly by simulation assuming the contact case IV as in [481], since the transfer characteristics would not saturate. As can be seen in Fig. 6.28, assuming contact situation II results in agreement for lower V_{GS}. However, for higher fields

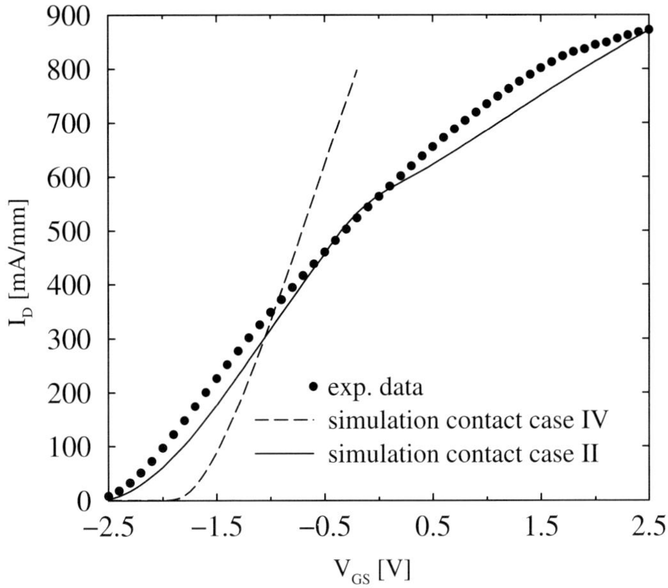

Fig. 6.28. Simulated and measured transfer characteristics of an AlGaN/GaN HEMT on SiC substrate with l_g= 200 nm.

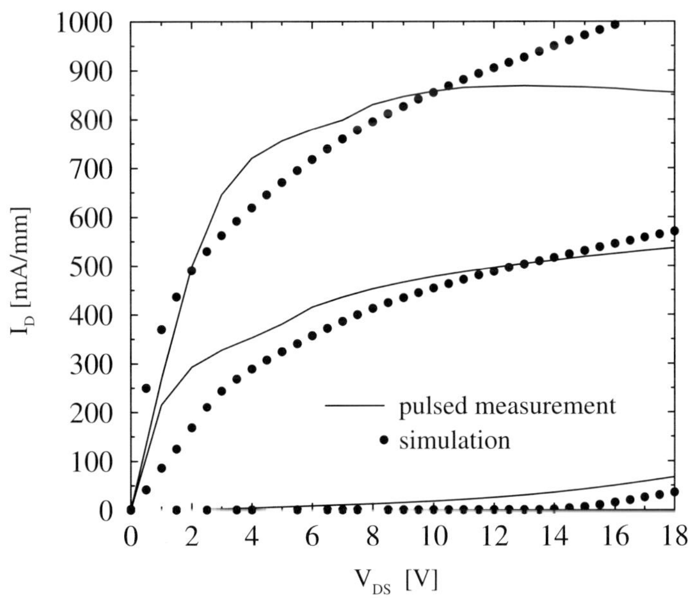

Fig. 6.29. Measured pulsed output characteristics of an AlGaN/GaN HEMT on SiC substrate and comparison to isothermal simulations with a gate length of l_g = 150 nm.

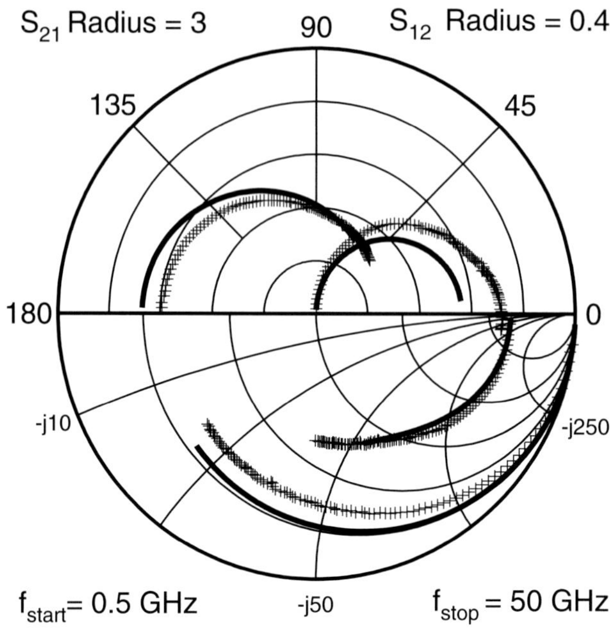

Fig. 6.30. Simulated (–) and measured (+) S-parameters S_{ij} between 0.5 GHz and 50 GHz at V_{DS}= 10 V.

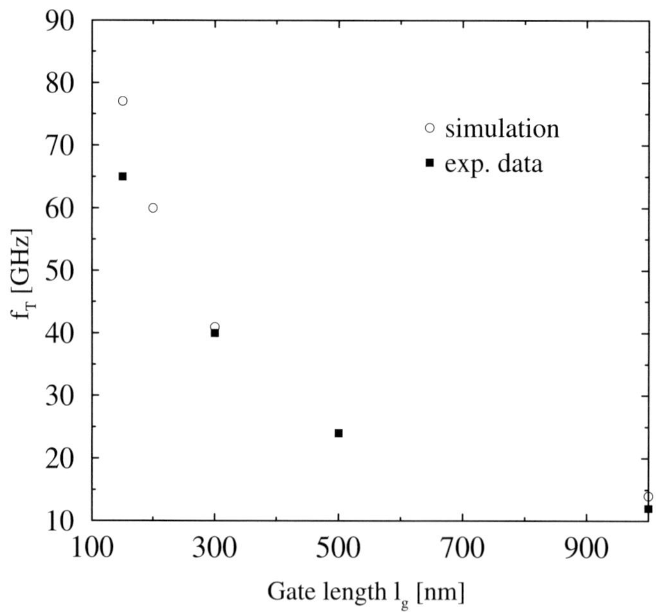

Fig. 6.31. Cut-off frequency f_{T} of an AlGaN/GaN HEMT on SiC substrate as a function of gate length l_{g}.

the simulation shows a rapid change, which occurs for a different V_{GS} in the measurement. The transconductance is strongly affected by the barrier conduction in AlGaN, and as shown in Fig. 6.28, the simulation underestimates the g_m characteristic of the measured device for higher V_{GS}.

High-Field Transport. Fig. 6.29 compares the measured pulsed and simulated isothermal output characteristics of an AlGaN/GaN HEMT on SiC. The pulse width is reduced to 1 μs, and the pulse repetition is 1 kHz, which leads to a significant reduction of thermal effects. It can be seen that the simulation agrees well with the measurements. Since pulsed operation would also reveal any dispersive effect, the cw measurements are compared to the pulsed measurements with the result that dispersive effects can be neglected in this case. For the mobility the model of (3.137) is used. It can be seen that for high power densities thermal effects cannot be completely excluded in the pulsed measurements, so that the simulation overestimates the currents for high power levels.

6.4.3 RF Simulation and Device Optimization

S-Parameter Simulation. The extraction of S-parameters by simulation and comparison to the measured values is performed as a more elaborate method of the charge control. Fig. 6.30 compares simulated and measured S-parameters up to 50 GHz. The parasitic elements were obtained in a similar fashion as used in the previous examples.

Choosing optimized contact conditions is a prerequisite for high-power, high-efficiency operation. In [508] it is shown that the barrier thickness has a critical impact on the device power performance: a larger barrier layer thickness improves the power performance. At the same time, scaling rules suggest the reduction of the barrier thickness for high-speed operation. An optimum must be found. Fig. 6.27 illustrates the optimization potential for the contacts. At the same time, device simulation can predict the overall device speed and the influence of the contact resistance on the cut-off frequency and gate delay times. Fig. 6.31 compares the simulated f_T versus gate length l_g with measurements when applying the scaling rules in the simulation and reducing the gate-to-channel separation with the gate length. This leads to higher f_T values in the simulations than in the measurements.

The examples in this section demonstrate the usefulness of TCAD simulations for GaN HEMTs to further exploit the optimization potential of the GaN HEMT technology.

7. Novel Devices

This chapter provides a glimpse of emerging approaches for new heterostructure high-speed and high-power devices.

7.1 InP DHBTs with GaAsSb Bases

Currently, two limitations prevail for InP HBTs (see Section 5.4). First of all, the active p-doping in $In_{0.53}Ga_{0.47}As$ base layers is limited to about 4×10^{19} cm^{-3}. Second, the $In_{0.53}Ga_{0.47}As$/InP interface requires the introduction of quaternary materials with reduced material quality and more complicated material growth. Thus, InP DHBTs with GaAsSb bases have recently been introduced to overcome these limitations. Fig. 7.1 gives a layer structure of such an HBT, similar to that suggested in [56].
Using the model for the InP/InGaAs/InGaAsP/InP HBT given in Section 5.4, the following study is performed. In this device, the $In_{0.53}Ga_{0.47}As$ base layer is replaced by $GaAs_{0.49}Sb_{0.51}$, which is also lattice-matched to the InP substrate, and the InGaAsP launcher is removed. Fig. 7.2 compares the simulated forward Gummel plots of the conventional InP/InGaAs/InGaAsP/ InP DHBT (Dev. 1) and the novel InP/GaAsSb/InP DHBT (Dev. 2). Measured data for the conventional device is included as a reference.
The following assumptions apply in the simulation:

- Identical activation of the p-doping in GaAsSb and InGaAs although higher activation can be expected in GaAsSb
- Higher electron minority in p-doped GaAsSb than in p-doped InGaAs
- Valence band discontinuity of 0.78 eV at the GaAsSb/InP interface [56].

The other material parameters for GaAsSb are described in Chapter 3.

As can be seen in Fig. 7.2, the replacement of InGaAs by GaAsSb in the base leads to a reduction in the turn-on voltage of the device and to an enhancement of the saturation current for the same device geometry. The advantage of the type II transition at the base/collector interface for low-power applications is visible. In addition, quaternary materials, such as InGaAsP, with reduced material quality

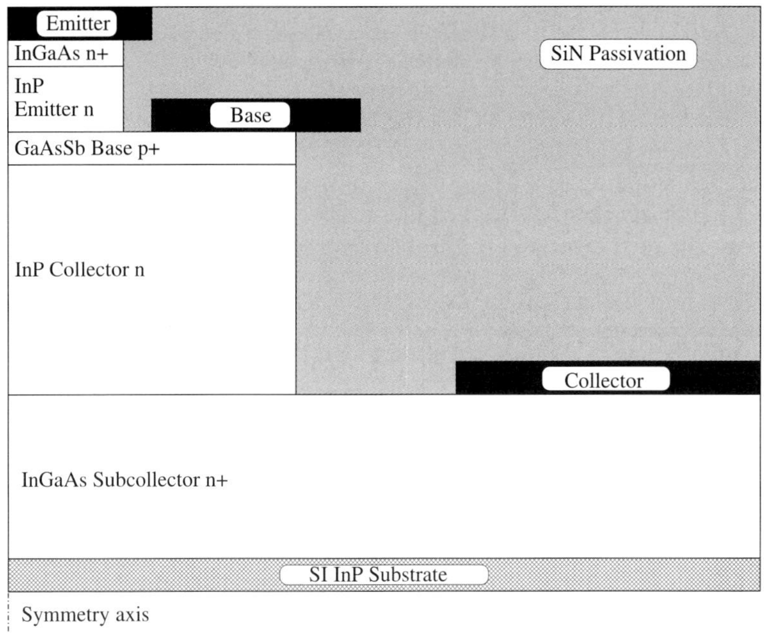

Fig. 7.1. Layer structure of an InP/GaAsSb/InP double heterojunction bipolar transistor.

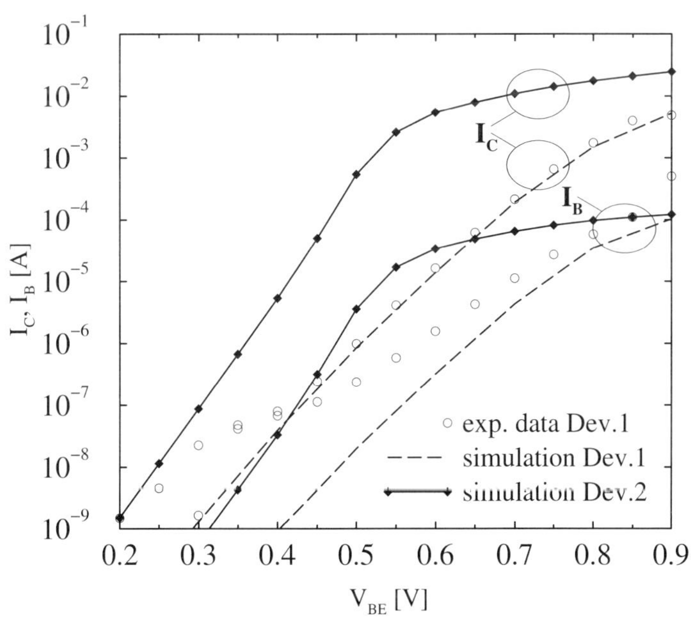

Fig. 7.2. Forward Gummel plot of $1 \times 8 \mu m$ InP/GaAsSb/InP DHBT in comparison to a conventional InP/InGaAs/InGaAsP/InP design.

and thus reduced transport properties, can be avoided. Reliability concerns in InGaAsP/InP are reduced by using GaAsSb, since near-ballistic transport is avoided.
An optimization issue remaining for the GaAsSb/InP HBTs is the improvement of the GaAsSb material quality for epitaxial growth using MOCVD and MBE. In addition, higher doping activations can be pursued to enable further scaling of the devices. Finally, process issues, such as surface passivation and the development of reproducible etch process, need to be resolved.

7.2 AlGaN/GaN HBTs

Both suffering and profiting from the wide bandgap material properties, AlGaN/GaN HBTs are very promising devices for high-voltage and high-power applications [332]. However, the highest gain which has been achieved so far remains $\beta \leq 10$.
A number of simulation studies of GaN HBTs have been presented to explain the prevailing shortcomings of the devices processed so far. The results were obtained with the device simulators ATLAS [346] and MEDICI [11, 284]. Overviews of remaining device issues of GaN HBTs are given in Refs. [333, 418].

A. Appendix: Benchmark Structures

This appendix contains a set of four realistic examples with the complete set of information for benchmarking heterostructure device simulators which we would like to suggest. The devices are realized and found operational in real world applications.

GaAs HBT

In Section 5.3, industrial AlGaAs/GaAs and InGaP/GaAs HBTs are given with layer and doping specifications. These devices can be used for benchmarking different device simulators, or simulations performed by the reader can be verified against experimental data, for example in Fig. 5.28, Fig. 5.30, and Fig. 5.33.

A Simple SiGe or GaAs HBT Structure

Vertical Structure. A device with a conservative design is shown in Fig. A.1. This structure has been used as a reference device in [398]. The user can choose between several emitter/base/collector layer configurations, such as $Si/Si_{1-x}Ge_x/Si$ ($x \leq 0.3$), $Al_xGa_{1-x}As/GaAs/GaAs$ ($x \leq 0.3$), or $In_xGa_{1-x}P$/ GaAs/GaAs ($x = 0.5$). The typical insulating material for SiGe HBTs is SiO_2 and Si_3N_4 for III-V HBTs.

Doping. The emitter doping can vary between 10^{19} cm^{-3} at the emitter contact and 3×10^{17} cm^{-3} at the base-emitter junction. The base doping can be chosen in the typical range 1-4$\times10^{19}$ cm^{-3} (default 2×10^{19} cm^{-3}). The collector doping can be 1-5$\times10^{16}$ cm^{-3} (default 2×10^{16} cm^{-3}) in the upper part of the collector and be as high as 10^{19} cm^{-3} in the lower subcollector region.

Contacts and Interfaces. Typically, all contacts are assumed to be non-alloyed ohmic contacts. In Si BJT or SiGe HBT, a poly-silicon emitter contact can also be chosen.

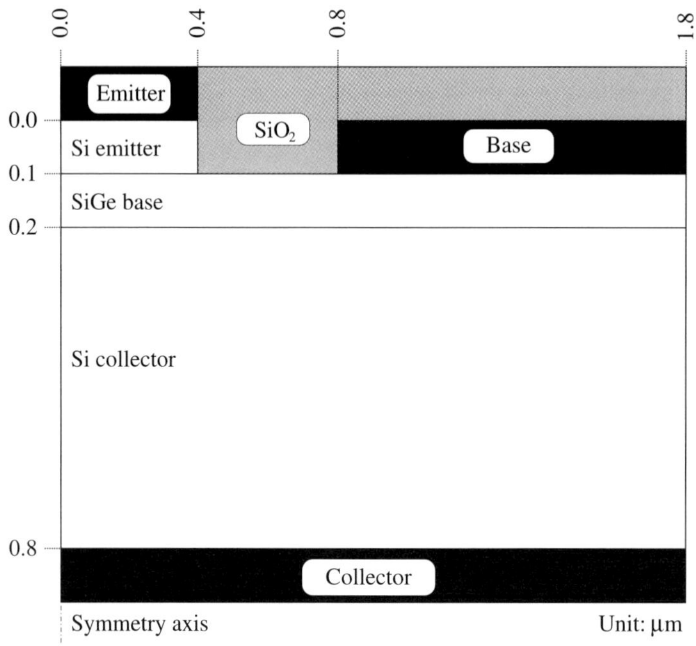

Fig. A.1. Layer structure of a 1×0.8 μm^2 HBT for benchmark.

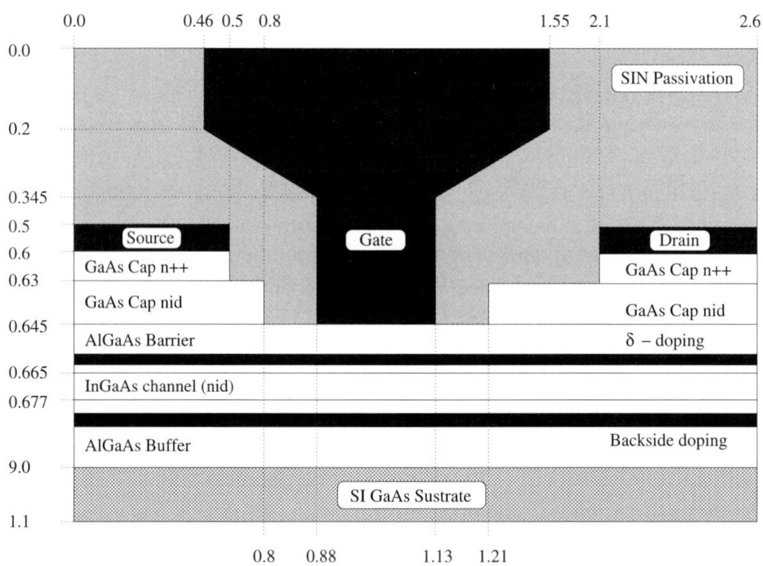

Fig. A.2. Layer structure of a double-recess double-heterojunction power AlGaAs/InGaAs/GaAs HEMT with l_g = 250 nm for benchmark (dimensions not to scale).

GaAs PHEMT

Vertical Layers. The device consists of a semi-insulating GaAs substrate (used as a thermal contact), a 223 nm thick $Al_{0.2}Ga_{0.8}As$ buffer, a 12 nm $In_{0.25}Ga_{0.75}As$ channel layer, and a 20 nm $Al_{0.2}Ga_{0.8}As$ barrier layer. The two GaAs cap layers have a thickness of 15 nm and 30 nm. The exact metal thicknesses are given in Fig. A.2. The exact recess structure is process-dependent and a suggestion is denoted in Fig. A.2.

Doping. The silicon doping in the GaAs caps amounts to 10^{16}cm^{-3} in the lower and to 3×10^{18} cm^{-3} in the upper layer. The δ-doping amounts to 5×10^{12} cm^{-2} in the upper layers with a thickness of 1.6 nm. The backside doping concentration is bulk type and amounts to 2.5×10^{18} cm^{-3} and assumed to be of rectangular shape with a thickness of 4 nm.

Contacts and Interfaces. The contact separation between source and drain is 1600 nm; the overall device width amounts to 2600 nm. A typical passivation is amorphous SiN with a relative permittivity of $\varepsilon_r = 7$. The gate shape is characterized by a 250 nm wide stem (nominal gate length) with a height of 300 nm. The exact slope of the gate shape is defined as given in Fig. A.2. The Schottky contact is characterized by a barrier height of 0.7 eV between the AlGaAs barrier and typical gate metalizations. The overall width of the gate head is 1090 nm; the ohmic contact thickness is about 100 nm. The SiN/AlGaAs and the SiN/GaAs are characterized by a degenerate surface charge density of 10^{12} cm^{-2}.

InAlAs/InGaAs HEMT on InP

Layer Structure. The device proposed is given in Fig. A.3. It consists of a semi-insulating InP substrate (used as a thermal contact), an $In_{0.52}Ga_{0.48}As$ buffer, a 12 nm $In_{0.53}Ga_{0.47}As$ channel layer, and a 12 nm $In_{0.52}Al_{0.48}As$ barrier layer. The contact cap is composed of a 30 nm nid-doped $In_{0.53}Ga_{0.47}As$ layer and a 50 nm $In_{0.53}Ga_{0.47}As$ layer silicon doped at 3×10^{18} cm^{-3}.

Doping. The background non-intentional doping (nid) is n-type and found to be 10^{15}cm^{-3} in all layers, if not stated differently. The silicon doping in the cap amounts to 10^{15}cm^{-3} and 3×10^{18} cm^{-3}. The device is single-side doped in the barrier layer and the δ-doping amounts to 5×10^{12} cm^{-2}. It is assumed to be of rectangular shape with a thickness of 1.6 nm and the distance of its center to the channel interface amounts to 5 nm.

Contacts and Interfaces. The contact separation amounts to 250 nm and the overall device width is 1250 nm. A typical passivation is amorphous SiN with a relative permittivity of $\varepsilon_r = 7$. The gate shape is characterized by a 150 nm wide stem with a of height 300 nm and with the slope as given in Fig. A.3. The Schottky contact is characterized by a barrier height of 0.6 eV. The thickness of the ohmic contacts amounts to 100 nm. The SiN surface is characterized by a negative surface charge density of 10^{11} cm^{-2}.

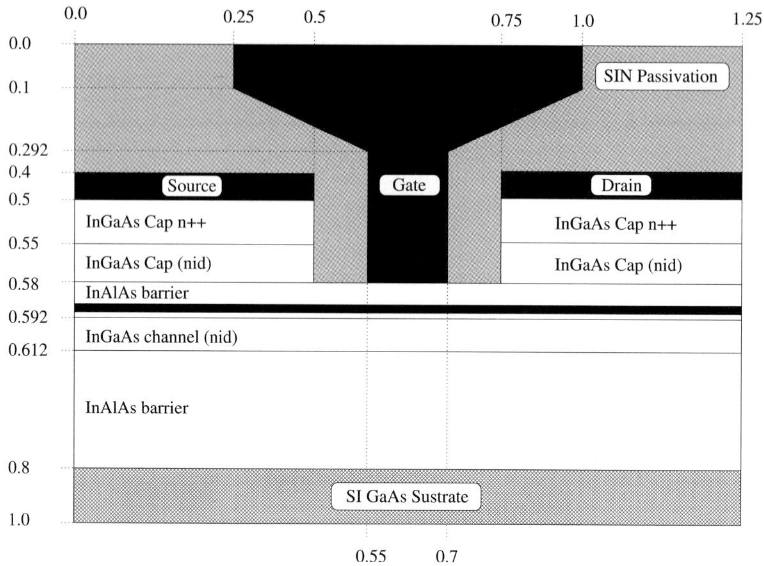

Fig. A.3. Layer structure of a double heterojunction single side doped InAlAs/InGaAs/InP HEMT (dimensions not to scale).

References

1. S. Adachi, "Material Parameters of $In_{1-x}Ga_xAs_yP_{1-y}$ and Related Binaries," *J.Appl.Phys.*, vol. 53, no. 12, pp. 8775–8792, 1982.
2. S. Adachi, "GaAs, AlAs, and $Al_xGa_{1-x}As$: Material Parameters for Use in Research and Device Applications," *J.Appl.Phys.*, vol. 58, no. 3, pp. R1–R29, 1985.
3. S. Adachi, "Band Gaps and Refractive Indices of AlGaAsSb, GaInAsSb, and InPAsSb: Key Properties for a Variety of the 2–4-μm Optoelectronic Device Applications," *J.Appl.Phys.*, vol. 61, no. 10, pp. 4869–4876, 1987.
4. S. Adachi, *Physical Properties of III-V Semiconductor Compounds.* Wiley, 1992.
5. S. Adachi, ed., *Properties of Aluminium Gallium Arsenide.* No. 7 in EMIS Datareviews Series, IEE INSPEC, 1993.
6. V. Agostinelli, T. Bordelon, X. Wang, C. Yeap, C. Maziar, and A. Tasch, "An Energy-Dependent Two-Dimensional Substrate Current Model for the Simulation of Submicrometer MOSFETs," *IEEE Electron Device Lett.*, vol. 13, no. 11, pp. 554–556, 1992.
7. D. Ahlgren, G. Freeman, S. Subbanna, R. Groves, D. Greenberg, J. Malinowski, D. Nguyen-Ngoc, S. Jeng, K. Stein, K. Schonenberg, D. Kiesling, B. Martin, S. Wu, D. Harame, and B. Meyerson, "A SiGe HBT BiCMOS Technology for Mixed Signal RF Applications," in *Proc. BCTM*, (Minneapolis), pp. 195–197, 1997.
8. R. Ahrenkiel, "Minority-Carrier Lifetime of III-V Compound Semiconductors," in *Proc. Intl. Symp. GaAs and Related Compounds*, (Freiburg), pp. 685–690, 1993.
9. R. Ahrenkiel and D. Dunlavy, "Minority-Carrier Lifetime in $Al_xGa_{1-x}As$," *J.Vac.Sci.Technol.A*, vol. 7, no. 3, pp. 822–826, 1989.
10. I. Akasaki and H. Amano, "Crystal Structure, Mechanical Properties and Thermal Properties of GaN," in Edgar [135], section 1.4, pp. 30–34.
11. E. Alekseev and D. Pavlidis, "DC and High-Frequency Performance of AlGaN/GaN Heterojunction Bipolar Transistors," *Solid-State Electron.*, vol. 44, no. 2, pp. 245–252, 2000.
12. E. Alekseev and D. Pavlidis, "Large-Signal Microwave Performance of GaN-Based NDR Diode Oscillators," *Solid-State Electron.*, vol. 44, no. 6, pp. 941–947, 2000.
13. Y. Ando, W. Contrata, N. Samoto, H. Miyamoto, K. Matsunaga, M. Kuzuhara, K. Kunihiro, K. Kasahara, T. Nakayama, Y. Takahashi, N. Hayama, and Y. Ohno, "Gate Length Scaling for $Al_{0.2}Ga_{0.8}N$/GaN HJFETs: Two-Dimensional Full Band Monte Carlo Simulation Including Polarization Effect," *IEEE Trans.Electron Devices*, vol. 47, no. 10, pp. 1965–1972, 2000.
14. Y. Ando, Y. Okamoto, H. Miyamoto, N. Hayama, T. Nakayama, K. Kasahara, and M. Kuzuhura, "A 110-W AlGaN/GaN Heterojunction FET on Thinned Sapphire Substrate," in *IEDM Tech.Dig.*, (Washington), pp. 381–384, 2001.
15. R. Anholt, "Dependence of GaAs Fringe Capacitances on the Fabrication Technologies," *Solid-State Electron.*, vol. 34, no. 5, pp. 515–520, 1991.

16. R. Anholt, "Distributive Effects on HBT S-Parameters," *Solid-State Electron.*, vol. 44, no. 4, pp. 729–737, 2000.
17. R. Anholt, "Simulating AlGaN/GaN HEMTs using G-PISCES-2B and PETs," *GaAs Simulation and Analysis News*, no. 1, Feb. 2000.
18. A. Anwar, S. Wu, and R. Webster, "Temperature Dependent Transport Properties in GaN, $Al_xGa_{1-x}N$, and $In_xGa_{1-x}N$ Semiconductors," *IEEE Trans.Electron Devices*, vol. 48, no. 3, pp. 567–572, Mar. 2001.
19. APSYS. http://www.crosslight.com/downloads/downloads.html.
20. V. Arora, D. Mui, and H. Morkoc, "High-Field Electron Drift Velocity and Temperature in Gallium Phosphide," *J.Appl.Phys.*, vol. 61, no. 9, pp. 4703–4704, 1987.
21. R. Asauskas, Z. Dobrovolskis, and A. Krotkus, "Maximum Drift Velocity of Electrons in Indium Antimonide at 77 K," *Sov.Phys.Semicond.*, vol. 14, no. 12, pp. 1377–1380, 1980.
22. P. Ashburn, *Design and Realization of Bipolar Transistors.* Chichester: Wiley, 1988.
23. U. Auer, R. Reuter, P. Ellrodt, W. Prost, and F. Tegude, "Characterization and Analysis of a New Gate Leakage Mechanism at High Drain Bias in InAlAs/InGaAs Heterostructure Field-Effect Transistors," in *Proc. Intl.Conf.InP and Rel.Mat.*, (Schwäbisch Gmünd), pp. 650–653, 1996.
24. Y. Baeyens, *Monolithic Microwave Integrated Circuits Using GaAs and InP Based Heterojunction Field-Effect Transistors.* Dissertation, Katholieke Universiteit Leuven, 1996.
25. S. Bahl, J. del Alamo, J. Dickmann, and S. Schildberg, "Off-State Breakdown in InAlAs/InGaAs MODFET's," *IEEE Trans.Electron Devices*, vol. 42, no. 1, pp. 15–22, 1995.
26. G. Bai, "High k Gate Stack for Sub-0.1 μm CMOS Technology," in *Proc. Symp. on Advances in Rapid Thermal Processing*, (Pennington), pp. 39–44, Electrochem.Soc., 1999.
27. R. Bank, D. Rose, and W. Fichtner, "Numerical Methods for Semiconductor Device Simulation," *IEEE Trans.Electron Devices*, vol. ED-30, no. 9, pp. 1031–1041, 1983.
28. G. Baraff, "Distribution Functions and Ionization Rates for Hot Electrons in Semiconductors," *Phys.Rev.*, vol. 128, no. 6, pp. 2507–2517, 1962.
29. S. Beebe, F. Rotella, Z. Sahul, D. Yergeau, G. McKenna, L. So, Z. Yu, K. Wu, E. Kan, J. McVittie, and R. Dutton, "Next Generation Stanford TCAD—PISCES 2ET and SUPREM 007," in *IEDM Tech.Dig.*, (San Francisco), pp. 213–216, 1994.
30. E. Bellotti, B. Doshi, K. Brennan, J. Albrecht, and P. Ruden, "Ensemble Monte Carlo Study of Electron Transport in Wurtzite InN," *J.Appl.Phys.*, vol. 85, no. 2, pp. 916–923, 1999.
31. H. Bennet and C. Wilson, "Statistical Comparisons of Data on Band-Gap Narrowing in Heavily Doped Silicon: Electrical and Optical Measurements," *J.Appl.Phys.*, vol. 55, no. 10, pp. 3582–3587, 1984.
32. J. Bergman, C. Hallin, and E. Janzen, "Temperature Dependence of the Minority Carrier Lifetime in GaAs/AlGaAs Double Heterostructures," *J.Appl.Phys.*, vol. 78, no. 7, pp. 4808–4810, 1995.
33. J. Berntgen, A. Schueppen, P. Maier, M. Tortschanoff, W. Kraus, and M. Averweg, "SiGe Technology Bears Fruit," *Mat.Sci.Eng.B*, vol. 89, no. 1-3, pp. 13–20, 2002.
34. M. Berroth and M. Bosch, "High-Frequency Equivalent Circuit of GaAs FETs," *IEEE Trans.Microwave Theory and Techniques*, vol. 39, no. 2, pp. 224–229, 1991.
35. C. Besikci and M. Razeghi, "Electron Transport Properties of $Ga_{0.51}In_{0.49}P$ for Device Applications," *IEEE Trans.Electron Devices*, vol. 41, no. 6, pp. 1066–1069, 1994.
36. A. Bessemoulin, J. Dishong, G. Clark, D. White, P. Quentin, H. Thomas, and D. Geiger, "1-W Broad Ka-Band Ultra Small High Power Amplifier MMICs Using 0.25 μm GaAs PHEMTs," in *GaAs IC Symp.Tech.Dig.*, (Monterey), pp. 40–43, 2002.

37. U. Bhapkar and M. Shur, "Monte Carlo Calculation of Velocity-Field Characteristics of Wurtzite GaN," *J.Appl.Phys.*, vol. 82, no. 4, pp. 1649–1655, 1997.
38. P. Bhattacharya, ed., *Properties of Lattice-Matched and Strained Indium Gallium Arsenide.* No. 8 in EMIS Datareviews Series, IEE INSPEC, 1993.
39. P. Bhattacharya, ed., *Semiconductor Optoelectronic Devices.* Upper Saddle River: Prentice Hall, 2nd ed., 1997.
40. P. Bhattacharya, "Deep Levels in InGaAs/GaAs and InGaAs/InP," in Bhattacharya [38], section 6.3, pp. 174–179.
41. T. Binder, J. Cervenka, K. Dragosits, A. Gehring, T. Grasser, R. K. M. Gritsch, M. Knaipp, H. Kosina, R. Mlekus, V. Palankovski, M. Rottinger, R. Rodriguez-Torres, G. Schrom, S. Selberherr, M. Stockinger, and S. Wagner, *MINIMOS-NT Device and Circuit Simulator, User's Guide, Release 2.0.* Institut für Mikroelektronik, Technische Universität Wien, 2002. http://www.iue.tuwien.ac.at/software/minimos-nt.
42. BIPOLE3. http://www.bipsim.com/mainframe.html.
43. Y. Biryulin, V. Karyaev, I. Novikova, T. Polyanskaya, V. Chaldyshev, and Y. Shmartsev, "Singularities in the Dependences of the Mobility and Quantum Efficiency of the Photoluminescence on the Composition of $GaAs_{1-x}Sb_x$ $(0.01 < x < 0.13)$ Solid Solutions," *Sov.Phys.Semicon.*, vol. 15, no. 11, pp. 1330–1332, 1981.
44. J. Blakemore, *Gallium Arsenide.* American Institute of Physics, 1987.
45. H. Blanck, K. Riepe, W. Doser, P. Auxemery, and D. Pons, "Industrial GaInP/GaAs Power HBT MMIC Process," in *Proc. European GaAs Rel.Comp.Appl.Symp. GAAS*, (Paris), pp. 4–7, 2000.
46. D. Blanks, G. Klimeck, R. Lake, D. Jovanovic, R. Bowen, C. Fernando, W. Frensley, and M. Leng, "NEMO: General Release of a New Comprehensive Quantum Device Simulator," in *Proc. Intl. Symp. Compound Semiconductors*, (San Diego), pp. 639–642, 1997.
47. K. Blotekjaer, "Transport Equations for Electrons in Two-Valley Semiconductors," *IEEE Trans.Electron Devices*, vol. ED-17, no. 1, pp. 38–47, 1970.
48. J. Böck, H. Knapp, K. Aufinger, M. Wurzer, S. Boguth, R. Schreiter, T. Meister, M. Rest, M. Ohnemus, and L. Treitinger, "12 ps Implanted Base Silicon Bipolar Technology," in *IEDM Tech.Dig.*, (Washington), pp. 553–556, 1999.
49. J. Böck, T. Meister, H. Knapp, D. Zöschg, H. Schäfer, K. Aufinger, M. Wurzer, S. Boguth, M. Franosch, R. Stengl, R. Schreiter, M. Rest, and L. Treitinger, "SiGe Bipolar Technology for Mixed Digital and Analogue RF Applications," in *IEDM Tech.Dig.*, (San Francisco), pp. 745–748, 2000.
50. J. Böck, H. Schäfer, H. Knapp, D. Zöschg, K. Aufinger, M. Wurzer, S. Boguth, M. Rest, R. Schreiter, R. Stengl, and T. Meister, "Sub 5 ps SiGe Bipolar Technology," in *IEDM Tech.Dig.*, (San Francisco), pp. 763–763, 2002.
51. J. Böck, H. Schäfer, H. Knapp, D. Zöschg, K. Aufinger, M. Wurzer, S. Boguth, R. Stengl, R. Schreiter, and T. Meister, "High-Speed SiGe:C Bipolar Technology," in *IEDM Tech.Dig.*, (Washington), pp. 344–347, 2001.
52. S. Bollaert, Y. Cordier, M. Zaknoune, H. Happy, V. Hoel, S. Lepilliet, D. Théron, and A. Cappy, "The Indium Content in Metamorphic $In_xAl_{1-x}As/In_xGa_{1-x}As$ HEMTs on GaAs Substrate: a New Structure Parameter," *Solid-State Electron.*, vol. 44, no. 6, pp. 1021–1027, 2000.
53. C. Bolognesi, "Antimonide-Based High-Speed Electronics: a Transistor Perspective," in *Proc. Intl.Conf.InP and Rel.Mat.*, (Stockholm), pp. 55–58, 2002.
54. C. Bolognesi, M. Dvorak, O. Pitts, and S. Watkins, "200 GHz $InP/GaAs_xSb_{1-x}/$ InP Double Heterojunction Bipolar Transistors," in *GaAs IC Symp.Tech.Dig.*, (Seattle), pp. 233–236, 2000.
55. C. Bolognesi, M. Dvorak, and S. Watkins, "InP/InGaAs/InP Double Heterojunction Bipolar Transistors," in *GaAs IC Symp.Tech.Dig.*, (Monterey), pp. 265–268, 2002.

56. C. Bolognesi, N. Matine, M. Dvorak, P. Yeo, X. Xu, and S. Watkins, "InP/GaAsSb/InP Double HBTs: A New Alternative for InP-Based DHBTs," *IEEE Trans.Electron Devices*, vol. 48, no. 11, pp. 2631–2639, 2001.
57. M. Borgarino, R. Plana, S. Delage, F. Fantini, and J. Graffeuil, "Influence of Surface Recombination on the Burn-In Effect in Microwave GaInP/GaAs HBT's," *IEEE Trans.Electron Devices*, vol. 46, no. 1, pp. 10–16, 1999.
58. G. Borghs, K. Bhattacharya, K. Deneffe, P. V. Mieghem, and R. Mertens, "Band-Gap Narrowing in Highly Doped n- and p-type GaAs Studied by Photoluminescence Spectroscopy," *J.Appl.Phys.*, vol. 66, no. 9, pp. 4381–4386, 1989.
59. D. Bose and B. Nag, "Electron Velocity in Indium Phosphide Single-Heterojunction Quantum Wells," *Semicond.Sci.Technol.*, vol. 6, no. 12, pp. 1135–1140, 1991.
60. B. Bouhafs, H. Aourag, M. Ferhat, A. Zaoui, and M. Certier, "Theoretical Analysis of Disorder Effects on Electronic and Optical Properties in InGaAsP Quaternary Alloy," *J.Appl.Phys.*, vol. 82, no. 10, pp. 4923–4930, Nov. 1997.
61. N. Bovolon, P. Baureis, J.-E. Müller, P. Zwicknagl, R. Schultheis, and E. Zanoni, "A Simple Method for the Thermal Resistance Measurement of AlGaAs/GaAs Heterojunction Bipolar Transistors," *IEEE Trans.Electron Devices*, vol. 45, no. 8, pp. 1846–1848, Aug. 1998.
62. N. Bovolon, R. Schultheis, J.-E. Müller, P. Zwicknagl, and E. Zanoni, "A Short-Term High-Current-Density Reliability Investigation of AlGaAs/GaAs Heterojunction Bipolar Transistors," *IEEE Electron Device Lett.*, vol. 19, no. 12, pp. 469–471, 1998.
63. H. Brand, *Thermoelektrizität und Hydrodynamik.* Dissertation, Technische Universität Wien, 1994. http://www.iue.tuwien.ac.at/phd/brand.
64. H. Brech, *Optimization of GaAs Based High Electron Mobility Transistors by Numerical Simulation.* Dissertation, Technische Universität Wien, 1998. http://www.iue.tuwien.ac.at/phd/brech.
65. H. Brech, T. Grave, T. Simlinger, and S. Selberherr, "Optimization of Pseudomorphic HEMT's Supported by Numerical Simulations," *IEEE Trans.Electron Devices*, vol. 44, no. 11, pp. 1822–1828, Nov. 1997.
66. K. Brennan, "Theory of the Steady-State Hole Drift Velocity in InGaAs," *Appl.Phys.Lett.*, vol. 51, no. 13, pp. 995–997, 1987.
67. K. Brennan and K. Hess, "High Field Transport in GaAs, InP, and InAs," *Solid-State Electron.*, vol. 27, no. 4, pp. 347–357, 1984.
68. K. Brennan, E. Bellotti, M. Farahmand, J. Haralson, P. Ruden, J. Albrecht, and A. Sutandi, "Materials Theory Based Modeling of Wide Band Gap Semiconductors: from Basic Properties to Devices," *Solid-State Electron.*, vol. 44, no. 2, pp. 195–204, 2000.
69. K. Brennan and A. Brown, eds., *Theory of Modern Electronic Semiconductor Devices.* New York: Wiley, 2002.
70. K. Brennan and P. Chiang, "Calculated Electron and Hole Steady-State Drift Velocities in Lattice Matched GaInP and AlGaInP," *J.Appl.Phys.*, vol. 71, no. 2, pp. 1055–1057, 1992.
71. K. Brennan, D. Park, K. Hess, and M. Littlejohn, "Theory of the Velocity-Field Relation in AlGaAs," *J.Appl.Phys.*, vol. 63, no. 10, pp. 5004–5008, 1988.
72. J. Bude, "MOSFET Modeling into the Ballistic Regime," in *Proc. Intl.Conf. on Simulation of Semiconductor Processes and Devices*, (Seattle), pp. 23–26, Sept. 2000.
73. F. Bufler, *Full-Band Monte Carlo Simulation of Electrons and Holes in Strained Si and SiGe.* Munich: Herbert Utz Verlag, 1998.
74. F. Bufler, P. Graf, B. Meinerzhagen, B. Adeline, M. Rieger, H. Kibbel, and G. Fischer, "Low- and High-Field Electron-Transport Parameters for Unstrained and Strained $Si_{1-x}Ge_x$," *IEEE Electron Device Lett.*, vol. 18, no. 6, pp. 264–266, 1997.

75. M. Bugajski, A. Kontkiewicz, and H. Mariette, "Energy Bands of Ternary Alloy Semiconductors: Coherent-Potential-Approximation Calculations," *Phys.Rev.B*, vol. 28, no. 12, pp. 7105–7114, 1983.
76. M. Bugajski and W. Lewandowski, "Concentration-Dependent Absorption and Photoluminescence of n-Type InP," *J.Appl.Phys.*, vol. 57, no. 2, pp. 521–530, 1985.
77. G. Bulman, V. Robbins, K. Brennan, K. Hess, and G. Stillman, "Experimental Determination of Impact Ionization Coefficients in (100) GaAs," *IEEE Electron Device Lett.*, vol. 4, no. 6, pp. 181–185, 1983.
78. G. Bulman, V. Robbins, and G. Stillman, "The Determination of Impact Ionization Coefficients in (100) Gallium Arsenide Using Avalanche Noise and Photocurrent Multiplication Measurements," *IEEE Trans.Electron Devices*, vol. 32, no. 11, pp. 2454–2466, 1985.
79. C. Bulutay, "Electron Initiated Impact Ionization in AlGaN Alloys," *Semicond.Sci.Technol.*, vol. 17, no. 10, pp. L59–L62, 2002.
80. E. Buturla, P. Cottrell, B. Grossman, and K. Salsburg, "Finite-Element Analysis of Semiconductor Devices: The FIELDAY Program," *IBM J.Res.Develop.*, vol. 44, no. 1-2, pp. 142–156, 2000.
81. C. Campell and S. Brown, "A Compact, 40 GHz, 0.5 W Power Amplifier MMIC," in *GaAs IC Symp.Tech.Dig.*, (Monterey), pp. 141–143, 1999.
82. C. Canali, C. Jacoboni, F. Nava, G. Ottaviani, and A. Quaranta, "Electron Drift Velocity in Silicon," *Phys.Rev.B*, vol. 12, no. 4, pp. 2265–2284, 1975.
83. X. Cao, J. van Hove, J. Klassen, C. Polley, A. Wowchak, P. Chow, D. King, A. Zhang, G. Dang, C. Monier, S. Pearton, and F. Ren, "Simulation of GaN/AlGaN Heterojunction Bipolar Transistors: Part I- NPN Structures," *Solid-State Electron.*, vol. 44, no. 4, pp. 1255–1259, 2000.
84. F. Capasso and G. Margaritondo, eds., *Heterojunction Band Discontinuities.* Elsevier, 1987.
85. F. Capasso, K. Mohammed, K. Alavi, A. Cho, and P. Foy, "Impact Ionization Rates for Electrons and Holes in $Al_{0.48}In_{0.52}As$," *Appl.Phys.Lett.*, vol. 45, no. 9, pp. 968–970, 1979.
86. M. Case, S. Maas, L. Larson, D. Rensch, D. Harame, and B. Meyerson, "An X-Band Monolithic Active Mixer in SiGe HBT Technology," in *IEEE Intl. Microwave Symp. Dig.*, (San Francisco), pp. 655–658, 1996.
87. H. Casey and M. Panish, *Heterostructure Lasers, Part B: Materials and Operating Characteristics.* Academic Press, 1978.
88. H. Casey-Jr. and M. Panish, "Composition Dependence of the $Ga_{1-x}Al_xAs$ Direct and Indirect Energy Gaps," *J.Appl.Phys.*, vol. 40, no. 12, pp. 4910–4912, 1969.
89. D. Caughey and R. Thomas, "Carrier Mobilities in Silicon Empirically Related to Doping and Field," *Proc.IEEE*, vol. 52, no. 12, pp. 2192–2193, 1967.
90. Y. Cengel, ed., *Heat Transfer: A Practical Approach.* Boston: McGraw-Hill, 1998.
91. C. Chan, Y. Chen, M. Chen, H. Lin, G. Jan, and Y. Chen, "Photoreflectance Spectroscopy of Strained-Layer (111)B InGaAs/GaAs Quantum Well Diodes," *J.Appl.Phys.*, vol. 84, no. 3, pp. 1595–1601, 1998.
92. N. Chand, "Carrier Diffusion Lengths and Lifetimes in InGaAs," in Bhattacharya [38], section 4.4, pp. 127–129.
93. N. Chand, T. Henderson, J. Klem, W. Masselink, R. Fischer, Y. Chang, and H. Morkoc, "Comprehensive Analysis of Si-Doped $Al_xGa_{1-x}As$ (x=0 to 1): Theory and Experiments," *Phys.Rev.B*, vol. 30, no. 8, pp. 4481–4492, 1984.
94. P. Chang, N. Li, A. Baca, H. Hou, C. Monier, J. Laroche, F. Ren, and S. Pearton, "Device Characteristics of the GaAs/InGaAsN/GaAs P-N-P Double Heterojunction Bipolar Transistor," *IEEE Electron Device Lett.*, vol. 22, no. 3, pp. 113–115, 2001.

95. H. Chau and D. Pavlidis, "A Physics-Based Fitting and Extrapolation Method for Measured Impact Ionization Coefficients in III-V Semiconductors," *J.Appl.Phys.*, vol. 72, no. 2, pp. 531–538, 1992.
96. C. Chelli, D. Cui, S. Hubbard, A. Eisenbach, D. Pavlidis, S. Krawczyk, and B. Sermage, "Minority Carrier Lifetimes in MOCVD-Grown C- and Zn-Doped InGaAs," in *Proc. Intl.Conf.InP and Rel.Mat.*, (Davos), pp. 127–130, 1999.
97. J. Chen and A. Cho, "Characterization of Te-Doped GaSb Grown by Molecular Beam Epitaxy Using SnTe," *J.Appl.Phys.*, vol. 70, no. 1, pp. 277–281, 1991.
98. B. Cheng, M. Cao, R. Rao, A. Inani, P. Voorde, W. Greene, J. Stork, Z. Yu, P. Zeitzoff, and J. Woo, "The Impact of High-k Gate Dielectrics and Metal Gate Electrodes on Sub-100 nm MOSFET's," *IEEE Trans.Electron Devices*, vol. 46, no. 7, pp. 1537–1544, July 1999.
99. M. Cherng, G. Stringfellow, and R. Cohen, "Organometallic Vapor Phase Epitaxial Growth of $GaAs_{0.5}Sb_{0.5}$," *Appl.Phys.Lett.*, vol. 44, no. 7, pp. 677–79, 1984.
100. M. Chertouk, W. Chang, C. Yuan, H. Chen, L. Lo, C. Chen, D. Tu, J. Liu, N. Draidia, and P. Chao, "The First 0.15 μm MHEMT 6" GaAs Foundry Service: Highly Reliable Process for 3 V Drain Bias Operations," in *Proc. Intl.Conf. GaAs Manufacturing Technology*, (Scottsdale), pp. 153–155, 2003.
101. M. Chertouk, H. Heiß, D. Xu, S. Kraus, W. Klein, G. Böhm, G. Tränkle, and G. Weimann, "Metamorphic InAlAs/InGaAs HEMTs on GaAs Substrates with Composite Channel and f_{max} of 350 GHz," in *Proc. Intl.Conf.InP and Rel.Mat.*, (Sapporo), pp. 737–740, 1995.
102. M. Chertouk, F. Steinhagen, H. Massler, M. Dammann, W. Haydl, K. Köhler, and G. Weimann, "W-Band High Gain Passivated 0.15 μm InP-Based HEMTs MMIC Technology with High Thermal Stability on InP Substrates," in *Proc. Intl.Conf.InP and Rel.Mat.*, (Tsukuba), pp. 227–230, 1998.
103. M. Chertouk, D. Tu, P. Meng, C. Yuan, W. Chang, C. Kuo, C. Chang, A. Chang, H. Chen, C. Chen, and P. Chao, "Manufacturable 0.15 μm PHEMT Process for High Volume and Low Cost on 6" GaAs Substrates: The First 0.15 μm PHEMT 6 " GaAs Foundry Fab," in *Proc. Intl.Conf. GaAs Manufacturing Technology*, (San Diego), pp. 138–141, 2002.
104. K. Christianson, "Reliability of III–V Based Heterojunction Bipolar Transistors," *Microelectron.Reliab.*, vol. 38, no. 1, pp. 153–161, 1997.
105. L. Cook, G. Bulman, and G. Stillman, "Electron and Hole Impact Ionization Coefficients in InP Determined by Photomultiplication Measurements," *Appl.Phys.Lett.*, vol. 40, no. 7, pp. 589–591, 1982.
106. J. Cressler and G. Niu, *Silicon-Germanium Heterojunction Bipolar Transistors*. Boston, London: Artech, 2003.
107. C. Crowell and S. Sze, "Temperature Dependence of Avalanche Multiplication in Semiconductors," *Appl.Phys.Lett.*, vol. 9, no. 6, pp. 242–244, 1966.
108. A. da Silva, C. Persson, M. Marcussen, E. Veje, and A. de Oliveira, "Band-Gap Shift in Heavily Doped n-type $Al_{0.3}Ga_{0.7}As$ Alloys," *Phys.Rev.B*, vol. 60, no. 4, pp. 2463–2467, 1999.
109. P. Damayanthi, R. Joshi, and J. McAdoo, "Electron Mobility and Drift Velocity Calculations for Bulk GaSb Material," *J.Appl.Phys.*, vol. 86, no. 9, pp. 5060–5064, 1999.
110. P. Damayanthi, R. Joshi, and J. McAdoo, "Calculations of Hole Transport Characteristics in Bulk GaSb with Comparisons to GaAs," *J.Appl.Phys.*, vol. 88, no. 2, pp. 817–821, 2000.
111. M. Dammann, M. Chertouk, W. Jantz, K. Köhler, K. H. Schmitt, and G. Weimann, "Reliability of Passivated 0.15 μm InAlAs/InGaAs HEMTs with Pseudomorphic Channel," in *Intl. Reliability Phys. Symp.*, (San Diego), pp. 99–102, 1999.

112. M. Dammann, A. Leuther, H. Konstanzer, and W. Jantz, "Effect of Gate Metal on Reliability of Metamorphic HEMTs," in *Proc. Intl. Symp. GaAs and Related Compounds*, (Baltimore), pp. 87–88, 2001.
113. S. Datta, K. Roenker, M. Cahay, and W. Stanchina, "Implications of Hole vs. Electron Transport Properties for High Speed PNP Heterojunction Bipolar Transistors," *Solid-State Electron.*, vol. 43, no. 1, pp. 73–79, Jan. 1999.
114. I. Daumiller, M. Seyboth, C. Kirchner, M. Kamp, and E. Kohn, "GaN/InGaN/GaN DHFETs Heterostructure FET," in *Proc. Device Research Conf.*, (Denver), pp. 1–2, 2000.
115. G. Davies, ed., *Properties and Growth of Diamond.* No. 9 in EMIS Datareviews Series, IEE INSPEC, 1994.
116. D. Decker and C. Dunn, "Determination of Germanium Ionization Coefficients from Small-Signal IMPATT Diode Characteristics," *IEEE Trans.Electron Devices*, vol. ED-17, no. 4, pp. 290–299, 1970.
117. D. Decker and C. Dunn, "Temperature Dependence of Carrier Ionization Rates and Saturated Velocities in Silicon," *J.Electr.Materials*, vol. 4, no. 3, pp. 527–547, 1975.
118. E. Dekel, E. Ehrenfreund, D. Gershoni, P. Boucaud, I. Sagnes, and Y. Campidelli, "Recombination Processes in SiGe/Si Quantum Wells Measured by Photoinduced Absorption Spectroscopy," *Phys.Rev.B*, vol. 56, no. 24, pp. 15734–15739, 1997.
119. J. del Alamo, S. Swirhun, and R. Swanson, "Simultaneous Measurement of Hole Lifetime, Hole Mobility and Bandgap Narrowing in Heavily Doped n-type Silicon," in *IEDM Tech.Dig.*, (Washington), pp. 290–293, 1985.
120. J. del Alamo and M. Sommerville, "Breakdown in Millimeter-Wave Power InP HEMTs: A Comparison with GaAs PHEMTs," *IEEE J.Solid-State Circuits*, vol. 34, no. 9, pp. 1204–1211, 1999.
121. J. del Alamo and R. Swanson, "Modelling of Minority-Carrier Transport in Heavily Doped Silicon Emitters," *Solid-State Electron.*, vol. 30, no. 11, pp. 1127–1136, 1987.
122. R. Deutschmann, *Entwicklung eines physikalischen HFET-Modells: Parameterextraktion und Verifikation.* Dissertation, Technische Universität Wien, 1995. http://www.iue.tuwien.ac.at/phd/deutschmann.
123. DEVICE. http://www.uv.ruhr-uni-bochum.de/.
124. R. Dietrich, A. Wieszt, A. Vescan, H. Leier, R. Stenzel, and W. Klix, "Power Handling Limits and Degradation of Large Area AlGaN/GaN RF-HEMTs," *Solid-State Electron.*, vol. 47, no. 1, pp. 123–125, 2003.
125. F. Diette, D. Langrez, J. Codron, E. Delos, D. Theron, and G. Salmer, "1510 mS/mm 0.1 μm Gate Length Pseudomorphic HEMTs with Intrinsic Current Gain Cutoff Frequency of 220 GHz," *Electron.Lett.*, vol. 32, no. 9, pp. 848–850, 1996.
126. C. Dimitriadis, "Carrier Recombination at Dislocations in Epitaxial Gallium Phosphide Layers," *Solid-State Electron.*, vol. 26, no. 7, pp. 633–637, 1983.
127. R. Dittrich and W. Schroeder, "Empirical Pseudopotential Band Structure of $In_{0.53}Ga_{0.47}As$ and $In_{0.52}Al_{0.48}As$," *Solid-State Electron.*, vol. 43, no. 2, pp. 403–407, 1999.
128. A. Dmitriev and A. Oruzheinikov, "Radiative Recombination Rates in GaN, InN, AlN and their Solid Solutions," in *Proc. Material Research Society Symp.* (D. Gaskill, C. Brandt, and R. Nemanich, eds.), vol. 423, (Pittsburgh), pp. 69–73, 1996.
129. K. Dragosits, V. Palankovski, and S. Selberherr, "Two-Dimensional Modeling of Quantum Mechanical Effects in Ultra-Short CMOS Devices," in *Intl.Conf. on Nanoelectronics and Electromagnetic Compatibility* (N. Mastorakis, V. Kluev, and D. Koruga, eds.), (Skiathos), pp. 113–116, WSEAS Press, 2002.
130. M. Drechsler, D. Hofmann, B. Meyer, T. Detchprohm, H. Amano, and I. Akasaki, "Determination of the Conduction Band Electron Effective Mass in Hexagonal GaN," *Jpn.J.Appl.Phys.*, vol. 34, no. 9, pp. L1178–L1179, 1995.

131. H. Duran, L. Ren, M. Py, O. Homan, U. Lott, and W. Bächtold, "High Performance InP-based HEMTs with Dry Etched Gate Recess for the Fabrication of Low-Noise Microwave Oscillators," in *Proc. Intl.Conf.InP and Rel.Mat.*, (Davos), pp. 319–322, 1999.
132. L. Eastman, V. Tilak, J. Smart, B. Green, E. Chumbes, R. Dimitrov, H. Kim, O. Ambacher, N. Weimann, T. Prunty, M. Murphy, W. J. Schaff, and J. Shealy, "Undoped AlGaN/GaN HEMTs for Microwave Power Amplification," *IEEE Trans.Electron Devices*, vol. 48, no. 3, pp. 479–485, 2001.
133. J. Eberhardt and E. Kasper, "Bandgap Narrowing in Strained SiGe on the Basis of Electrical Measurements on Si/SiGe/Si Hetero Bipolar Transistors," *Mat.Sci.Eng.B*, vol. 89, no. 1–3, pp. 93–96, 2002.
134. K. Eberl, O. Schmidt, and R. Duschl, "Structural Properties of SiC and SiGeC Alloy Layers on Si," in Kasper and Lyutovich [252], section 2.5, pp. 75–88.
135. J. Edgar, ed., *Properties of Group III Nitrides.* No. 11 in EMIS Datareviews Series, IEE INSPEC, 1994.
136. EMIS-Group, ed., *Properties of Gallium Arsenide.* No. 2 in EMIS Datareviews Series, IEE INSPEC, 2 ed., 1990.
137. EMIS-Group, ed., *Properties of Indium Phosphide.* No. 6 in EMIS Datareviews Series, IEE INSPEC, 1991.
138. W. Engl, A. Emunds, B. Meinerzhagen, H. Peifer, and T. Thoma, "Bridging the Gap between the Hydrodynamic and the Monte Carlo Model – An Attempt," in *VLSI Process/Device Modeling Workshop*, (Osaka), pp. 32–33, 1989.
139. W. Engl, R. Laur, and H. Dirks, "MEDUSA - A Simulator for Modular Circuits," *IEEE Trans.Computer-Aided Design of Integrated Circuits and Systems*, vol. 1, no. 2, pp. 85–93, 1982.
140. M. Ershov and V. Ryzhii, "High-Field Electron Transport in SiGe Alloy," *Jpn.J.Appl.Phys.*, vol. 33, no. 3A, pp. 1365–1371, 1994.
141. J. Fan, "Heterostructure Device Wafer Manufacturing for Telecom Applications for 4" and 6" Wafer Fabs," in *Proc. Intl.Conf. GaAs Manufacturing Technology*, (St. Louis), pp. 193–197, 1999.
142. D. Fanning, L. Witkowski, J. Stidham, H. Tserng, M. Muir, and P. Saunier, "Dielectrically Defined Optical T-gate for High Power GaAs pHEMTs," in *Proc. Intl.Conf. GaAs Manufacturing Technology*, (San Diego), pp. 83–86, 2002.
143. M. Farahmand, C. Garetto, E. Bellotti, K. Brennan, M. Goano, E. Ghillino, G. Ghione, J. Albrecht, and P. Ruden, "Monte Carlo Simulation of Electron Transport in the III-Nitride Wurtzite Phase Materials System: Binaries and Ternaries," *IEEE Trans.Electron Devices*, vol. 48, no. 3, pp. 535–542, Mar. 2001.
144. D. Ferry, *Semiconductors.* New York: Macmillan, 1991.
145. C. Fischer, *Bauelementsimulation in einer computergestützten Entwurfsumgebung.* Dissertation, Technische Universität Wien, 1994.
http://www.iue.tuwien.ac.at/phd/fischer.
146. C. Fischer, P. Habaš, O. Heinreichsberger, H. Kosina, P. Lindorfer, P. Pichler, H. Pötzl, C. Sala, A. Schütz, S. Selberherr, M. Stiftinger, and M. Thurner, *MINIMOS 6 User's Guide.* Inst. für Mikroelektronik, Technische Universität Wien, 1994.
147. M. Fischetti, "Monte Carlo Simulation of Transport in Technologically Significant Semiconductors of the Diamond and Zinc-Blende Structures–Part I: Homogeneous Transport," *IEEE Trans.Electron Devices*, vol. 38, no. 3, pp. 634–649, 1991.
148. M. Fischetti and S. Laux, "Band Structure, Deformation Potentials, and Carrier Mobility in Strained Si, Ge, and SiGe Alloys," *J.Appl.Phys.*, vol. 80, no. 4, pp. 2234–2252, Aug. 1996.

149. M. Fischetti, S. Laux, and E. Crabbé, "Understanding Hot-Electron Transport in Silicon Devices: Is There a Shortcut?," *J.Appl.Phys.*, vol. 78, no. 2, pp. 1058–1085, 1995.
150. M. Fischetti, N. Sano, S. E. Laux, and K. Natori, "Full-Band-Structure Theory of High-field Transport and Impact Ionization of Electrons and Holes in Ge, Si, and GaAs," *IEEE J. Technology Computer Aided Design*, 1997. http://www.ieee.org/products/online/journal/tcad/accepted/fischetti-feb97/.
151. FLOODS and FLOOPS. http://www.tec.ufl.edu/flooxs/.
152. M. Foisy, *A Physical Model for the Bias Dependence of the Modulation-Doped Field-Effect Transistor's High Frequency Performance.* PhD thesis, Cornell University, 1990.
153. J. Fossum and D. Lee, "A Physical Model for the Dependence of Carrier Lifetime on Doping Density in Nondegenerate Silicon," *Solid-State Electron.*, vol. 25, no. 8, pp. 741–747, 1982.
154. B. Foutz, S. O'Leary, M. Shur, and L. Eastman, "Transient Electron Transport in Wurtzite GaN, InN, and AlN," *J.Appl.Phys.*, vol. 85, no. 11, pp. 7727–7734, 1999.
155. G. Freeman, D. Ahlgren, D. Greenberg, R. Groves, F. Huang, G. Hugo, B. Jagannathan, J. Jeng, J. Johnson, K. Schonenberg, K. Stein, R. Volant, and S. Subbanna, "A 0.18 μm 90 GHz f_T SiGe HBT BiCMOS, ASIC-Compatible, Copper Interconnect Technology for RF and Microwave Applications," in *IEDM Tech.Dig.*, (Washington), pp. 569–572, 1999.
156. W. Frensley, "Boundary Conditions for Open Quantum Systems Driven Far From Equilibrium," *Reviews of Modern Physics*, vol. 62, no. 3, pp. 745–789, 1990.
157. S. Fu, T. Chin, M. Ho, C. Tu, and P. Aspeck, "Impact Ionization Coefficients in (100) GaInP," *Appl.Phys.Lett.*, vol. 66, no. 25, pp. 3507–3509, 1995.
158. F. Fuchs, K. Kheng, P. Koidl, and K. Schwarz, "Fermi-Edge Singularity in Degenerate n-Type Bulk InAs," *Phys.Rev.B*, vol. 48, no. 11, pp. 7884–7888, 1993.
159. A. Fujihara, Y. Ikenaga, H. Takahashi, M. Kawanaka, and S. Tanaka, "High-Speed InP/InGaAs DHBTs with Ballistic Collector Launcher Structure," in *IEDM Tech.Dig.*, (Washington), pp. 772–775, 2001.
160. G-PISCES-2B. http://www.gateway-modeling.com/products.htm.
161. F. Gaensslen and R. Jaeger, "Temperature Dependent Threshold Behavior of Depletion Mode MOSFET's," *Solid-State Electron.*, vol. 22, no. 4, pp. 423–430, 1979.
162. F. Gaensslen, R. Jaeger, and J. Walker, "Low-Temperature Threshold Behavior of Depletion Mode Devices - Characterization and Simulation," in *IEDM Tech.Dig.*, (Washington), pp. 520–524, 1976.
163. A. Galvanauskas, A. Gorelenok, Z. Dobrovol'skis, S. Kershulis, Y. Pozhela, A.Reklaitis, and N. Shmidt, "Transport Phenomena and Alloy Scattering in $In_{1-x}Ga_xAs_yP_{1-y}$ Compounds," *Sov.Phys.Semicond.*, vol. 22, no. 9, pp. 1055–1058, 1988.
164. D. Gaskill, L. Rowland, and K. Doverspike, "Electrical Properties of AlN, GaN, and AlGaN," in Edgar [135], section 3.2, pp. 101–116.
165. W. Geerts, J. Mackenzie, C. Abernathy, S. Pearton, and T. Schmiedel, "Electrical Transport in p-GaN, n-InN, and n-InGaN," *Solid-State Electron.*, vol. 39, no. 9, pp. 1289–1294, 1996.
166. A. Geissberger, M. Fresina, L. Kapitan, C. Barratt, T. Kin, M. Hoppe, D. Streit, T. Block, M. Lammert, and A. Oki, "Transfer and Qualification of a Layout-Compatible Second Source HBT Technology for Mobile Phone Applications," in *Proc. Intl.Conf. GaAs Manufacturing Technology*, (St. Louis), pp. 245–248, 1999.
167. B. Gelmont and Z. Sokolova, "Auger Recombination in Direct-Gap N-type Semiconductors," *Sov.Phys.Semicond.*, vol. 16, no. 9, pp. 1067–1068, 1982.

168. B. Gelmont, Z. Sokolova, and I. Yassievich, "Auger Recombination in Direct-Gap p-type Semiconductors," *Sov.Phys.Semicond.*, vol. 16, no. 4, pp. 382–387, 1982.
169. W. Gerlach, H. Schlangenotto, and H. Maeder, "On the Radiative Recombination Rate in Silicon," *Phys.stat.sol.(a)*, vol. 13, no. 1, pp. 277–283, 1972.
170. M. Gershenzon and R. Mikulyak, "Radiative Pair Recombination and Surface Recombination in GaP Photoluminescence," *Appl.Phys.Lett.*, vol. 8, no. 10, pp. 245–247, 1966.
171. J. Geßner, F. Schwierz, H. Mau, D. Nuernbergk, M. Roßberg, and D. Schipanski, "Simulation of the Frequency Limits of SiGe HBTs," in *Proc. Intl.Conf. on Modeling and Simulation of Microsystems*, (San Juan, Puerto Rico), pp. 407–410, Apr. 1999.
172. R. Ghin, J. David, S. Plimmer, M. Hopkinson, G. Rees, D. Herbert, and D. Wight, "Avalanche Multiplication and Breakdown in $Ga_{0.52}In_{0.48}P$ Diodes," *IEEE Trans.Electron Devices*, vol. 45, no. 10, pp. 2096–2101, 1998.
173. M. Goano, E. Belloti, E. Ghillino, C. Garetto, G. Ghione, and K. Brennan, "Band Structure Nonlocal Pseudopotential Calculation of the III-Nitride Wurtzide Phase Materials System. Part II: Ternary Alloys $Al_xGa_{1-x}N$, $In_xGa_{1-x}N$, and $In_xAl_{1-x}N$," *J.Appl.Phys.*, vol. 88, no. 11, pp. 6476–6482, 2000.
174. M. Goano, E. Belloti, E. Ghillino, G. Ghione, and K. Brennan, "Band Structure Nonlocal Pseudopotential Calculation of the III-Nitride Wurtzide Phase Materials System. Part I: Binary Compounds GaN, AlN, and InN," *J.Appl.Phys.*, vol. 88, no. 11, pp. 6467–6475, 2000.
175. J. Godin, M. Riet, S. Blayac, P. Berdaguer, V. Dhalluin, F. Alexandre, M. Kahn, A. Pinquier, A. Kasbari, J. Moulu, and A. Konczykowska, "InP DHBT Technology and Design for 40 Gbit/s Full-Rate-Clock Communication Circuits," in *GaAs IC Symp.Tech.Dig.*, (Monterey), pp. 215–218, 2002.
176. B. Gonzales, V. Palankovski, H. Kosina, A. Hernandez, and S. Selberherr, "An Energy Relaxation Time Model for Device Simulation," *Solid-State Electron.*, vol. 43, no. 9, pp. 1791–1795, 1999.
177. T. González Sánchez, J. Velázques Pérez, P. Gutiérrez Conde, and D. Pardo Collantes, "Five-Valley Model for the Study of Electron Transport Properties at Very High Electric Fields in GaAs," *Semicond.Sci.Technol.*, vol. 6, no. 9, pp. 862–871, 1991.
178. T. González Sánchez, J. Velázques Pérez, P. Gutiérrez Conde, and D. Pardo Collantes, "Electron Transport in InP Under High Electric Field Conditions," *Semicond.Sci.Technol.*, vol. 7, no. 1, pp. 31–36, 1992.
179. M. Gospodinova, V. Mollov, R. Arnaudov, and P. Philippov, "Study of the DC Biasing Effect on Insertion Losses in High-Frequency Interconnections," *Microelectronics Journal*, vol. 31, no. 11-12, pp. 1009–1014, 2000.
180. S. Goto, T. Ueda, T. Ohshima, and H. Kakinuma, "Effect of the Growth Conditions on Electrical Properties of Si-Doped $In_{0.52}Al_{0.48}As$ Grown by Metalorganic Vapor Phase Epitaxy," *Jpn.J.Appl.Phys.*, vol. 38, no. 2, pp. 1048–1051, 1999.
181. W. Grant, "Electron and Hole Ionization Rates in Epitaxial Silicon at High Electric Fields," *Solid-State Electron.*, vol. 16, no. 10, pp. 1189–1203, 1973.
182. T. Grasser, *Mixed-Mode Device Simulation.* Dissertation, Technische Universität Wien, 1999. http://www.iue.tuwien.ac.at/phd/grasser.
183. T. Grasser, H. Kosina, C. Heitzinger, and S. Selberherr, "Accurate Impact Ionization Model which Accounts for Hot and Cold Carrier Populations," *Appl.Phys.Lett.*, vol. 80, no. 4, pp. 613–615, 2002.
184. T. Grasser, H. Kosina, and S. Selberherr, "Investigation of Spurious Velocity Overshoot Using Monte Carlo Data," *Appl.Phys.Lett.*, vol. 79, no. 12, pp. 1900–1903, 2001.
185. T. Grasser, V. Palankovski, G. Schrom, and S. Selberherr, "Hydrodynamic Mixed-Mode Simulation," in *Proc. Intl.Conf. on Simulation of Semiconductor Processes and Devices* (K. De Meyer and S. Biesemans, eds.), pp. 247–250, Leuven: Springer, 1998.

186. T. Grasser and S. Selberherr, "Fully Coupled Electrothermal Mixed-Mode Device Simulation of SiGe HBT Circuits," *IEEE Trans.Electron Devices*, vol. 48, no. 7, pp. 1421–1427, 2001.
187. T. Grasser, T. Tang, H. Kosina, and S. Selberherr, "A Review of Hydrodynamic and Energy-Transport Models for Semiconductor Device Simulation," *Proc.IEEE*, vol. 91, no. 2, pp. 251–274, 2003.
188. M. Green, "Intrinsic Concentration, Effective Densities of States, and Effective Mass in Silicon," *J.Appl.Phys.*, vol. 67, no. 6, pp. 2944–2954, 1990.
189. C. Grein, K. A. El-Rub, M. Flatte, and H. Ehrenreich, "Band Structure Engineering of Superlattice-Based Short-;Mid-, and Long-Wavelength Infrared Avalanche Photodiodes for Improved Impact Ionization Rates," *J.Appl.Phys.*, vol. 92, no. 7, pp. 3771–3777, Oct. 2002.
190. I. Gresham, N. Jain, T. Budka, A. Alexanian, N. Kinayman, B. Ziegner, S. Brown, and P. Staecker, "A Compact Manufacturable 76–77 GHz Radar Module for Commercial ACC Applications," *IEEE Trans.Microwave Theory and Techniques*, vol. 49, no. 1, pp. 44–58, 2001.
191. E. Grilli, M. Guzzi, R. Zamboni, and L. Pavesi, "High-Precision Determination of the Temperature Dependence of the Fundamental Energy Gap in Gallium Arsenide," *Phys.Rev.B*, vol. 45, no. 4, pp. 1638–1644, 1992.
192. H. Grubin, "High Field Transport and the Gunn Effect in AlGaAs/GaAs Structures," in Adachi [5], section 7.5, pp. 197–213.
193. M. Grupen, K. Hess, and G. Song, "Simulation of Transport over Heterojunctions," in *Simulation of Semiconductor Devices and Processes* (W. Fichtner and D. Aemmer, eds.), vol. 4, (Konstanz), pp. 303–310, Hartung-Gorre, 1991.
194. J. Gunn, "Microwave Oscillations of Current in III-V Semiconductors," *Solid-State Comm.*, vol. 1, no. 4, pp. 88–91, 1963.
195. M. Gupta, "Power Gain in Feedback Amplifiers," *IEEE Trans.Microwave Theory and Techniques*, vol. 40, no. 5, pp. 864–879, 1992.
196. P. Habaš, *Analysis of Physical Effects in Small Silicon MOS Devices.* Dissertation, Technische Universität Wien, 1993.
197. Z. Hang, H. Shen, and F. Pollak, "Temperature Dependence of the E_0 and $E_0+\Delta_0$ Gaps of InP up to 600 degrees ^{o}C," *Solid-State Comm.*, vol. 73, no. 1, pp. 15–18, 1990.
198. W. Hänsch, *The Drift Diffusion Equation and its Application in MOSFET Modeling.* Wien, New York: Springer, 1991.
199. W. Hänsch, M. Orlowski, and W. Weber, "The Hot-Electron Problem in Submicron MOSFET," in *18th European Solid State Device Research Conference - ESSDERC 88* (J.-P. Nougier and D. Gasquet, eds.), vol. 49 of *Journal de Physique*, (Les Ulis Cedex, France), pp. 597–606, les éditions de physique, 1988.
200. W. Hänsch, T. Vogelsang, R. Kircher, and M. Orlowski, "Carrier Transport Near the Si/SiO_2 Interface of a MOSFET," *Solid-State Electron.*, vol. 32, no. 10, pp. 839–849, 1989.
201. H. Happy, S. Bollaert, H. Foure, and A. Cappy, "Numerical Analysis of Device Performance of Metamorphic $In_yAl_{1-y}As/In_xGa_{1-x}As$ ($0.3\leq x \leq 0.6$) HEMTs on GaAs Substrate," *IEEE Trans.Electron Devices*, vol. 45, no. 10, pp. 2089–2095, 1998.
202. E. Harmon, M. Lovejoy, M. Melloch, M. Lundstrom, T. de Lyon, , and J. Woodall, "Experimental Observation of a Minority Electron Mobility Enhancement in Degenerately Doped p-Type GaAs," *Appl.Phys.Lett.*, vol. 63, no. 4, pp. 5597–5602, 1993.
203. E. Harmon, M. Melloch, and M. Lundstrom, "Effective Band-Gap Shrinkage in GaAs," *Appl.Phys.Lett.*, vol. 64, no. 4, pp. 502–504, Jan. 1994.
204. E. Harmon, M. Melloch, M. Lundstrom, and M. Lovejoy, "Experimental Determination of the Effects of Degenerate Fermi Statistics on Heavily p-Doped GaAs," *Appl.Phys.Lett.*, vol. 58, no. 15, pp. 1647–1649, 1991.

205. G. Harris, ed., *Properties of Silicon Carbide.* No. 13 in EMIS Datareviews Series, IEE INSPEC, 1995.
206. T. Hashimoto, Y. Nonaka, T. Saito, K. Sasahara, T. Tominari, K. Sakai, K. Tokunaga, T. Fujiwara, S. Wada, T. Udo, T. Jinbo, K. Washio, and H. Hosoe, "Integration of a 0.13-μm CMOS and a High Performance Self-Aligned SiGe HBT Featuring Low Base Resistance," in *IEDM Tech.Dig.*, (San Francisco), pp. 779–782, 2002.
207. M. Haupt, *Molekularstrahl-Epitaxie und Charakterisierung von* $Ga_xIn_{1-x}As/Al_xIn_{1-x}As$ *Potentialtopfstrukturen (in German).* Diplomarbeit, Universität Freiburg, 1994.
208. M. Haupt, K. Köhler, P. Ganser, S. Emminger, S. Müller, and W. Rothemund, "Growth of High Quality $Al_{0.48}In_{0.52}As/Ga_{0.47}In_{0.53}As$ Heterostructures Using Strain Relaxed $Al_xGa_yIn_{1-x-y}As$ Buffer Layers on GaAs," *Appl.Phys.Lett.*, vol. 69, no. 3, pp. 412–414, 1996.
209. J. Hauser, T. Glisson, and M. Littlejohn, "Negative Resistance and Peak Velocity in the Central (000) Valley of III-V Semiconductors," *Solid-State Electron.*, vol. 22, no. 5, pp. 487–493, 1979.
210. J. Haynes and W. Shockley, "The Mobility and Life of Injected Holes and Electrons in Germanium," *Phys.Rev.*, vol. 81, no. 5, pp. 835–843, 1951.
211. B. Heinemann, D. Knoll, R. Barth, D. Bolze, K. Blum, J. Drews, K. Ehwald, G. Fischer, K. Köpke, D. Krüger, R. Kurps, H. Rücker, P. Schley, W. Winkler, and H. Wulf, "Cost-Effective High-Performance High-Voltage SiGe:C HBTs with 100 GHz f_T and $BV_{CE0} \times f_T$ Products Exceeding 220 VGHz," in *IEDM Tech.Dig.*, (Washington), pp. 348–351, 2001.
212. C. Henry, R. Logan, F. Merrit, and C. Bethea, "Radiative and Nonradiative Lifetimes in N-Type and P-Type 1.6 μm InGaAs," *Electron.Lett.*, vol. 20, no. 9, pp. 358–359, 1984.
213. K. Hess, ed., *Monte Carlo Device Simulation: Full Band and Beyond.* Boston, Dordrecht, London: Kluwer, 1991.
214. K. Hess, *Theory of Semiconductor Devices.* IEEE Press, 2000.
215. O. Hildebrand, W. Kuebart, and M. Pilkuhn, "Resonant Enhancement of Impact in $Ga_{1-x}Al_xSb$," *Appl.Phys.Lett.*, vol. 37, no. 9, pp. 801–804, 1980.
216. P. Hill, J. Schlafer, W. Powazinik, M. Urban, E. Eichen, and R. Olshansky, "Measurement of Hole Velocity in n-Type InGaAs," *Appl.Phys.Lett.*, vol. 50, no. 18, pp. 1260–1262, 1987.
217. M. Hirose, J. Yoshida, and N. Toyoda, "An Improved Two-Dimensional Simulation Model (MEGA) for GaAs MESFET Applicable to LSI Design," *IEEE Trans.Computer-Aided Design*, vol. 7, no. 2, pp. 225–230, 1988.
218. W. Hoke, C. Whelan, P. Marsh, P. Lemonias, P. Lyman, S. Lardizabal, R. Leoni, T. Kazior, A. Torabi, J. Jang, D. Dumka, I. Adesida, K. Chang, and K. Hsieh, "Progress in GaAs-Based Metamorphic Technology," in *Proc. Intl.Conf. GaAs Manufacturing Technology*, (Las Vegas), pp. 225–229, 2001.
219. A. Hössinger and S. Selberherr, "Accurate Three-Dimensional Simulation of Damage Caused by Ion Implantation," in *Proc. Intl.Conf. on Modeling and Simulation of Microsystems*, (San Juan, Puerto Rico), pp. 363–366, Apr. 1999.
220. M.-K. Hudait, P. Modak, S. Hardikar, and S. Krupanidhi, "Zn Incorporation and Band Gap Shrinkage in p-type GaAs," *J.Appl.Phys.*, vol. 82, no. 10, pp. 4931–4937, 1997.
221. P. Huguet, P. Auxemery, G. Pataut, P. Fellon, D. Geiger, and H. Jung, "Space Evaluation of P-HEMT MMIC Process PH15," in *Proc. European Space Components Conf.*, (Noordwijk), pp. 199–204, 2000.

222. L. Huldt, N. Nilsson, and K. Svantesson, "The Temperature Dependence of Band-to-Band Auger Recombination in Silicon," *Appl.Phys.Lett.*, vol. 35, no. 10, pp. 776–777, 1979.
223. G. Hurkx, D. Klaassen, and M. Knuvers, "A New Recombination Model for Device Simulation Including Tunneling," *IEEE Trans.Electron Devices*, vol. 39, no. 2, pp. 331–338, 1992.
224. M. Ida, K. Kurishima, N. Watanabe, and T. Enoki, "InP/InGaAs DHBTs with 341-GHz f_T at High Current Density of over 800 kA/cm^2," in *IEDM Tech.Dig.*, (Washington), pp. 776–779, 2001.
225. M. Ieong, P. Solomon, S. Laux, H. Wong, and D. Chidambarrao, "Comparison of Raised and Schottky Source/Drain MOSFETs Using a Novel Tunneling Contact Model," in *IEDM Tech.Dig.*, (San Francicso), pp. 733–736, 1998.
226. M. Ilegems, "InP-Based Lattice-Matched Heterostructures," in Bhattacharya [38], section 1.3, pp. 16–25.
227. ISE Integrated Systems Engineering AG, Zürich, Switzerland, *DESSIS-ISE, ISE TCAD Release 6.0*, July 1999.
228. H. Ito and T. Ishibashi, "Minority-Electron Mobility in *p*-Type GaAs," *J.Appl.Phys.*, vol. 65, no. 12, pp. 5197–5199, 1989.
229. S. Iyer, G. Patton, S. Delage, S. Tiwari, and J. Stork, "Silicon-Germanium Base Heterojunction Bipolar Transistors by Molecular Beam Epitaxy," in *Proc. Si-MBE Symp.*, (Hawaii), pp. 114–125, Oct. 1987.
230. C. Jacoboni, C. Canali, G. Ottaviani, and A. Quaranta, "A Review of Some Charge Transport Properties of Silicon," *Solid-State Electron.*, vol. 20, no. 2, pp. 77–89, 1977.
231. C. Jacoboni and P. Lugli, *The Monte Carlo Method for Semiconductor Device Simulation.* Wien-New York: Springer, 1989.
232. C. Jacoboni, F. Nava, C. Canali, and G. Ottaviani, "Electron Drift Velocity and Diffusivity in Germanium," *Phys.Rev.B*, vol. 24, no. 2, pp. 1014–1026, 1981.
233. B. Jagannathan, M. Khater, F. Pagette, J. Rieh, D. Angell, H. Chen, J. Florkey, F. Golan, D. Greenberg, R. Groves, S. Jeng, J. Johnson, E. Mengistu, K. Schonenberg, C. Schnabel, P. Smith, A. Stricker, D. Ahlgren, G. Freeman, K. Stein, and S. Subbanna, "Self-Aligned SiGe NPN Transistor With 285 GHz $f_{\max}$ and 207 GHz f_T in a Manufacturable Technology," *IEEE Electron Device Lett.*, vol. 23, no. 5, pp. 258–260, 2002.
234. S. Jain, *Germanium-Silicon Strained Layers and Heterostructures*, vol. 24 of *Advances in Electronics and Electron Physics.* Academic Press, 1994.
235. S. Jain, J. McGregor, and D. Roulston, "Band-Gap Narrowing in Novel III-V Semiconductors," *J.Appl.Phys.*, vol. 68, no. 7, pp. 3747–3749, 1990.
236. S. Jain and D. Roulston, "A Simple Expression for Band Gap Narrowing (BGN) in Heavily Doped Si, Ge, GaAs, and Ge_xSi_{1-x} Strained Layers," *Solid-State Electron.*, vol. 34, no. 5, pp. 453–465, 1991.
237. B. Jalali and S. Pearton, eds., *InP HBTs: Growth,Processing,and Applications.* Boston: Artech House, 1995.
238. E. Johnson, "Physical Limitations on Frequency and Power Parameters of Transistors," *RCA Review*, vol. 26, no. pt. 5, pp. 163–177, June 1965.
239. J. Johnson, A. Stricker, A. Joseph, and J. Slinkman, "A Technology Simulation Methodology for AC-Performance Optimization of SiGe HBTs," in *IEDM Tech.Dig.*, (Washington), pp. 489–492, 2001.
240. R. Johnson, Z. Kachwalla, and C. Snowden, "Multi-Tone Microwave Simulation of HEMTs Using a Physics-Based Electro-Thermal CAD Model," in *Proc. European Microwave Conf.*, (Munich), pp. 107–110, 1999.
241. R. Johnson and O. Eknoyan, "High-field Electron Drift Velocity Measurements in Gallium Phosphide," *J.Appl.Phys.*, vol. 58, no. 3, pp. 1402–1403, 1985.

242. J. Jongsma, "AMS Low Power Bluetooth Solution in 0.35μm SiGe-BiCMOS," in *Proc. Informationstagung Mikroelektronik* (G. Fiedler, ed.), (Vienna), pp. 249–254, 2003.
243. C. Jungemann, B. Neinhüs, S. Decker, and B. Meinerzhagen, "Hierarchical 2-D DD and HD Noise Simulations of Si and SiGe Devices–Part II: Results," *IEEE Trans.Electron Devices*, vol. 49, no. 7, pp. 1258–1264, July 2002.
244. C. Jungemann, B. Neinhüs, and B. Meinerzhagen, "Full-Band Monte Carlo Device Simulation of a SiGe/Si HBT with a Realistic Ge Profile," *IEICE Trans.Electron.*, vol. E83-C, no. 8, pp. 1228–1234, 2000.
245. C. Jungemann, B. Neinhüs, and B. Meinerzhagen, "Hierarchical 2-D DD and HD Noise Simulations of Si and SiGe Devices–Part I: Theory," *IEEE Trans.Electron Devices*, vol. 49, no. 7, pp. 1250–1257, July 2002.
246. C. Jungemann, S. Yamaguchi, and H. Goto, "Is There Experimental Evidence for a Difference between Surface and Bulk Impact Ionization in Silicon?," in *IEDM Tech.Dig.*, (San Francisco), pp. 383–386, 1996.
247. C. Jungemann, B. Neinhüs, and B. Meinerzhagen, "Comparative Study of Electron Transit Times Evaluated by DD, HD, and MC Device Simulation for a SiGe HBT," *IEEE Trans.Electron Devices*, vol. 48, no. 10, pp. 2216–2220, Oct. 2001.
248. M. Kahn, S. Blayac, M. Riet, P. Berdaguer, V. Dhalluin, F. Alexandre, F. Aniel, and J. Godin, "Effect of Base Thickness Reduction on High Speed Characteristics and Base Resistance of InGaAs/InP Heterojunction Bipolar Transistors," in *Proc. Intl.Conf.InP and Rel.Mat.*, (Santa Barbara), pp. 134–137, 2003.
249. K. Kalna, S. Roy, A. Asenov, K. Elgaid, and I. Thayne, "RF Analysis of Aggressively Scaled PHEMTs," in *Proc. European Solid-State Device Research Conf.* (W. Lane, G. Crean, F. McCabe, and H. Grünbacher, eds.), (Cork), pp. 156–159, Frontier Group, 2000.
250. Y. Kao and O. Eknoyan, "Electron and Hole Carrier Mobilities for Liquid Phase Epitaxially Grown GaP in the Temperature Range 200–550 K," *J.Appl.Phys.*, vol. 54, no. 5, pp. 2468–2471, 1983.
251. K. Kasahara, H. Miyamoto, Y. Ando, Y. Okamoto, T. Nakayama, and M. Kuzuhara, "Ka-Band 2.3 W Power AlGaN/GaN Heterojunction FET," in *IEDM Tech.Dig.*, (San Francisco), pp. 677–680, 2002.
252. E. Kasper and K. Lyutovich, eds., *Silicon Germanium and SiGe-Carbon.* No. 24 in EMIS Datareviews Series, IEE INSPEC, 2000.
253. A. Katz, *Indium Phosphide and Related Materials.* Boston: Artech House, 1992.
254. M. Kerr, J. Schmidt, A. Cuevas, and J. Bultman, "Surface Recombination Velocity of Phosphorus-Diffused Silicon Solar Cell Emitters Passivated with Plasma Enhanced Chemical Vapor Deposited Silicon Nitride and Thermal Silicon Oxide," *J.Appl.Phys.*, vol. 89, no. 7, pp. 3821–3826, 2001.
255. K. Kim, M. Lee, J. Bahng, K. Shim, and B. Choe, "Optical Constants and Electronic Interband Transitions of Disordered $GaAs_{1-x}P_x$ Alloys," *J.Appl.Phys.*, vol. 84, no. 7, pp. 3696–3699, Oct. 1998.
256. J. King, ed., *Material Handbook for Hybrid Microelectronics.* Artech House, 1988.
257. C. Kittel, *Introduction to Solid State Physics.* New York: Wiley, 6 ed., 1986.
258. D. Klaassen, J. Slotboom, and H. de Graaff, "Unified Apparent Bandgap Narrowing in n- and p-Type Silicon," *Solid-State Electron.*, vol. 35, no. 2, pp. 125–129, 1992.
259. M. Knaipp, *Modellierung von Temperatureinflüssen in Halbleiterbauelementen.* Dissertation, Technische Universität Wien, 1998. http://www.iue.tuwien.ac.at/phd/knaipp.

260. K. Kobayashi, L. Yang, A. Guitierrez-Aitken, E. Kaneshiro, F. Yamada, H. Yen, T. Block, A. Oki, and D. Streit, "A 25 GHz InGaAs/InAlAs-InP Power MMIC with 48 % Power Added Efficiency," in *IEDM Tech.Dig.*, (Washington), pp. 409–412, 1999.
261. K. Köhler, S. Müller, N. Rollbühler, R. Kiefer, R. Quay, and G. Weimann, "Multi-wafer Epitaxy of AlGaN/GaN Heterostructures for Power Applications," in *Proc. Intl. Symp. Compound Semiconductors* (M. Ilegems, ed.), (Lausanne), pp. 235–238, IOP Publishing, 2003.
262. J. Kolnik, I. Oguzman, K. Brennan, R. Wang, and P. Ruden, "Monte Carlo Calculation of Electron Initiated Impact Ionization in Bulk Zinc-blende and Wurtzite GaN," *J.Appl.Phys.*, vol. 81, no. 2, pp. 726–733, 1997.
263. K. Kong, D. Boone, M. King, B. Nguyen, M. Vernon, E. Reese, and G. Brehm, "A Compact 30 GHz MMIC High Power Amplifier (3 W CW) in Chip and Packaged Form," in *GaAs IC Symp.Tech.Dig.*, (Monterey), pp. 37–39, 2002.
264. C. Köpf, *Modellierung des Elektronentransports in Verbindungshalbleiterlegierungen*. Dissertation, Technische Universität Wien, 1997. (in German) http://www.iue.tuwien.ac.at/phd/koepf.
265. C. Köpf, G. Kaiblinger-Grujin, H. Kosina, and S. Selberherr, "Influence of Dopant Species on Electron Mobility in InP," in *Proc. Intl.Conf.InP and Rel.Mat.*, (Hyannis), pp. 280–283, 1997.
266. C. Köpf, G. Kaiblinger-Grujin, H. Kosina, and S. Selberherr, "Reexamination of Electron Mobility Dependence on Dopants in GaAs," in *Proc. European Solid-State Device Research Conf.* (H. Grünbacher, ed.), (Stuttgart), pp. 304–307, Editions Frontieres, 1997.
267. C. Köpf, H. Kosina, and S. Selberherr, "Physical Models for Strained and Relaxed GaInAs Alloys: Band Structure and Low-Field Transport," *Solid-State Electron.*, vol. 41, no. 8, pp. 1139–1152, 1997.
268. H. Kosina and G. Kaiblinger-Grujin, "Ionized-Impurity Scattering of Majority Electrons in Silicon," *Solid-State Electron.*, vol. 42, no. 3, pp. 331–338, 1998.
269. H. Kosina, G. Klimeck, M. Nedjalkov, and S. Selberherr, "Comparison of Numerical Quantum Device Models," in *Proc. Intl.Conf. on Simulation of Semiconductor Processes and Devices*, (Boston), pp. 171–174, 2003.
270. H. Kosina, M. Nedjalkov, and S. Selberherr, "Theory of the Monte Carlo Method for Semiconductor Device Simulation," *IEEE Trans.Electron Devices*, vol. 47, no. 10, pp. 1898–1908, 2000.
271. H. Kosina and S. Selberherr, "A Hybrid Device Simulator that Combines Monte Carlo and Drift-Diffusion Analysis," *IEEE Trans.Computer-Aided Design*, vol. 13, no. 2, pp. 201–210, 1994.
272. H. Kosina and C. Troger, "SPIN — A Schrödinger-Poisson Solver Including Non-parabolic Bands," *VLSI Design*, vol. 8, no. 1-4, pp. 489–493, 1998.
273. K. Kramer and W. Hitchon, *Semiconductor Devices – A Simulation Approach*. Prentice Hall, 1997.
274. H. Kressel and J. Butler, *Semiconductor Lasers and Heterojunction LEDs*. New York: Academic Press, 1977.
275. M. Krijn, "Heterojunction Band Offsets and Effective Masses in III-V Quaternary Alloys," *Semicond.Sci.Technol.*, vol. 6, no. 1, pp. 27–31, 1991.
276. S. Krishnan, G. Yeap, B. Yu, Q. Xiang, and M.-R. Lin, "High-k Scaling of Gate Insulators: an Insightful Study," in *Proc. Conf. on Microelectronic Device Technology*, vol. 3506, (Santa Clara), pp. 65–72, SPIE, 1998.
277. H. Kroemer, "Theory of a Wide-Gap Emitter for Transistors," *Proc. IRE*, vol. 45, no. 11, pp. 1535–1537, Nov. 1957.

278. S. Krukowski, A. Witek, J. Adamczyk, J. Jun, M. Bockowski, I. Grzegory, B. Lucznik, G. Nowak, M. Wroblewski, A. Presz, S. Gierlotka, S. Stelmach, B. Palosz, S. Porowski, and P. Zinn, "Thermal Properties of Indium Nitride," *J.Phys.Chem.Solids*, vol. 59, no. 3, pp. 289–295, 1998.
279. S. Krupenin, R. Blanchard, M. Sommerville, J. del Alamo, K. Duh, and P. Chao, "Physical Mechanisms Limiting the Manufacturing Uniformity of Millimeter-Wave Power InP HEMTs," *IEEE Trans.Electron Devices*, vol. 47, no. 8, pp. 1560–1565, 2000.
280. K. Kunihiro, K. Kasahara, Y. Takahashi, and Y. Ohno, "Experimental Evaluation of Impact Ionization Coefficients in GaN," *IEEE Electron Device Lett.*, vol. 20, no. 12, pp. 608–610, 1999.
281. P. Kurpas, A. Maaßdorf, W. Doser, P. Heymann, B. Janke, F. Schnieder, H. Blanck, P. Auxemery, D. Pons, W. Heinrich, and J. Würfl, "10 W GaInP/GaAs Power HBTs for Base Station Applications," in *IEDM Tech.Dig.*, (San Francisco), pp. 681–684, 2002.
282. R. Lai, M. Barsky, R. Grundbacher, P. Liu, T. Chin, M. Nishimoto, R. Elmajarian, R. Rodriguez, L. Tran, A. Gutierrez, A. Oki, and D. Streit, "InP HEMT Amplifier Development for G-Band (140-220 GHz) Applications," in *IEDM Tech.Dig.*, (San Francisco), pp. 175–177, 2000.
283. R. Lai, P. H. Liu, J. Scarpulla, R. T. D. Leung, D. Eng, R. Grundbacher, M. Aust, J. Lee, and M. Hoppe, "0.1 μm InGaAs/AlGaAs/GaAs HEMT MMIC Production Process for High Performance Commercial Ka-Band LNAs," in *Proc. Intl.Conf. GaAs Manufacturing Technology*, (Washington), pp. 249–250, 2000.
284. D. Lambert, D. Lin, and R. Dupuis, "Simulation of the Characteristics of AlGaN/GaN Heterojunction Bipolar Transistors," *Solid-State Electron.*, vol. 44, no. 2, pp. 253–257, 2000.
285. W. Lambrecht and B. Segall, "Band Structure of AlGaN, $(AlN)_x(SiC)_{1-x}$ and Diamond/c-BN Alloys," in Edgar [135], section 5.2, pp. 163–166.
286. M. Landolt and J. Börnstein, *Zahlenwerte und Funktionen aus Naturwissenschaften und Technik*, vol. 17/A of *Neue Serie, Gruppe III.* Berlin: Springer, 1982.
287. M. Landolt and J. Börnstein, *Numerical Data and Functional Relationships in Science and Technology*, vol. 22/A of *New Series, Group III.* Berlin: Springer, 1987.
288. M. Landolt and J. Börnstein, *Numerical Data and Functional Relationships in Science and Techology*, vol. 41 of *New Series, Group III.* Berlin: Springer, 2001.
289. P. Landsberg, *Recombination in Semiconductors.* Cambridge University Press, 1991.
290. P. Lautenschlager, M. Garriga, and M. Cardona, "Temperature Dependence of the Interband Critical-Point Parameters of InP," *Phys.Rev.B*, vol. 36, no. 9, pp. 4813–4820, 1987.
291. S. Laux, "Techniques for Small-Signal Analysis of Semiconductor Devices," *IEEE Trans.Electron Devices*, vol. ED-32, no. 10, pp. 2028–2037, 1985.
292. S. Laux and M. Fischetti, "The DAMOCLES Monte Carlo Device Simulation Program," in *Computational Electronics* (K. Hess, J. Leburton, and U. Ravaioli, eds.), pp. 87–92, Kluwer, 1991.
293. C. Lee, R. Logan, R. Batdorf, J. Kleimack, and W. Wiegmann, "Ionization Rates of Holes and Electrons in Silicon," *Phys.Rev.*, vol. 134, no. 3A, pp. A761–773, 1964.
294. J. Lee, A. Gutierrez-Aitken, S. Li, and P. Bhattacharya, "Impact Ionization Coefficients in $Si_{1-x}Ge_x$," *Appl.Phys.Lett.*, vol. 66, no. 2, pp. 204–205, 1995.
295. I. Leu and A. Neugroschel, "Minority-Carrier Transport Parameters in Heavily Doped p-type Silicon at 296 and 77 K," *IEEE Trans.Electron Devices*, vol. 40, no. 10, pp. 1872–1875, 1993.

296. D. Leung, Y. Chou, J. Scarpulla, R. Kono, R. Lai, D. Eng, P. Liu, and D. Streit, "High-Reliabiltiy Deep Submicron GaAs Pseudomorphic HEMT MMIC Amplifiers," in *Proc. Intl.Conf. GaAs Manufacturing Technology*, (Las Vegas), pp. 115–118, 2001.
297. M. Levinstein, S. Rumyantsev, and M. Shur, *Si, Ge, C(Diamond), GaAs, GaP, GaSb, InAs, InP, InSb*, vol. 1. Singapore: World Scientific, 1996.
298. M. Levinstein, S. Rumyantsev, and M. Shur, *Ternary and Quaternary III-V Compounds*, vol. 2. Singapore: World Scientific, 1999.
299. X. Li, Q. Du, J. Heroux, and W. Wang, "n-Channel AlSb/GaSb Modulation-Doped Field-Effect Transistors," *Solid-State Electron.*, vol. 41, no. 12, pp. 1853–1856, 1997.
300. M. Libezny, S. Jain, J. Poortmans, M. Caymax, J. Nijs, R. Mertens, K. Werner, and P. Balk, "Photoluminiscence Determination of the Fermi Energy in Heavily Doped Strained $Si_{1-x}Ge_x$ Layers," *Appl.Phys.Lett.*, vol. 64, no. 15, pp. 1953–1955, 1994.
301. P. Lim, B. Schineller, O. Schön, K. Heime, and M. Heuken, "Photoluminescence of GaN: Mg Grown by Metalorganic Vapor-Phase Epitaxy (MOVPE)," *J.Cryst.Growth*, vol. 205, no. 1-2, pp. 1–10, 1999.
302. A. Lindell, M. Pessa, A. Salokatve, F. Bernardini, R. Nieminen, and M. Paalanen, "Band Offsets at the GaInP/GaAs Heterojunction," *J.Appl.Phys.*, vol. 82, no. 7, pp. 3374–3380, Oct. 1997.
303. J. Liou, *Principles&Analysis of AlGaAs/GaAs Heterojunction Bipolar Transistors.* Boston: Artech House, 1996.
304. M. Littlejohn, K. Kim, and H. Tian, "High-Field Transport in InGaAs and Related Heterostructures," in Bhattacharya [38], section 4.2, pp. 107–116.
305. W. Liu, *Fundamentals of III-V Devices: HBTs, MESFETs, and HFETs/HEMTs.* New York: John Wiley & Sons, 1999.
306. C. Lombardi, S. Manzini, A. Saporito, and M. Vanzi, "A Physically Based Mobility Model for Numerical Simulation of Nonplanar Devices," *IEEE Trans.Computer-Aided Design*, vol. 7, no. 11, pp. 1164–1171, 1988.
307. D. Look, D. Lorance, J. Sizelove, C. Stutz, K. Evans, and D. Whitson, "Alloy Scattering in p-Type $Al_xGa_{1-x}As$," *J.Appl.Phys.*, vol. 71, no. 1, pp. 260–266, 1992.
308. D. Look, J. Sizelove, S. Keller, Y. Wu, U. Mishra, and S. DenBaars, "Accurate Mobility and Carrier Concentration Analysis for GaN," *Solid-State Comm.*, vol. 102, no. 4, pp. 297–300, 1997.
309. J. López-González and L. Prat, "The Importance of Bandgap Narrowing Distribution Between the Conduction and Valence Bands in Abrupt HBT's," *IEEE Trans.Electron Devices*, vol. 44, no. 7, pp. 1046–1051, 1997.
310. T. Low, C. Hutchison, P. Canfield, T. Shirley, R. Yeats, J. Chang, G. Essilfie, M. Culver, W. Whiteley, D. D'Avanzo, N. Pan, J. Elliot, and C. Lutz, "Migration from an AlGaAs to an InGaP Emitter HBT IC Process for Improved Reliability," in *GaAs IC Symp.Tech.Dig.*, (Atlanta), pp. 153–157, Nov. 1998.
311. J. Lowney and H. Bennett, "Majority and Minority Electron and Hole Mobilities in Heavily Doped GaAs," *J.Appl.Phys.*, vol. 69, no. 10, pp. 7102–7110, 1991.
312. Z. Lu, M. Hanna, and A. Majerfeld, "Determination of Band Gap Narrowing and Hole Density for Heavily C-doped GaAs by Photoluminiscence Spectroscopy," *Appl.Phys.Lett.*, vol. 64, no. 1, pp. 88–90, 1994.
313. M. Lundstrom, *Fundamentals of Carrier Transport.* Cambridge University Press, 2000.
314. M. Lundstrom, M. Klausmeier-Brown, M. Melloch, R. Ahrenkiel, and B. Keyes, "Device-Related Material Properties of Heavily Doped Gallium Arsenide," *Solid-State Electron.*, vol. 33, no. 6, pp. 693–704, 1990.

315. E. Lyumkis, R. Mickevicius, O. Penzin, B. Polsky, and K. E. Sayed, "Numerical Analysis of Electron Tunneling through Hetero-Interfaces and Schottky Barriers in Heterostructure Devices," in *GaAs IC Symp.Tech.Dig.*, (Seattle), pp. 129–132, 2000. Updated version.
316. E. Lyumkis, R. Mickevicius, O. Penzin, B. Polsky, K. E. Sayed, A. Wettstein, and W. Fichtner, "Simulations of Quantum Transport in HEMT Using Density Gradient Model," in *GaAs IC Symp.Tech.Dig.*, (Monterey), pp. 233–236, 2002.
317. P. Ma, P. Zampardi, L. Zhang, and M. Chang, "Determining the Effectiveness of HBT Emitter Ledge Passivation by Using an On-Ledge Schottky Diode Potentiometer," *IEEE Electron Device Lett.*, vol. 20, no. 9, pp. 460–462, 1999.
318. P. Mackowiak and W. Nakwaski, "Threshold Currents of Nitride Vertical-Cavity Surface-Emitting Lasers with Various Active Regions," *MRS Internet J. Nitride Semicond.Res.*, vol. 3, 1998. http://nsr.mij.mrs.org/3/35/.
319. Y. Mamontov and M. Willander, "Simulation of Bandgap Narrowing and Incomplete Ionization in Strained $Si_{1-x}Ge_x$ Alloys on ⟨001⟩ Si Substrate (for Temperatures from 40K up to 400K)," *Solid-State Electron.*, vol. 38, no. 3, pp. 599–607, 1995.
320. W. Marsetz, *Thermische und elektrische Konzeption von GaAs-HEMTs für Leistungsverstärker bis 80 GHz (in German).* Dissertation, Universität Karlsruhe, 1999.
321. A. Marshak, "Transport Equations for Highly Doped Devices and Heterostructures," *Solid-State Electron.*, vol. 30, no. 11, pp. 1089–1093, 1987.
322. A. Marshak, M. Shibib, J. Fossum, and F. Lindholm, "Rigid Band Analysis of Heavily Doped Semiconductor Devices," *IEEE Trans.Electron Devices*, vol. ED-28, no. 3, pp. 293–298, 1981.
323. M. Martinez, D. Look, J. Sizelove, and F. Schuermeyer, "Monte Carlo Simulation of Bulk Hole Transport in $Al_xGa_{1-x}As$, $In_{1-x}Al_xAs$, and $GaAs_xSb_{1-x}$," *J.Appl.Phys.*, vol. 77, no. 2, pp. 661–664, 1995.
324. G. Masetti, M. Severi, and S. Solmi, "Modeling of Carrier Mobility Against Carrier Concentration in Arsenic-, Phosphorus- and Boron-Doped Silicon," *IEEE Trans.Electron Devices*, vol. ED-30, no. 7, pp. 764–769, 1983.
325. N. Matine, M. Dvorak, S. Lam, C. Bolognesi, Y. Houng, and N. Moll, "Demonstration of GSMBE Grown $InP/GaAs_{0.51}Sb_{0.49}/InP$ DHBTs," in *GaAs IC Symp.Tech.Dig.*, (Washington), pp. 239–242, 2000.
326. T. Matsuoka, H. Okamoto, M. Nakao, H. Harima, H. Takahata, H. Mitate, S. Mizuno, Y. Uchiyama, and T. Makimoto, "MOVPE Growth of Wurtzide InN and its Characteristics," in *Int. Conf. Solid-State Devices Materials (SSDM)*, (Tokyo), pp. 132–133, 2003.
327. Z. Matutinović-Krstelj, V. Venkataraman, E. Prinz, J. Sturm, and C.W.Magee, "A Comprehensive Study of Lateral and Vertical Current Transport in $Si/Si_{1-x}Ge_x/Si$ HBT's," in *IEDM Tech.Dig.*, (Washington), pp. 87–90, 1993.
328. Matweb, "MatWeb Materials Property Database." http://www.matweb.com.
329. K. Matzuzawa, K. Uchida, and A. Nishiyama, "A Unified Simulation of Schottky and Ohmic Contacts," *IEEE Trans.Electron Devices*, vol. 47, no. 1, pp. 103–108, 2000.
330. P. Maycock, "Thermal Conductivity of Silicon, Germanium, III-V Compounds and III-V Alloys," *Solid-State Electron.*, vol. 10, no. 3, pp. 161–168, 1967.
331. Ferro Ceramic Grinding Inc., "Properties of Ceramic Materials: Physical, Mechanical, Thermal and Electrical." http://ferroceramic.com/ceramic-materials/.
332. L. McCarthy, P. Kozodoy, M. Rodwell, S. DenBaars, and U. Mishra, "AlGaN/GaN Heterojunction Bipolar Transistor," *IEEE Electron Device Lett.*, vol. 20, no. 6, pp. 277–279, 1999.
333. L. McCarthy, I. Smorchkova, H. Xing, P. Kozodoy, P. Fini, J. Limb, D. Pulfrey, J. Speck, M. Rodwell, S. DenBaars, and U. Mishra, "GaN HBT: Toward an RF Device," *IEEE Trans.Electron Devices*, vol. 48, no. 3, pp. 543–551, 2001.

334. B. McDermott, E. Gernter, S. Pittman, C. Seabury, and M. Chang, "Growth and Doping of GaAsSb via Metalorganic Chemical Vapor Deposition for InP Heterojunction Bipolar Transistors," *Appl.Phys.Lett.*, vol. 68, no. 10, pp. 1386–1388, 1996.
335. J. McMacken and S. Chamberlain, "CHORD: A Modular Semiconductor Device Simulation Development Tool Incorporating External Network Models," *IEEE Trans.Computer-Aided Design*, vol. 8, no. 8, pp. 826–836, 1989.
336. MEDICI. http://www.synopsys.com/products/avmrg/device_sim_ds.html.
337. B. Meinerzhagen and W. Engl, "The Influence of the Thermal Equilibrium Approximation on the Accuracy of Classical Two-Dimensional Numerical Modeling of Silicon Submicrometer MOS Transistors," *IEEE Trans.Electron Devices*, vol. ED-35, no. 5, pp. 689–697, 1988.
338. W. Meng, "Crystal Structure, Mechanical Properties and Thermal Properties and Refractive Index of AlN," in Edgar [135], section 1.3, pp. 22–29.
339. A. Milnes and A. Polyakov, "Gallium Antimonide Device Related Properties," *Solid-State Electron.*, vol. 36, no. 6, pp. 803–818, 1993.
340. T. Mimura, S. Hiyamizu, T. Fujii, and K. Nanbu, "A New Field-Effect Transistor with Selectively Doped GaAs/n-$Al_xGa_{1-x}As$ Heterojunctions," *Japanese J.Appl.Phys.*, vol. 19, no. 5, pp. L225–L227, 1980.
341. J. Minch, S. Park, T. Keating, and S. Chuang, "Theory and Experiment of $In_{1-x}Ga_xAs_yP_{1-y}$ and $In_{1-x-y}Ga_xAl_yAs$ Long-Wavelength Strained Quantum-Well Lasers," *IEEE J.Quantum Electronics*, vol. 35, no. 5, pp. 771–782, 1999.
342. U. Mishra, P. Parikh, and Y. Wu, "AlGaN/GaN HEMTs - An Overview of Device Operation and Applications," *Proc.IEEE*, vol. 90, no. 6, pp. 1022–1031, 2002.
343. T. Mnatsakanov, M. Levinshtein, L. Pomortseva, S. Yurkov, G. Simin, and M. Khan, "Carrier Mobility Model for GaN," *Solid-State Electron.*, vol. 47, no. 1, pp. 111–115, 2003.
344. K. Mochizuki, T. Oka, K. Ouchi, and T. Tanoue, "Reliability Investigation of Heavily C-doped InGaP/GaAs HBTs Operated under a Very High Current-Density Condition," *Solid-State Electron.*, vol. 43, no. 8, pp. 1425–1428, 1999.
345. M. Mokhtari, C. Fields, and R. Rajavel, "100+ GHz Static Divide-By-2 Circuit in InP-DHBT Technology," in *GaAs IC Symp.Tech.Dig.*, (Monterey), pp. 291–294, 2002.
346. C. Monier, F. Ren, J. Han, P. Chang, R. Shul, K. Lee, A. Zhang, A. Baca, and S. Pearton, "Simulation of NPN and PNP AlGaN/GaN Heterojunction Bipolar Transistors Performances: Limiting Factors and Optimum Design," *IEEE Trans.Electron Devices*, vol. 48, no. 3, pp. 427–432, 2001.
347. C. Morton and C. Snowden, "Comparison of Quasi-2D and Ensemble Monte Carlo Simulations for Deep Submicron HEMTs," in *IEEE Intl. Microwave Symp. Dig.*, (Boston), pp. 153–156, 1998.
348. C. Morton, C. Snowden, and M. Howes, "HEMT Physical Model for MMIC CAD," in *Proc. European Microwave Conf.*, (Bologna), pp. 199–204, 1995.
349. S. Mottet and J. Viallet, "Thermionic Emission in Heterojunctions," in *Proc. Intl.Conf. on Simulation of Semiconductor Devices and Processes* (G. Baccarani and M. Rudan, eds.), (Bologna), pp. 97–108, Tecnoprint, 1988.
350. J.-E. Müller, P. Baureis, O. Berger, T. Boettner, N. Bovolon, R. Schultheis, G. Packeiser, and P. Zwicknagl, "A Small Chip Size 2 W, 62% Efficient HBT MMIC for 3 V PCN Applications," *IEEE J.Solid-State Circuits*, vol. 33, no. 9, pp. 1277–1283, Sept. 1998.
351. M. Murti, B. Grietens, C. van Hoof, and G. Borghs, "1.7-1.9 μm $In_xGa_{1-x}As$/ $In_yAl_{1-y}As$ Light-Emitting Diodes Lattice-Mismatched Grown on GaAs," *J. Appl. Phys.*, vol. 78, no. 1, pp. 578–580, 1995.

352. J. Muth, J. Lee, I. Shmagin, R. Kolbas, H. Casey, B. Keller, U. Mishra, and S. DenBaars, "Absorption Coeffient, Energy Gap, Exciton Binding Energy, and Recombination Lifetime of GaN Obtained from Transmission Measurements," *Appl.Phys.Lett.*, vol. 71, no. 18, pp. 2572–2574, 1997.
353. B. Nag and M. Das, "Electron Mobility in $In_{0.5}Ga_{0.5}P$," *J.Appl.Phys.*, vol. 83, no. 11, pp. 5862–5864, June 1998.
354. B. Nag, *Electron Transport in Compound Semiconductors*, vol. 11 of *Springer Series in Solid-State Sciences*. Springer, 1980.
355. M. Nawaz, "The Role of Background Concentration in AlInAs/GaInAs/InP Based Double Heterojunction HEMTs," *Solid-State Electron.*, vol. 42, no. 10, pp. 1849–1854, 1998.
356. M. Nedjalkov, H. Kosina, S. Selberherr, and I. Dimov, "A Backward Monte Carlo Method for Simulation of the Electron Quantum Kinetics in Semiconductors," *VLSI Design*, vol. 13, no. 1-4, pp. 405–411, 2001.
357. M. Nedjalkov and P. Vitanov, "Iteration Approach for Solving the Boltzmann Equation with the Monte Carlo Method," *Solid-State Electron.*, vol. 32, no. 10, pp. 893–896, 1989.
358. B. Neinhüs, S. Decker, P. Graf, F. Bufler, and B. Meinerzhagen, "Consistent Hydrodynamic and Monte-Carlo Simulation of SiGe HBTs Based on Table Models for the Relaxation Times," *VLSI Design*, vol. 8, no. 1-4, pp. 387–391, 1998.
359. B. Neinhüs, P. Graf, S. Decker, and B. Meinerzhagen, "Examination of Transient Drift-Diffusion and Hydrodynamic Modeling Accuracy for SiGe HBTs by 2D Monte-Carlo Device Simulation," in *Proc. European Solid-State Device Research Conf.* (H. Grünbacher, ed.), (Stuttgart), pp. 188–191, Editions Frontieres, 1997.
360. NEMO. http://www.cfdrc.com/nemo/.
361. A. Neukermans and G. Kino, "Absolute Measurement of the Electron Velocity-Field Characteristics of InSb," *Phys.Rev.B*, vol. 7, no. 6, pp. 2703–2709, 1973.
362. A. Neviani, G. Meneghesso, E. Zanoni, M. Hafizi, and C. Canali, "Positive Temperature Dependence of the Electron Impact Ionization Coefficient in $In_{0.53}Ga_{0.47}As$/InP HBT's," *IEEE Electron Device Lett.*, vol. 18, no. 12, pp. 619–621, 1997.
363. nextnano3. http://www.webplexity.de/nextnano3.php.
364. K. Ng, *Complete Guide to Semiconductor Devices*. McGraw-Hill, 1995.
365. L. Nguyen, *Realization of Ultra High Speed Modulation-Doped Field Effect Transistors*. PhD thesis, Cornell University, 1989.
366. NIST. National Institute of Standards and Technology. http://www.nist.gov.
367. C. Nuese, H. Kressel, and I. Ladany, "The Future for LEDs," *IEEE Spectrum*, vol. 9, no. 5, pp. 28–38, 1972.
368. R. Oberhuber, G. Zandler, and P. Vogl, "Mobility of Two-Dimensional Electrons in AlGaN/GaN Modulation-Doped Field Effect Transistors," *Appl.Phys.Lett.*, vol. 73, no. 6, pp. 818–820, 1998.
369. K. Oda, E. Ohue, I. Suzumura, R. Hayami, A. Kodama, H. Shimamoto, and K. Washio, "Self-Aligned Selective-Epitaxial-Growth $Si_{1-x-y}Ge_xC_y$ HBT Technology Featuring 170-GHz $f_{\max}$," in *IEDM Tech.Dig.*, (Washington), pp. 332–335, 2001.
370. I. Oguzman, E. Belotti, K. Brennan, J. Kolnik, R. Wang, and P. Ruden, "Theory of Hole Initiated Impact Ionization in Bulk Zincblende and Wurtzite GaN," *J.Appl.Phys.*, vol. 81, no. 12, pp. 7827–7834, 1997.
371. T. Oka, K. Hirata, K. Ouchi, H. Uchiyama, T. Taniguchi, K. Mochizuki, and T. Nakamura, "Advanced Performance of Small-Scaled InGaP/GaAs HBT's with f_T over 150 GHz and $f_{\max}$ over 250 GHz," in *IEDM Tech.Dig.*, (San Francisco), pp. 653–656, 1998.

372. Y. Okuto and C. Crowell, "Threshold Energy Effect on Avalanche Breakdown Voltage in Semiconductor Junctions," *Solid-State Electron.*, vol. 18, no. 2, pp. 161–168, 1975.
373. S. O'Leary, B. Foutz, M. Shur, U. Bhapkar, and L. Eastman, "Electron Transport in Wurtzide Indium Nitride," *J.Appl.Phys.*, vol. 83, no. 2, pp. 826–829, 1998.
374. S. O'Leary, B. Foutz, M. Shur, U. Bhapkar, and L. Eastman, "Monte Carlo Simulation of Electron Transport in Wurtzite Aluminum Nitride," *Solid-State Comm.*, vol. 105, no. 10, pp. 621–626, 1998.
375. J. Olson, R. Ahrenkiel, D. Dunlavy, B. Keyes, and A. Kibbler, "Ultralow Recombination Velocity at $Ga_{0.5}In_{0.5}P$/GaAs Heterointerfaces," *Appl.Phys.Lett.*, vol. 55, no. 12, pp. 1208–1210, 1989.
376. F. Osaka, T. Mikawa, and T. Kaneda, "Electron and Hole Ionization Coefficients in (100) Oriented $Ga_{0.33}In_{0.67}As_{0.70}P_{0.30}$," *Appl.Phys.Lett.*, vol. 45, no. 3, pp. 292–293, 1984.
377. F. Osaka, T. Mikawa, and T. Kaneda, "Impact Ionization Coefficients of Electrons and Holes in (100)-Oriented $Ga_{1-x}In_xAs_yP_{1-y}$," *IEEE J.Quantum Electronics*, vol. 21, no. 9, pp. 1326–1336, 1985.
378. K. Osamura, S. Naka, and Y. Murakami, "Preparation and Optical Properties of $Ga_{1-x}In_xN$ Thin Films," *J.Appl.Phys.*, vol. 46, no. 8, pp. 3432–3437, 1975.
379. V. Palankovski, N. Belova, T. Grasser, H. Puchner, S. Aronowitz, and S. Selberherr, "A Methodology for Deep Sub-0.25μm CMOS Technology Prediction," *IEEE Trans.Electron Devices*, vol. 48, no. 10, pp. 2331–2336, Oct. 2001.
380. V. Palankovski, G. Kaiblinger-Grujin, and S. Selberherr, "Study of Dopant-Dependent Band Gap Narrowing in Compound Semiconductor Devices," *Materials Science & Engineering B*, vol. 66, no. 1-3, pp. 46–49, 1999.
381. V. Palankovski, R. Quay, and S. Selberherr, "Industrial Application of Heterostructure Device Simulation," *IEEE J.Solid-State Circuits*, vol. 36, no. 9, pp. 1365–1370, Sept. 2001.
382. V. Palankovski, G. Röhrer, E. Wachmann, J. Kraft, B. Löffler, J. Cervenka, R. Quay, T. Grasser, and S. Selberherr, "Optimization of High-Speed SiGe HBTs," in *Proc. Intl.Symp. Elec.Dev. for Microwave and Optoelectronic Applications EDMO*, (Vienna), pp. 187–191, 2001.
383. V. Palankovski, R. Schultheis, A. Bonacina, and S. Selberherr, "Effectiveness of Silicon Nitride Passivation in III-V Based Heterojunction Bipolar Transistors," *J. Radiation Effects and Defects in Solids*, vol. 156, no. 1-4, pp. 261–265, 2001.
384. V. Palankovski, R. Schultheis, and S. Selberherr, "Simulation of Power Heterojunction Bipolar Transistors on Gallium Arsenide," *IEEE Trans.Electron Devices*, vol. 48, no. 6, pp. 1264–1269, June 2001.
385. D. Palmer, "The Semiconductors Information: Web-Site for Semiconductor Scientists, Engineers and Technologists Across the World."
http://www.semiconductors.co.uk.
386. J. Palmour, S. Sheppard, R. Smith, S. Allen, W. Pribble, T. Smith, Z. Ring, J. Sumakeris, A. Saxler, and J. Milligan, "Wide Bandgap Semiconductor Devices and MMICs for RF Microwave Applications," in *IEDM Tech.Dig.*, (Washington), pp. 385–388, 2001.
387. N. Pan, J. Elliott, M. Knowles, D. Vu, K. Kishimoto, J. Twynam, H. Sato, M. Fresina, and G. Stillman, "High Reliability InGaP/GaAs HBT," *IEEE Electron Device Lett.*, vol. 19, no. 4, pp. 115–117, 1998.
388. A. Parker and D. Root, "Pulse Measurements to Quantity Dispersion in PHEMTs," in *IEEE Conf. URSI Intl. Symp. Signals Systems Electronics*, (New York), pp. 444–449, 1998.

389. R. Passler and G. Oelgart, "Appropriate Analytical Description of the Temperature Dependence of Exciton Peak Positions in GaAs/$Al_xGa_{1-x}As$ Multiple Quantum Wells and the $\Gamma_{8v} - \Gamma_{6c}$ Gap of GaAs," *J.Appl.Phys.*, vol. 82, no. 5, pp. 2611–2616, 1997.
390. M. Patil, U. Ravaioli, and K. Hess, "Real-Space Transfer in AlGaAs/GaAs Heterojunctions," in Adachi [5], section 7.6, pp. 214–220.
391. L. Pavesi, "Photoluminescence Spectra of AlGaAs," in Adachi [5], section 9.1, pp. 245–268.
392. L. Pavesi and M. Guzzi, "Photoluminescence of $Al_xGa_{1-x}As$ Alloys," *J.Appl.Phys.*, vol. 75, no. 10, pp. 4779–4838, 1994.
393. A. Peaker, "Defect Energy Levels in LPE InP," in EMIS-Group [137], section 11.7, pp. 251–252.
394. A. Peaker and J. Nicholas, "Defect Energy Levels in MBE GaAs," in EMIS-Group [136], section 13.10, pp. 334–336.
395. A. Peaker and F. Saleemi, "Defect Energy Levels in AlGaAs," in Adachi [5], section 9.2, pp. 269–277.
396. T. Pearsall, "Impact Ionization Rates for Electron and Holes in $Ga_{0.47}In_{0.53}As$," *Appl.Phys.Lett.*, vol. 36, no. 3, pp. 218–220, 1980.
397. T. Pearsall, R. Nahory, and M. Pollack, "Impact Ionization Coefficients for Electrons and Holes in $In_{0.14}Ga_{0.86}As$," *Appl.Phys.Lett.*, vol. 27, no. 6, pp. 330–332, 1975.
398. B. Pejčinović, L. Kay, T.-W. Tang, and D. Navon, "Numerical Simulation and Comparison of Si BJTs and $Si_{1-x}Ge_x$ HBTs," *IEEE Trans.Electron Devices*, vol. 36, no. 10, pp. 2129–2137, 1989.
399. B. Pereiaslavets, K. Bachem, J. Braunstein, and L. Eastman, "GaInP/InGaAs/-GaAs Graded Barrier MODFET Grown by OMVPE: Design, Fabrication, and Device Results," *IEEE Trans.Electron Devices*, vol. 43, no. 10, pp. 1659–1664, 1996.
400. D. Peters, W. Daumann, W. Brockerhoff, R. Reuter, E. Koenig, and F. Tegude, "Direct Calculation of the HBT Small-Signal Equivalent Circuit with Special Emphasis to the Feedback Capacitance," in *Proc. European Microwave Conference*, (Bologna, Italy), pp. 1032–1036, 1995.
401. C. Pigorsch, *Mikroskopisch/makroskopische Simulation von Nanobauelementen (in German).* Dissertation, Technische Universität Dresden, 2000.
402. S. Pinel, A. Marty, J. Tasselli, J. Bailbe, E. Beyne, R. V. Hoof, S. Marco, J. Morante, O. Vendier, and M. Huan, "Thermal Modeling and Management in Ultrathin Chip Stack Technology," *IEEE Trans. Components and Packaging Technologies*, vol. 25, no. 2, pp. 244–253, 2002.
403. PISCES-ET. http://www-tcad.stanford.edu/tcad.html.
404. S. Plimmer, J. David, G. Rees, and P. Robson, "Ionization Coefficients in $Al_xGa_{1-x}As$ ($x = 0 - 0.60$)," *Semicond.Sci.Technol.*, vol. 15, no. 7, pp. 692–699, 2000.
405. K. Ploog and O. Brandt, "Doping of Group III Nitrides," *J.Vac.Sci.Technol.A*, vol. A16, no. 3, pp. 1609–1614, 1998.
406. F. Pollak, "Energy Gaps of AlGaAs," in Adachi [5], section 4.6, pp. 79–81.
407. W. Press, B. Flannery, S. Teukolski, and W. Vetterling, *Numerical Recipees in C.* Cambridge University Press, 1988.
408. E. Prinz, P. Garone, P. Schwartz, X. Xiao, and J. Sturm, "The Effect of Base-Emitter Spacers and Strain-Dependent Densities of States in Si/$Si_{1-x}Ge_x$/Si Heterojunction Bipolar Transistors," in *IEDM Tech.Dig.*, (Washington), pp. 639–642, 1989.
409. D. Pulfrey and S. Fathpour, "Performance Predictions for N-P-N $Al_xGa_{1-x}N$/GaN HBTs," *IEEE Trans.Electron Devices*, vol. 48, no. 3, pp. 597–602, 2001.

410. W. Quade, E. Schöll, and M. Rudan, "Impact Ionization within the Hydrodynamic Approach to Semiconductor Transport," *Solid-State Electron.*, vol. 36, no. 10, pp. 1493–1505, 1993.
411. R. Quay, K. Hess, R. Reuter, M. Schlechtweg, T. Grave, V. Palankovski, and S. Selberherr, "Nonlinear Electronic Transport and Device Performance of HEMTs," *IEEE Trans.Electron Devices*, vol. 48, no. 2, pp. 210–217, 2001.
412. R. Quay, C. Moglestue, V. Palankovski, and S. Selberherr, "A Temperature Dependent Model for the Saturation Velocity in Semiconductor Materials," *Materials Science in Semiconductor Processing*, vol. 3, no. 1-2, pp. 149–155, 2000.
413. R. Quay, C. Moglestue, V. Palankovski, and S. Selberherr, "A Temperature Dependent Model for the Saturation Velocity in Semiconductor Materials," *Materials Science in Semiconductor Processing*, vol. 3, no. 1-2, pp. 149–155, 2000.
414. R. Quay, V. Palankovski, M. Chertouk, A. Leuther, and S. Selberherr, "Simulation of InAlAs/InGaAs High Electron Mobility Transistors with a Single Set of Physical Parameters," in *IEDM Tech.Dig.*, (San Fransisco), pp. 186–189, 2000.
415. R. Quay, V. Palankovski, R. Reuter, M. Schlechtweg, W. Kellner, and S. Selberherr, "III-V Device Optimization by Physics-Based S-Parameter Simulation," in *Proc. Intl. Symp. Compound Semiconductors*, (Berlin), pp. 325–328, 1999.
416. R. Quay, R. Reuter, V. Palankovski, and S. Selberherr, "S-Parameter Simulation of RF-HEMTs," in *Proc. Intl.Symp. Elec.Dev. for Microwave and Optoelectronic Applications EDMO*, (Manchester), pp. 13–18, 1998.
417. L. Reggiani, C. Canali, F. Nava, and G. Ottaviani, "Hole Drift Velocity in Germanium," *Phys.Rev.B*, vol. 16, no. 6, pp. 2781–2791, 1977.
418. F. Ren, J. Han, R. Hickman, J. V. Hove, P. Chow, J. Klaasen, J. LaRoche, K. Jung, H. Cho, X. Cao, S. Donovan, R. Kopf, R. Wilson, A. Baca, R. Shul, L. Zhang, C. Willison, C. Abernathy, and S. Pearton, "GaN/AlGaN HBT Fabrication," *Solid-State Electron.*, vol. 44, no. 2, pp. 239–244, 2000.
419. R. Reuter, M. Agethen, U. Auer, S. van Waasen, D. Peters, W. Brockerhoff, and F. Tegude, "Investigation and Modeling of Impact Ionization with Regard to the RF and Noise Behavior of HFET," *IEEE Trans.Microwave Theory and Techniques*, vol. 45, no. 6, pp. 977–983, 1997.
420. D. Richey, J. Cressler, and A. Joseph, "Scaling Issues and Ge Profile Optimization in Advanced UHV/CVD SiGe HBT's," *IEEE Trans.Electron Devices*, vol. 44, no. 3, pp. 431–440, 1997.
421. B. Ridley, "A Note on the Origin of the Yellow Luminescence in GaN," *J.Phys.:Condensed Matter*, vol. 10, no. 27, pp. L461–L463, 1998.
422. I. Riech, P. Diaz, and E. Marin, "Study of Nonradiative Recombination Mechanisms in Semiconductors by Photoacoustic Measurements," *Phys.stat.sol.(b)*, vol. 220, no. 1, pp. 305–308, 2000.
423. J. Rieh, P. Bhattacharya, and E. Croke, "Temperature Dependent Minority Electron Mobilities in Strained $Si_{1-x}Ge_x$ ($0.2 \leq x \leq 0.4$) Layers," *IEEE Trans.Electron Devices*, vol. 47, no. 4, pp. 883–890, Apr. 2000.
424. J. Rieh, B. Jagannathan, H. Chen, K. Schonenberg, D. Angell, A. Chinthakindi, J. Florkey, F. Golan, D. Greenberg, S.-J. Jeng, M. Khater, F. Pagette, C. Schnabel, P. Smith, A. Stricker, K. Vaed, R. Volant, D. Ahlgren, G. Freeman, K. Stein, and S. Subbanna, "SiGe HBTs with Cut-off Frequency Near 300 GHz," in *IEDM Tech.Dig.*, (San Francisco), pp. 771–774, 2002.
425. V. Robbins, S. Smith, and G. Stillman, "Impact Ionization in $Al_xGa_{1-x}As$ for $x=$ 0.1-0.4," *Appl.Phys.Lett.*, vol. 52, no. 4, pp. 296–298, 1988.
426. H. Rohdin, A. Wakita, A. Nagy, V. Robbins, N. Moll, and C. Su, "A 0.1 μm MHEMT Millimeter-Wave IC Technology Designed for Manufacturability," *Solid-State Electron.*, vol. 43, no. 8, pp. 1645–1654, 1999.

427. J. Roldan, F. Gamiz, J. Lopez-Villanueva, and J. Carceller, "A Monte Carlo Study on the Electron-Transport Properties of High-Performance Strained-Si on Relaxed $Si_{1-x}Ge_x$ Channel MOSFETs," *J.Appl.Phys.*, vol. 80, no. 9, pp. 5121–5128, 1996.
428. J. Rollins and J. Choma, "Mixed-Mode PISCES-SPICE Coupled Circuit and Device Solver," *IEEE Trans.Computer-Aided Design*, vol. 7, no. 8, pp. 862–867, 1988.
429. F. Rotella, G. Ma, Z. Yu, and R. Dutton, "Modeling, Analysis, and Design of RF LDMOS Devices Using Harmonic-Balance Device Simulation," *IEEE Trans.Microwave Theory and Techniques*, vol. 48, no. 6, pp. 991–999, 2000.
430. J. Ruch, "High-Field Transport Properties of GaSb," *Appl.Phys.Lett.*, vol. 20, no. 7, pp. 246–249, 1972.
431. M. Rudolph, R. Doerner, K. Beilenhoff, and P. Heymann, "Scalable GaInP/GaAs HBT Large-Signal Model," *IEEE Trans.Microwave Theory and Techniques*, vol. 48, no. 12, pp. 2370–2376, 2000.
432. M. Rudolph, R. Doerner, and P. Heymann, "Direct Extraction of HBT Equivalent-Circuit Elements," *IEEE Trans.Microwave Theory and Techniques*, vol. 47, no. 1, pp. 82–84, 1999.
433. A. Rusani, J. Kuchenbecker, M. Borgarino, R. Plana, J. Graffeuil, and M. Vanzi, "Investigation of the Burn-In Effect in Microwave GaInP/GaAs HBTs by Means of Numerical Simulations," in *Proc. Intl.Symp. Elec.Dev. for Microwave and Optoelectronic Applications EDMO*, (London), pp. 260–265, 1999.
434. N. Sano, T. Aoki, M. Tomizava, and A. Yoshii, "Electron Transport and Impact Ionization in Si," *Phys.Rev.B*, vol. 41, no. 17, pp. 12122–12128, 1990.
435. J. Sato-Iwanaga, K. Fujimoto, H. Masato, Y. Ota, K. Inoue, B. Trojyanovsky, Z. Yu, and R. Dutton, "Distorsion Analysis of GaAs MESFETs Based on Physical Model Using PISCES-HB," in *IEDM Tech.Dig.*, (San Francisco), pp. 163–166, 1996.
436. D. Sawdai, X. Zhang, D. Pavlidis, and P. Bhattacharya, "Performance Optimization of PNP InAlAs/InGaAs HBTs," in *Proc. IEEE Cornell Conf. Advanced Concepts High Speed Semiconductor Devices and Circuits*, (Ithaca), pp. 269–277, 1997.
437. M. Schatzmayr, E. Wachmann, M. Mühlberger, C. Schelling, and F. Schäffler, "A Fully Certified SiGe-BiCMOS Process for ASICS and Multiproduct Wafers," in *Int. Semicond. Dev. Res. Symp. (ISDRS)*, (Charlottesville, Virginia), p. 299, 1999.
438. A. Schenk, "A Model for the Field and Temperature Dependence of Shockley-Read-Hall Lifetimes in Silicon," *Solid-State Electron.*, vol. 35, no. 11, pp. 1585–1596, 1992.
439. A. Schenk, "Rigorous Theory and Simplified Model of the Band-to-Band Tunneling in Silicon," *Solid-State Electron.*, vol. 36, no. 1, pp. 19–34, 1993.
440. A. Schenk, *Advanced Physical Models for Silicon Device Simulation.* Computational Microelectronics, Wien, New York: Springer, 1998.
441. I. Schmale, *Entwicklung eines konsistenten elektrothermischen Grossignalmodells für Feldeffekttransistoren als Grundlage für den zuverlässigen Entwurf monolithischer Frequenzverdoppler (in German).* Dissertation, Universität-Gesamthochschule Kassel, 1999.
442. D. Schroeder, *Modelling of Interface Carrier Transport for Device Simulation.* Wien, New York: Springer, 1994.
443. D. Schroeder, *Boundary and Interface Conditions of Transport Equations for Device Simulation*, vol. 36 of *Festkörperprobleme/Advances in Solid State Physics*, pp. 265–283. Vieweg, 1996.
444. E. Schubert, *Doping in III-V Semiconductors.* Cambridge University Press, 1993.
445. R. Schultheis, P. Baureis, J. Müller, N. Bovolon, and P. Zwicknagl, "Small Signal Modeling of HBTs for High Power Application," in *Intl.Workshop on Integrated Nonlinear Microwave and Millimeterwave Circuits*, (Duisburg), pp. 118–126, 1998.

446. R. Schultheis, N. Bovolon, J.-E. Müller, and P. Zwicknagl, "Modelling of Heterojunction Bipolar Transistors (HBTs) Based on Gallium Arsenide (GaAs)," *Intl.J. of RF and Microwave Computer-Aided Engineering*, vol. 10, no. 1, pp. 33–42, 2000.
447. V. Schwarz, *A Clock Recovery Circuit for 40 Gb/s using InP/InGaAs Heterojunction Bipolar Transistors.* PhD thesis, Swiss Federal Insitute of Technology, 2002.
448. S. Selberherr, *Analysis and Simulation of Semiconductor Devices.* Wien, New York: Springer, 1984.
449. S. Selberherr, "MOS Device Modeling at 77K," *IEEE Trans.Electron Devices*, vol. 36, no. 8, pp. 1464–1474, 1989.
450. S. Selberherr, "Device Modeling and Physics," *Physica Scripta*, vol. 35, pp. 293–298, 1991.
451. S. Selberherr, W. Hänsch, M. Seavey, and J. Slotboom, "The Evolution of the MINIMOS Mobility Model," *Solid-State Electron.*, vol. 33, no. 11, pp. 1425–1436, 1990.
452. Z. Shafi, P. Ashburn, I. Post, D. Robbins, W. Leong, C. Gibbings, and S. Nigrin, "Analysis and Modeling of the Base Currents of $Si/Si_{1-x}Ge_x$ Heterojunction Bipolar Transistors Fabricated in High and Low Oxygen Content Material," *J.Appl.Phys.*, vol. 78, no. 4, pp. 2823–2829, 1995.
453. J. Shah, "Auger Recombination Coefficients in $In_{0.53}Ga_{0.47}As$," in Bhattacharya [38], section 6.4, pp. 180–183.
454. W. Shan, J. Ager, K. Yu, W. Walukiewicz, E. Haller, M. Martin, W. McKinney, and W. Yang, "Dependence of the Fundamental Band Gap of $Al_xGa_{1-x}N$ on Alloy Composition and Pressure," *J.Appl.Phys.*, vol. 85, no. 12, pp. 8505–8507, 1999.
455. A. Sharma, N. Ravindra, S. Auluck, and V. Srivastava, "Temperature-Dependent Effective Masses in III-V Compound Semiconductors," *Phys.stat.sol.(b)*, vol. 120, no. 2, pp. 715–721, 1983.
456. J. Shealy, J. Smart, M. Poulton, R. Sadler, D. Grider, S. Gibb, B. Hosse, B. Sousa, D. Halchin, V. Steel, P. Garber, P. Wilkerson, B. Zaroff, J. Dick, T. Mercier, J. Bonaker, M. Hamilton, C. Greer, and M. Isenhour, "Gallium Nitride (GaN) HEMT's: Progress and Potential for Commercial Applications," in *GaAs IC Symp.Tech.Dig.*, (Monterey), pp. 243–246, 2002.
457. H. Shen, S. Pan, Z. Hang, Jing-Leng, F. Pollak, J. Woodall, and R. Sacks, "Photoreflectance of GaAs and $Ga_{0.82}Al_{0.18}As$ at Elevated Temperatures up to 600 ^oC," *Appl.Phys.Lett.*, vol. 53, no. 12, pp. 1080–1082, 1988.
458. B. Sheu, D. Scharfetter, P. Ko, and M. Jeng, "BSIM: Berkeley Short-Channel IGFET Model for MOS Transistors," *IEEE J.Solid-State Circuits*, vol. SC-22, no. 4, pp. 558–566, 1987.
459. K. Shim and H. Rabitz, "Electronic and Structural Properties of the Pentanary Alloy $Ga_xIn_{1-x}P_ySb_zAs_{1-y-z}$," *J.Appl.Phys.*, vol. 85, no. 11, pp. 7705–7715, 1999.
460. M. Shur, *GaAs Devices and Circuits.* Plenum Press, 1987.
461. M. Shur, "Low Ballistic Mobility in Submicron HEMTs," *IEEE Electron Device Lett.*, vol. 23, no. 9, pp. 511–513, 2002.
462. M. Shur and M. Khan, "GaN/AlGaN Heterostructure Devices: Photodetectors and Field-Effect Transistors," *MRS Bulletin*, vol. 22, no. 2, pp. 44–50, 1997.
463. E. Sichel and J. Pankove, "Thermal Conductivity of GaN," *J.Phys.Chem.Solids*, vol. 38, no. 3, p. 333, 1977.
464. R. Sieg and S. Ringel, "Reabsorption, Band-Gap Narrowing, and the Reconciliation of Photoluminescence Spectra with Electrical Measurements for Epitaxial n-InP," *J.Appl.Phys.*, vol. 80, no. 1, pp. 448–458, 1996.
465. Silvaco International, Santa Clara, USA, *ATLAS, User's Manual, Ed. 6*, Nov. 1998.
466. T. Simlinger, *Simulation von Heterostruktur-Feldeffekttransistoren.* Dissertation, Technische Universität Wien, 1996. http://www.iue.tuwien.ac.at/phd/simlinger.

467. T. Simlinger, H. Brech, T. Grave, and S. Selberherr, "Simulation of Submicron Double-Heterojunction High Electron Mobility Transistors with MINIMOS-NT," *IEEE Trans.Electron Devices*, vol. 44, no. 5, pp. 700–707, 1997.
468. T. Simlinger, M. Rottinger, and S. Selberherr, "A Method for Unified Treatment of Interface Conditions Suitable for Device Simulation," in *Simulation of Semiconductor Processes and Devices*, (Cambridge, Massachusetts), pp. 173–176, 1997.
469. J. Singh, *Physics of Semiconductors and their Heterostructures.* McGraw Hill, 1993.
470. G. Slack, "Nonmetallic Crystals with High Thermal Conductivity," *J.Phys.Chem.-Solids*, vol. 34, no. 2, pp. 321–335, 1973.
471. G. Slack, R. Tanzilli, R. Pohl, and J. Vandersande, "The Intrinsic Thermal Conductivity of AlN," *J.Phys.Chem.Solids*, vol. 48, no. 7, pp. 641–647, 1987.
472. J. Slotboom and H. de Graaff, "Measurements of Bandgap Narrowing in Si Bipolar Transistors," *Solid-State Electron.*, vol. 19, no. 10, pp. 857–862, 1976.
473. J. Slotboom, G. Streutker, M. v. Dort, P. Woerlee, A. Pruijmboom, and D. Gravesteijn, "Non-Local Impact Ionization in Silicon Devices," in *IEDM Tech.Dig.*, (Washington), pp. 127–130, 1991.
474. S. Smirnov, H. Kosina, and S. Selberherr, "Investigation of the Electron Mobility in Strained $Si_{1-x}Ge_x$ at High Ge Composition," in *Proc. Intl.Conf. on Simulation of Semiconductor Processes and Devices*, (Kobe), pp. 29–32, 2002.
475. C. Snowden, ed., *Semiconductor Device Modeling.* London: Springer, 1989.
476. M. Sokolich, C. Fields, B. Shi, Y. Brown, M. Montes, R. Martinez, A. Kramer, S. Thomas, and M. Madhav, "A Low Power 72.8 GHz Static Frequency Divider Implemented in AlInAs/InGaAs HBT IC Technology," in *GaAs IC Symp.Tech.Dig.*, (Seattle), pp. 81–84, 2000.
477. M. Sokolich, S. Thomas III, and C. Fields, "High-Speed and Low Power InAlAs/InGaAs Heterojunction Bipolar Transistors for Dense Ultra High Speed Digital Applications," in *IEDM Tech.Dig.*, (Washington), pp. 780–783, 2001.
478. R. Sorge, B. Heinemann, H. Zeindl, G. Lippert, and D. Wolansky, "Determination of the Generation Lifetime in Si/SiGe/Si Epitaxial Layer Stacks Grown by MBE and LPCVD Using Non Equilibrium Simultaneous HF/LF MOS CV Measurements," in *Proc. European Solid-State Device Research Conf.* (G. Baccarani and M. Rudan, eds.), (Gif-sur-Yvette Cedex), pp. 237–240, Editions Frontieres, 1996.
479. M. Sotoodeh, A. Khalid, and A. Rezazadeh, "Numerical Modeling of AlGaAs/GaAs and InGaP/GaAs Single and Double HBTs," in *Proc. Intl.Symp. Elec.Dev. for Microwave and Optoelectronic Applications EDMO*, (London), pp. 223–228, 1997.
480. M. Sotoodeh, A. Khalid, and A. Rezazadeh, "Empirical Low-Field Mobility Model for III-V Compounds Applicable in Device Simulation Codes," *J.Appl.Phys.*, vol. 87, no. 6, pp. 2890–2900, Mar. 2000.
481. R. Stenzel, C. Pigorsch, W. Klix, A. Vescan, and H. Leier, "Simulation of AlGaN/GaN-HFETs Including Spontaneous and Piezoelectric Polarization Charges," in *Proc. Intl. Symp. Compound Semiconductors*, (Berlin), pp. 511–514, 1999.
482. T. Stoica and L. Vescan, "Quantum Efficiency of SiGe LEDs," *Semicond.Sci.Technol.*, vol. 18, no. 6, pp. 409–416, June 2003.
483. G. Stollwerk, O. Sulima, and A. Bett, "Characterisation and Simulation of GaSb Device-Related Properties," *IEEE Trans.Electron Devices*, vol. 47, no. 2, pp. 448–457, 2000.
484. R. Strasser and S. Selberherr, "Practical Inverse Modeling with SIESTA," *IEICE Trans.Electron.*, vol. E83-C, no. 8, pp. 1303–1310, 2000.
485. R. Stratton, "Diffusion of Hot and Cold Electrons in Semiconductor Barriers," *Phys.Rev.*, vol. 126, no. 6, pp. 2002–2014, 1962.

486. U. Strauss, A. Heberle, W. Rühle, H. Tews, T. Lauterbach, and K. Bachem, "Influence of the Annealing on Electron Lifetimes in Transistor Base-Layers on GaAs:C," in *Proc. Intl. Symp. GaAs and Related Compounds*, (Freiburg), pp. 739–742, 1993.
487. D. Streit, A. Oki, and M. Hoppe, "High Volume HBT Production for Commercial Applications," in *Proc. Intl.Symp. Elec.Dev. for Microwave and Optoelectronic Applications EDMO*, (Manchester), pp. 111–118, 1998.
488. J. Sturm and H. Yin, "Si/SiGe/Si Heterojunction Bipolar Transistors," in Kasper and Lyutovich [252], section 7.1, pp. 305–318.
489. S. Subbanna, G. Tuttle, and H. Kroemer, "N-type Doping of Gallium Antimonide and Aluminum Antimonide Grown by Molecular Beam Epitaxy Using Lead Telluride as a Tellurium Dopant Source," *J.Electron.Mat.*, vol. 17, no. 4, pp. 297–303, 1988.
490. T. Suemitsu, T. Enoki, N. Sano, M. Tomizawa, and Y. Ishii, "An Analysis of the Kink Phenomena in InAlAs/InGaAs HEMTs Using Two-Dimensional Device Simulation," *IEEE Trans.Electron Devices*, vol. 45, no. 12, pp. 2390–2399, 1998.
491. T. Suemitsu, H. Yokoyama, T. Ishii, T. Enoki, G. Meneghesso, and E. Zanoni, "30-nm Two-Step Recess Gate InP-Based InAlAs/InGaAs HEMTs," *IEEE Trans.Electron Devices*, vol. 49, no. 10, pp. 1694–1700, 2002.
492. G. Sullivan, C. Farley, W. Ho, R. Pierson, M. Szwed, M. Lind, and R. Bernescut, "High Gain AlInAs/GaAsSb/AlInAs NPN HBTs on InP," *J.Electron.Mat.*, vol. 21, no. 12, pp. 1123–1125, 1992.
493. J. Sutherland and J. Hauser, "A Computer Analysis of Heterojunction and Graded Composition Solar Cells," *IEEE Trans.Electron Devices*, vol. ED-24, no. 4, pp. 363–372, 1977.
494. S. Swirhun, D. Kane, and R. Swanson, "Measurements of Electron Lifetime, Electron Mobility and Band-Gap Narrowing in Heavily Doped p-Type Silicon," in *IEDM Tech.Dig.*, (San Francicso), pp. 24–27, 1996.
495. S. Sze, *Physics of Semiconductor Devices.* New York: Wiley, 1969.
496. S. Sze, *Physics of Semiconductor Devices.* New York: Wiley, second ed., 1981.
497. K. Taguchi, T. Torikai, Y. Sugimoto, K. Makita, and H. Ishihara, "Temperature Dependence of Impact Ionization Coefficients in InP," *J.Appl.Phys.*, vol. 59, no. 2, pp. 476–481, 1986.
498. M. Takeshima, "Effect of Auger Recombination on Laser Operation in $Ga_{1-x}Al_xAs$," *J.Appl.Phys.*, vol. 58, no. 10, pp. 3846–3850, 1985.
499. T. Tang, S. Ramaswamy, and J. Nam, "An Improved Hydrodynamic Transport Model for Silicon," *IEEE Trans.Electron Devices*, vol. 40, no. 8, pp. 1469–1476, 1993.
500. T. Tansley, "Crystal Structure, Mechanical Properties, Thermal Properties and Refractive Index of InN," in Edgar [135], section 1.5, pp. 35–40.
501. T. Tansley and C. Foley, "Optical Band Gap of Indium Nitride," *J.Appl.Phys.*, vol. 59, no. 9, pp. 3241–3244, 1986.
502. K. Tappura, "Electrical and Optical Properties of GaInAsP Grown by Gas-source Molecular Beam Epitaxy," *J.Appl.Phys.*, vol. 74, no. 7, pp. 4565–4570, 1993.
503. Technology Modeling Associates, Inc., Sunnyvale, California, *TMA Medici, Two-Dimensional Device Simulation Program, Version 4.0 User's Manual*, Oct. 1997.
504. Technology Modeling Associates, Inc., Sunnyvale, California, *TMA TSUPREM-4, Two-Dimensional Process Simulation Program, Version 6.5 User's Manual*, May 1997.
505. M. Telford, ed., *Advanced Semiconductors Buyers' Guide.* Cambridge: Elsevier, 2000.
506. R. Thoma, A. Emunds, B. Meinerzhagen, H. Peifer, and W. Engl, "Hydrodynamic Equations for Semiconductors with Nonparabolic Band Structure," *IEEE Trans.Electron Devices*, vol. 38, no. 6, pp. 1343–1353, 1991.

507. M. Thurner and S. Selberherr, "Comparison of Long- and Short-Channel MOSFETs Carried out by 3D-MINIMOS," in *17th European Solid State Device Research Conference - ESSDERC'87* (P. Calzolari and G. Soncini, eds.), (Bologna), pp. 409–412, Tecnoprint, 1987.
508. V. Tilak, B. Green, V. Kaper, H. Kim, T. Prunty, J. Smart, J. Shealy, and L. Eastman, "Influence of Barrier Thickness on the High-power Performance of AlGaN/GaN HEMTs," *IEEE Electron Device Lett.*, vol. 22, no. 11, pp. 504–506, 2001.
509. M. Timmons, J. Hutchby, and R. A. D. Dunlavy, "Minority-Carrier Lifetime Measurements in AlGaAs Alloys by Transient Photoluminescence," in *Intl.Symp. GaAs and Related Compounds*, no. 96 in Inst.Phys.Conf.Ser., (Bristol), pp. 289–294, IOP Publishing, 1989.
510. M. Tinston, "National Compound Semiconductor Roadmap," 2002. http://ncsr.csci-va.com/.
511. L. Tirino, M. Weber, K. Brennan, E. Bellotti, and M. Goano, "Dependence of the Impact Ionization Coefficients in GaAs, cubic SiC, and Zinc-Blende GaN," *J.Appl.Phys.*, vol. 94, no. 1, pp. 423–430, July 2003.
512. A. Titkov, E. Iluridze, I. Mironov, and V. Cheban, "Interband Auger Recombination Involving a Spin-orbit Split-off Valence Band in p-type GaSb Crystals," *Sov.Phys.Semicond.*, vol. 20, no. 1, pp. 14–19, 1986.
513. S. Tiwari, *Compound Semiconductor Device Physics*. Academic Press, 1992.
514. S. Tiwari and S. Wright, "Material Properties of p-type GaAs at Large Dopings," *Appl.Phys.Lett.*, vol. 56, no. 6, pp. 563–565, 1990.
515. S. Tiwari, S. Wright, and D. Frank, "Compound Semiconductor Heterostructure Bipolar Transistors," *IBM J.Res.Develop.*, vol. 34, no. 4, pp. 550–567, 1990.
516. J. Tomm, A. Maaßdorf, Y. Mazur, S. Gramlich, E. Richter, F. Brunner, M. Weyers, F. Tränkle, V. Malyarchuk, T. Günther, C. Lienau, and M. Jurisch, "The Impact of Defects to Minority-Carrier Dynamics in Heavily Doped GaAs:C Analyzed by Transient Photoluminescence Spectroscopy," *Mat.Sci.Eng.B*, vol. 91–92, no. 1, pp. 25–28, Apr. 2002.
517. R. Trew, "AlGaN/GaN HFET Amplifier Performance and Limitations," in *IEEE Intl. Microwave Symp. Dig.*, vol. 3, (Seattle), pp. 1811–1814, 2002.
518. H. Tserng and P. Saunier, "A Highly-Efficient 7-Watt 16 GHz Monolithic Pseudomorphic HEMT Amplifier," in *Microwave and Millimeter Wave Monolithic Circ. Symp. Dig. of Papers*, (Atlanta), pp. 105–108, 1993.
519. G. Turner, S. Eglash, and A. Strauss, "Molecular-Beam Epitaxial Growth of High-Mobility n-GaSb," *J.Vac.Sci.Technol.B*, vol. 11, no. 3, pp. 864–867, 1993.
520. M. Tyagi and R. VanOverstraeten, "Minority Carrier Recombination in Heavily Doped Silicon," *Solid-State Electron.*, vol. 26, no. 6, pp. 577–597, 1983.
521. J. Urquhart, D. Robbins, R. Taylor, and A. Moseley, "Impact Ionization Coefficients in $In_{0.53}Ga_{0.47}As$," *Semicond.Sci.Technol.*, vol. 5, no. 7, pp. 789–791, 1990.
522. M. Vaidyanathan and D. Pulfrey, "Effects of Quasi-Ballistic Base Transport on the High-Frequency Characteristics of Bipolar Transistors," *IEEE Trans.Electron Devices*, vol. 44, no. 4, pp. 618–627, 1997.
523. C. van de Walle, "Band Lineups and Deformation Potentials in the Model-Solid Theory," *Phys.Rev.B*, vol. 39, no. 3, pp. 1871–1883, 1989.
524. C. van de Walle and R. Martin, "Theoretical Study of Band Offsets at Semiconductor Interfaces," *Phys.Rev.B*, vol. 35, no. 15, pp. 8154–8165, 1987.
525. M. van Dort, J. Slotboom, G. Streutker, and P. Woerlee, "Lifetime Calculations of MOSFET's Using Depth-Dependent Non-Local Impact Ionization," in *Proc. Intl.Conf. on Simulation of Semiconductor Devices and Processes* (S. Selberherr, H. Stippel, and E. Strasser, eds.), (Vienna), pp. 469–472, Springer, 1993.

526. R. van Overstraeten and H. de Man, "Measurement of the Ionization Rates in Diffused Silicon p-n Junctions," *Solid-State Electron.*, vol. 13, no. 5, pp. 583–608, 1970.
527. W. van Roosbroeck and W. Shockley, "Photon-Radiative Recombination of Electrons and Holes in Germanium," *Phys.Rev.*, vol. 94, no. 6, pp. 1558–1560, 1954.
528. J. Vandersande, "Thermal Conductivity of Natural-Isotope Diamond," in Davies [115], section 1.7, pp. 33–35.
529. Y. Varshni, "Band-to-Band Radiative Recombination in Groups IV, VI, and III-V Semiconductors(I)," *Phys. Status Solidi*, vol. 19, no. 2, pp. 459–514, 1967.
530. Y. Varshni, "Temperature Dependence of the Energy Gap in Semiconductors," *Physica*, vol. 34, no. 1, pp. 149–154, 1967.
531. L. Vorobev, V. Stafeev, V. Shalygin, and A. Shturbin, "Light-Modulation Investigation of Impact Ionization in Intrinsic InSb," *Sov.Phys.Semicond.*, vol. 17, no. 1, pp. 81–84, 1983.
532. I. Vurgaftman, J. Meyer, and L. Ram-Mohan, "Band Parameters for III-V Compound Semiconductors and their Alloys," *J.Appl.Phys.*, vol. 89, no. 11, pp. 5815–5875, 2001.
533. J. Wagner, H. Obloh, M. Kunzer, P. Schlotter, W. Pletschen, R. Kiefer, U. Kaufmann, and K. Köhler, *Group III-Nitride Heterostructures: From Materials Research to Devices*, vol. 40 of *Festkörperprobleme*, pp. 641–656. Vieweg, 2000.
534. S. Wagner, V. Palankovski, T. Grasser, R. Schultheis, and S. Selberherr, "Small-Signal Analysis and Direct S-Parameter Extraction," in *Proc. Intl.Symp. Elec.Dev. for Microwave and Optoelectronic Applications EDMO*, (Manchester), pp. 50–55, 2002.
535. D. Walker, X. Zhang, A. Saxler, P. Kung, J. Xu, and M. Razeghi, "$Al_xGa_{1-x}N$ ($0 \leq x \leq 1$) Ultraviolet Photodetectors Grown on Sapphire by Metal-Organic Chemical Vapor Deposition," *Appl.Phys.Lett.*, vol. 70, no. 8, pp. 949–951, 1997.
536. W. Walukiewicz, "Low Field Mobilities in AlGaAs Alloys," in Adachi [5], section 7.1, pp. 167–172.
537. C.-H. Wang and A. Neugroschel, "Minority-Carrier Transport Parameters in Degenerate n-Type Silicon," *IEEE Electron Device Lett.*, vol. 11, no. 12, pp. 576–578, 1990.
538. P. Wang, S. Holmes, T. Le, R. Stradling, I. Ferguson, and A. de Oliveira, "Electrical and Magneto-Optical Studies of MBE InAs on GaAs," *Semicond.Sci.Technol.*, vol. 7, no. 6, pp. 767–786, 1992.
539. X. Wang, V. Chandramouli, C. Maziar, and A. Tasch, "An Efficient Multi-Band Monte Carlo Model with Anisotropic Impact Ionization," in *IEDM Tech.Dig.*, (San Francisco), pp. 725–728, 1992.
540. Y. Wang, H. Chou, T. Tsai, M. Tu, Y. Chen, P. Chao, C. Chen, S. Liu, and C. Wu, "High Yield InGaP HBT Manufacturing Technology on 150-nm GaAs Wafers," in *Proc. Intl.Conf. GaAs Manufacturing Technology*, (San Diego), pp. 247–250, 2002.
541. K. Washio, E. Ohue, R. Hayami, A. Kodama, H. Shimamoto, M. Miura, K. Oda, I. Suzumura, T. Tominari, and T. Hashimoto, "Ultra-High-Speed Scaled-down Self-Aligned SEG SiGe HBTs," in *IEDM Tech.Dig.*, (San Francisco), pp. 767–770, 2002.
542. K. Washio, E. Ohue, H. Shimamoto, K. Oda, R. Hayami, Y. Kiyota, M. Tanabe, M. Kondo, T. Hashimoto, and T. Harada, "A 0.2 μm 180-GHz-f_{max} 6.7-ps-ECL SOI/HRS Self-Aligned SEG SiGe HBT/CMOS Technology for Microwave and High-Speed Digital Applications," in *IEDM Tech.Dig.*, (San Francisco), pp. 741–744, 2000.
543. K. Washio, E. Ohue, M. Tanabe, and T. Onai, "Self-Aligned Meta/IDP Si Bipolar Technology with 12-ps ECL and 45-GHz Dynamic Frequency Divider," *IEEE Trans.Electron Devices*, vol. 44, no. 11, pp. 2078–2082, 1997.
544. I. Watanabe, T. Torikai, K. Makita, K. Fukushima, and T. Uji, "Impact Ionization Rates in (100) $Al_{0.48}In_{0.52}As$," *IEEE Electron Device Lett.*, vol. 11, no. 10, pp. 437–438, 1990.

545. S. Watkins, O. Pitts, C. Dale, X. Xu, M. Dvorak, N. Matine, and C. Bolognesi, "Heavily Carbon-Doped GaAsSb Grown on InP for HBT Applications," *J.Cryst.Growth*, vol. 221, no. 1-4, pp. 59–65, 2000.
546. R. C. Weast, ed., *Handbook of Chemistry and Physics*. CRC Press, 68th ed., 1987.
547. C. Whelan, P. Marsh, S. Lardizabal, W. Hoke, R. McTaggart, and T. Kazior, "Low Noise and Power Metamorphic HEMT Devices and Circuits with $x = 30\%$ to 60% $In_xGa_{1-x}As$ Channels on GaAs Substrates," in *Proc. European GaAs Rel.Comp.Appl.Symp. GAAS*, (Paris), pp. 189–192, 2000.
548. J. Wiley, *Semiconductors and Semimetals*, vol. 10. New York: Academic Press, 1975.
549. V. Wilkinson and A. Adams, "The Effect of Temperature and Pressure on InGaAs Band Structure," in Bhattacharya [38], section 3.2, pp. 70–75.
550. K. Wolfstirn, "Hole and Electron Mobilities in Doped Silicon from Radiochemical and Conductivity Measurements," *J.Phys.Chem.Solids*, vol. 16, no. 3-4, pp. 279–284, 1960.
551. C. Wu and E. Yang, "Carrier Transport Across Heterojunction Interfaces," *Solid-State Electron.*, vol. 22, no. 3, pp. 241–248, 1979.
552. C. Wu, E. Sovero, and B. Massey, "40 GHz Transimpedance Amplifier with Differential Outputs using InP/InGaAs Heterojunction Bipolar Transistors," in *GaAs IC Symp.Tech.Dig.*, (Monterey), pp. 63–66, 2002.
553. D. Xu, H. Heiß, S. Kraus, M. Sexl, G. Böhm, G. Tränkle, G. Weimann, and G. Abstreiter, "Design and Fabrication of Double Modulation Doped InAlAs/InGaAs/InAs Heterojunction FET's for High-Speed and Millimeter-Wave Applications," *IEEE Trans.Electron Devices*, vol. 45, no. 1, pp. 21–30, 1998.
554. D. Xu, T. Suemitsu, J. Osaka, Y. Umeda, Y. Yamane, Y. Ishii, T. Ishii, and T. Tamamura, "An 0.03-μm Gate-Length Enhancement-Mode InAlAs/InGaAs/InP MODFET's with 300 GHz f_T and 2 S/mm Extrinsic Transconductance," *IEEE Electron Device Lett.*, vol. 20, no. 5, pp. 206–208, 1999.
555. J. Xu and M. Shur, "Velocity-Field Dependence in GaAs," *IEEE Trans.Electron Devices*, vol. ED-34, no. 8, pp. 1831–1832, 1987.
556. T. Yamada, J. Zhou, H. Miyata, and D. Ferry, "In-Plane Transport Properties in $Si/Si_{1-x}Ge_x$ Structure and its FET Performance by Computer Simulation," *IEEE Trans.Electron Devices*, vol. 41, no. 9, pp. 1513–1522, 1994.
557. Y. Yamashita, A. Endoh, K. Shinohara, K. Hikosaka, T. Matsui, S. Hiyamizu, and T. Mimura, "Pseudomorphic $In_{0.52}Al_{0.48}As/In_{0.7}Ga_{0.3}As$ HEMTs with an Ultrahigh f_T of 562 GHz," *IEEE Electron Device Lett.*, vol. 23, no. 10, pp. 573–575, 2002.
558. B. Yan, J. Luo, and Q. Zhang, "Study of Band-Gap Narrowing Effect and Nonradiative Recombination Centers for Heavily C-doped GaAs by Photoluminiscence Spectroscopy," *J.Appl.Phys.*, vol. 77, no. 9, pp. 4822–4824, May 1995.
559. K. Yang, J. East, and G. Haddad, "Numerical Modeling of Abrupt Heterojunctions Using a Thermionic-Field Emission Boundary Condition," *Solid-State Electron.*, vol. 36, no. 3, pp. 321–330, 1993.
560. Y. Yang, J. Frackoviak, C. Liu, C. Chen, L. Chua, W. Sung, A. Tate, J. Tong, R. Reyes, R. Kopf, R. Ruel, D. Werder, V. Houtsma, G. Georgiou, J. Weiner, Y. Baeyens, and Y. Chen, "High-Performance and High-Uniformity InP/InGaAs/InP DHBT Technology for High-Speed Optical Communication Systems," in *GaAs IC Symp.Tech.Dig.*, (Monterey), pp. 269–272, 2002.
561. H. Yao and A. Compaan, "Plasmons, Photoluminiscence, and Band-Gap Narrowing in Very Heavily Doped *n*-GaAs," *Appl.Phys.Lett.*, vol. 57, no. 2, pp. 147–149, 1990.
562. R. Yarborough, B. Landini, R. Welser, J. Yang, and T. Henderson, "Enhanced CDMA Performance from an InGaP/InGaAsN/GaAs N-p-N Double Heterojunction Bipolar Transistor," in *GaAs IC Symp.Tech.Dig.*, (Monterey), pp. 273–276, 2002.

563. K. Yeom, J. Hinckley, and J. Singh, "Theoretical Study on Threshold Energy and Impact Ionization Coefficient for Electrons in $Si_{1-x}Ge_x$," *Appl.Phys.Lett.*, vol. 64, no. 22, pp. 2985–2987, 1994.
564. K. Yeom, J. Hinckley, and J. Singh, "Calculation of Electron and Hole Impact Ionization Coefficients in SiGe Alloys," *J.Appl.Phys.*, vol. 80, no. 12, pp. 6773–6782, 1996.
565. S. Yi, S. Chung, H. Rohdin, M. Hueschen, D. Bour, N. Moll, D. Chamberlin, and J. Amano, "Growth and Device Performance of InP/GaAsSb HBTs," in *Proc. Intl.Conf.InP and Rel.Mat.*, (Santa Barbara), pp. 380–384, 2003.
566. P. Yoder, U. Krumbein, K. Gärtner, N. Sasaki, and W. Fichtner, "Statistical Enhancement of Terminal Current Estimation for Monte Carlo Device Simulation," *VLSI Design*, vol. 6, no. 1-4, pp. 303–306, 1998.
567. I. Yoon, T. Ji, S. Oh, J. Choi, and H. Park, "Concentration Dependent Photoluminescence of Te-doped $In_{0.5}Ga_{0.5}P$ Layers Grown by Liquid Phase Epitaxy," *J.Appl.Phys.*, vol. 82, no. 8, pp. 4024–4027, Oct. 1997.
568. L. Yu, D. Qiao, Q. Xing, S. Lau, K. Butros, and J. Redwing, "Ni and Ti Schottky Barriers on n-AlGaN grown onm SiC Substrates," *Appl.Phys.Lett.*, vol. 73, no. 2, pp. 238–240, 1998.
569. Z. Yu, B. Riccó, and R. Dutton, "A Comprehensive Analytical and Numerical Model of Polysilicon Emitter Contacts in Bipolar Transistors," *IEEE Trans.Electron Devices*, vol. 31, no. 6, pp. 773–784, 1984.
570. J. Yuan, *SiGe, GaAs, and InP Heterojunction Bipolar Transistors.* New York: Wiley, 2002.
571. E. Yurova and E. Pel, "Electron Mobility in the Intermediate Range of Compositions of $GaAs_{1-x}P$ Solid Solutions," *Sov.Phys.Semicond.*, vol. 7, no. 2, pp. 273–274, 1973.
572. X. Zheng, P. Yuan, X. Sun, G. Kinsey, A. Holmes, B. Streetman, and J. Campbell, "Temperature Dependence of the Ionization Coefficients of $Al_xGa_{1-x}As$," *IEEE J.Quantum Electronics*, vol. 36, no. 10, pp. 1168–1173, 2000.
573. B. Zhou, K. Butcher, H. Zou, X. Li, and T. Tansley, "Photoconductive Decay in LCVD/PECVD Low Temperature Grown GaN," in *Proc. Topical Workshop III-V Nitrides*, (Nagoya), pp. 243–245, 1995.
574. E. Zielinski, H. Schweizer, K. Streubel, H. Eisele, and G. Weimann, "Excitonic Transitions and Exciton Damping Process in InGaAs/InP," *J.Appl.Phys.*, vol. 59, no. 6, pp. 2196–2204, 1986.

Index

List of Figures

List of Tables

Springer-Verlag
and the Environment

WE AT SPRINGER-VERLAG FIRMLY BELIEVE THAT AN international science publisher has a special obligation to the environment, and our corporate policies consistently reflect this conviction.

WE ALSO EXPECT OUR BUSINESS PARTNERS – PRINTERS, paper mills, packaging manufacturers, etc. – to commit themselves to using environmentally friendly materials and production processes.

THE PAPER IN THIS BOOK IS MADE FROM NO-CHLORINE pulp and is acid free, in conformance with international standards for paper permanency.